The maintenance of a stable acid–base status within biological tissue is a fundamental homeostatic process in all organisms, necessary to preserve the metabolic function of proteins and other macromolecules. The study of acid–base regulation has advanced enormously over recent decades due to the develoment of increasingly accurate and sensitive techniques for measuring acid–base variables. This volume brings together contributions from leading comparative physiologists working on factors affecting the acid–base status of the internal fluids of animals and plants. The result is a broad-ranging, authoritative and accessible review of the most recent and exciting discoveries in this area, together with a critical look at current techniques and tools. As such, it provides an unparalleled resource for researchers and their students in the fields of animal physiology, plant physiology, cell biology and biochemistry.

SOCIETY FOR EXPERIMENTAL BIOLOGY
SEMINAR SERIES: 68

REGULATION OF TISSUE pH IN PLANTS
AND ANIMALS
A REAPPRAISAL OF CURRENT TECHNIQUES

DAMAGED

SOCIETY FOR EXPERIMENTAL BIOLOGY SEMINAR SERIES

A series of multi-author volumes developed from seminars held by the Society for Experimental Biology. Each volume serves not only as an introductory review of a specific topic, but also introduces the reader to experimental evidence to support the theories and principles discussed, and points the way to new research.

REGULATION OF TISSUE pH IN PLANTS AND ANIMALS
A REAPPRAISAL OF CURRENT TECHNIQUES

Edited by

S. Egginton

Department of Physiology
University of Birmingham

E.W. Taylor

Department of Biological Sciences
University of Birmingham

J.A. Raven

Department of Biological Sciences
University of Dundee

CAMBRIDGE
UNIVERSITY PRESS

CAMBRIDGE UNIVERSITY PRESS
Cambridge, New York, Melbourne, Madrid, Cape Town, Singapore, São Paulo

Cambridge University Press
The Edinburgh Building, Cambridge CB2 8RU, UK

Published in the United States of America by Cambridge University Press, New York

www.cambridge.org
Information on this title: www.cambridge.org/9780521623179

First published 1999
This digitally printed version 2007

A catalogue record for this publication is available from the British Library

Library of Congress Cataloguing in Publication data

Regulation of tissue pH in plants and animals: a reappraisal of current
techniques / edited by S. Egginton, E.W. Taylor, J.A. Raven.
  p. cm.—(Society for Experimental Biology seminar series; 68)
Includes bibliographical references and index.
ISBN 0–521–62317–0 (hardcover)
1. Acid–base equilibrium. I. Egginton, S. II. Taylor, E. W.
III. Raven, John A. IV. Series: Seminar series (Society for Experimental
Biology (Great Britain)); 68.
QP90.7.R44 1999
581.7′5–dc21 98–34469 CIP

ISBN 978-0-521-62317-9 hardback
ISBN 978-0-521-03938-3 paperback

Contents

Contributors

BLATT, M.R.
Department of Biological Sciences, University of London, Wye College, Kent TN25 5AH, UK.

BULLOCK, A.J.
Department of Physiology, P.O. Box 147, University of Liverpool, Crown Street, Liverpool L69 3BX, UK.

BUTTELL, N.
Department of Physiology, P.O. Box 147, University of Liverpool, Crown Street, Liverpool L69 3BX, UK.

DUQUETTE, R.A.
Department of Physiology, P.O. Box 147, University of Liverpool, Crown Street, Liverpool L69 3BX, UK.

EGGINTON, S.
Department of Physiology, University of Birmingham, Edgbaston, Birmingham B15 2TT, UK.

FELLE, H.H.
Botanisches Institut 1, Universität Giessen, Seckenbergstrasse 17–21, D-35390 Giessen, Germany.

GRABOV, A.
Department of Biological Sciences, University of London, Wye College, Kent TN25 5AH, UK.

HEISLER, N.
Department of Animal Physiology, Humboldt Universität zu Berlin, Abderhaldenhaus, Philippstrasse 13, D-10115 Berlin, Germany.

JACKSON, D.C.
Division of Biomedical Sciences, Brown University, Box G, Providence RI 02912-0001, USA.

KINSEY, S.T.
Department of Biological Sciences, University of North Carolina at Wilmington, 601 South College Road, Wilmington, NC 28403-3297, USA.

MALAN, A.
University of Strasbourg 1, CNRS, URA 1332, 12 rue Université, F-67000, Strasbourg, France.

MOERLAND, T.S.

Department of Biological Sciences, Florida State University, Biology Unit 1, B-157, Tallahassee, FL 32306-2050, USA.

PÖRTNER, H.O.

Alfred-Wegener Institute for Polar and Marine Science, Postfach 120161, Columbusstrasse, D-27568 Bremerhaven, Germany.

RATCLIFFE, R.G.

Department of Plant Sciences, University of Oxford, South Parks Road, Oxford OX1 3RB, UK.

RAVEN, J.A.

Department of Biological Sciences, University of Dundee, Dundee DD1 4HN, UK.

SARTORIS, F.J.

Alfred-Wegener Institute for Polar and Marine Science, Postfach 120161, Columbusstrasse, D-27568 Bremerhaven, Germany.

SCHWIENING, C.J.

Physiological Laboratory, Downing Street, Cambridge CB2 3EG, UK.

TAYLOR, E.W.

School of Biological Sciences, University of Birmingham, Edgbaston, Birmingham B15 2TT, UK.

TUFTS, B.L.

Biology Department, Queens University, Kingston K7L 3N6, Canada.

TYLER-JONES, R.

Cripps Computing Centre, University of Nottingham, Nottingham NG7 2RD, UK.

WANG, Y.

Department of Biology, McMaster University, 1200 Main Street West, Hamilton, Ontario L85 4K1, Canada.

WHITELEY, N.M.

School of Biological Sciences, University of Wales, Bangor, Gwynedd LL57 2UW, UK.

WILSON, R.W.

Department of Biological Sciences, University of Exeter, Hatherly Laboratories, Prince of Wales Road, Exeter EX4 4PS, UK.

WOOD, C.M.

Department of Biology, McMaster University, 1200 Main Street West, Hamilton, Ontario L8S 4K1, Canada.

WRAY, S.

Department of Physiology, P.O. Box 147, University of Liverpool, Crown Street, Liverpool L69 3BX, UK.

ZAMMIT, V.A.

Hannah Research Institute, Ayr KA6 5HL, UK.

Preface

The regulation of [H$^+$] (commonly expressed as pH) in the internal
fluids, and consequently in the tissues of living organisms, is central to
the processes of homeostatic control. This is because pH determines the
charge state on proteins and other macromolecules, which in turn affects
vital metabolic processes such as enzyme–substrate binding and the
catalytic functions of enzymes. Consideration of pH regulation is, therefore, of fundamental importance to biology. Many factors affect pH,
including rates of production, accumulation and excretion of acid
metabolites. Regulation of acid–base status involves a number of other
physiological processes, most prominently ionoregulation and respiration. Comparative physiologists have been actively working in this
area for the last decade, considering new problems of environmental
pH variation (e.g. temperature, acid or alkaline waters, availability of
ions, respiratory gases or food) and physiological mechanisms (e.g.
rates of activity, buffering using skeletal calcium salts, role of the gut,
regulation in different tissues and in response to discrete bodily functions such as respiration, digestion or excretion). This volume considers
current work by leading plant and animal physiologists, all with international reputations for their work on pH regulation.

Accurate measurement of pH in tissues is complicated by their sub-
division into discrete compartments (extracellular fluid volume, intracellular cytosol, tonoplast, nucleoplasm, mitochondria etc.). The small
size of these compartments coupled with the highly labile nature of pH
variation, which can be critically affected by sampling technique, presents serious technical problems. Initial chapters consider the application of current techniques for measurement of pH in biological systems, the relative merits of which are being hotly debated. The use of
ion-sensitive eletrodes and fluorescent dyes (Felle, Schwiening) and of
non-invasive nuclear magnetic resonance (Kinsey and Moerland) in the
measurement of intracellular pH (pH$_i$) are considered. Next are discussions on the influence of environmental conditions on acid–base
regulation (Pörtner and Sartoris) and proton metabolism during exercise

(Wood and Wang). The importance of branchial ion movement in the maintenance of acid–base balance in fishes (Heisler) is complemented by discussion of the role of [H⁺] in controlling ionoregulation in plants (Blatt and Grabov). The consequences of specific CO_2-concentrating mechanisms (Raven) and periods of anoxia (Ratcliffe) for pH_i regulation in plants are explored. Contributions describing the role of skeletal elements (Jackson, Whiteley) and the gut (Wilson) detail how secondary mechanisms may be used to control pH. Constrasting studies into pH regulation of individual cells – smooth muscle (Wray *et al.*) and erythrocytes (Tufts) – lead to a detailed account of control at the organismal level during periods of starvation (Zammit) and hibernation (Malan). As André Malan pointed out, the 1977 SI committee headed by Siggaard-Andersen recommended the non-SI unit of pH be replaced by the chemical potential, in kJ/mol. Unfortunately, Henderson put the zero on the pH scale corresponding to a very strong rather than a very weak acid, resulting in the rather awkward figure of −44 kJ/mol for the arterial pH of a mammal at 37 °C! The volume therefore ends with a reappraisal of the use of [H⁺] rather than pH when defining acid–base status in biological systems (Tyler-Jones and Taylor). Authors have differed in their use of standard units and we have generally left them in their preferred format, so that concentration is given either as M or mol/l.

This volume follows the excellent *Acid–base regulation in animals* edited by Norbert Heisler (Elsevier, 1986) and *Comparative aspects of extracellular acid–base balance* edited by Jean-Paul Truchot (Springer-Verlag, 1987), which provide essential historical background to the comparative aspects of this topic. The present volume provides an update of information based on new technologies, recognising the universal influence of tissue pH on cellular function. Importantly, the specific technical difficulties associated with using plant or animal cells turn out to provide complementary insights into pH regulation, which means the book will be of equal interest to plant and animal physiologists, cell biologists and biochemists. Our aim was to provide a critical introduction targeted at advanced undergraduate and graduate students as well as researchers in this area, written by acknowledged experts in the field. It is based on a two-day symposium at the 1997 Annual Meeting of the Society for Experimental Biology that provided a forum for animated debate, and we hope the book also stimulates in its readers a re-evaluation of techniques and new directions.

SE
EWT
JAR

CHRISTOF J. SCHWIENING

Measurement of intracellular pH: a comparison between ion-sensitive microelectrodes and fluorescent dyes

Introduction

Intracellular pH (pH_i) has been studied extensively for over a century using a wide range of techniques. These techniques have been subject to constant improvements to the extent that useful measurements can now be made in even the smallest of cells. This chapter outlines the development of ion-sensitive microelectrodes and dyes and highlights some of the difficulties and dangers of both of these methods. It then concentrates on some recent advances in our understanding of how fluorescent pH-sensitive dyes might best be used. Finally, a tabulated comparison between the two techniques is presented.

Historical perspective on pH_i measurement

Although pH_i changes had been observed much earlier, the first measurements of what could loosely be described as intracellular pH were made around 1910 using cell extracts and platinum/hydrogen electrodes (for a review see Caldwell, 1956). For example, in 1912 Michaelis and Davidoff measured the pH of blood and noted that red cell lysis caused a change in bulk pH. The technique of cell lysis was perhaps most suited to the measurement of pH_i in non-nucleated erythrocytes where intracellular compartments did not complicate the measurements. At around the same time various workers were using naturally occurring pH indicators to visualise changes in pH_i (e.g. Crozier, 1918). The problems associated with such indirect techniques were recognised very early on and the search for better methods led in three directions: distribution of weak acids and bases; smaller electrodes to allow direct measurement of pH_i; and better indicators and techniques to load them into cells.

Weak acid and base distribution

Detailed discussion of the distribution of weak acids or bases is beyond the scope of this chapter but readers who are interested should consider the review by Roos and Boron (1981). In broad terms, the distribution of the weak acid or base across the plasma membrane is dependent upon the pH in both the extracellular and intracellular compartments. Fridericia (1920) was the first to use this technique to measure the pH_i of red blood cells using CO_2 and he obtained a pH_i of 7.30. Later dimethyloxazolidine-dione (DMO; Waddell & Butler, 1959) was introduced as a pH_i indicator. Although the technique is easy, accurate and can be applied to small cells, its use was limited by the fact that it requires the destruction of the tissue and therefore cannot easily be used to measure pH_i changes over time. The technique also has a major problem in that it can alter pH_i. However, the principle of weak acid or base permeation and equilibration across the cell membrane is of some interest. De Vries (1871) first showed colour changes in living beetroot when exposed to ammonia. The colour changes reflected changes in intracellular pH as the uncharged, lipid-soluble weak base diffused across the cell membrane and then associated with H^+ causing a profound alkalinisation. The movement of uncharged weak acids or bases therefore not only allowed pH_i to be estimated, but was also a powerful means of adding or removing acid equivalents from cells. The application of weak acids or bases to cells has become the ubiquitous method for challenging and thereby studying pH_i regulation.

pH-sensitive electrodes

Early electrodes

Intracellular pH-sensitive electrodes have been made out of numerous materials. They can be broadly divided into three groups:

> metals: platinum–hydrogen electrodes (Taylor & Whitaker, 1927); antimony (Buytendijk & Woerdeman, 1927); tungsten (Caldwell, 1954);
> glass: (Caldwell, 1954; Hinke, 1967; Thomas, 1974);
> liquid membranes: bicarbonate sensor; nigericin (Matsumura *et al.*, 1980); neutral H^+ exchanger tri-n-dodecylamine (Ammann *et al.*, 1981).

High electrical resistance, however, presented a major problem for the miniaturisation of metal minielectrodes. It was not until the 1950s that significant progress was made in producing microelectrodes that could be widely used. pH-sensitive glass, although discovered around 1900, became the material of choice. It has relatively low electrical resistance, no sensitivity to oxidising and reducing agents, dissolved

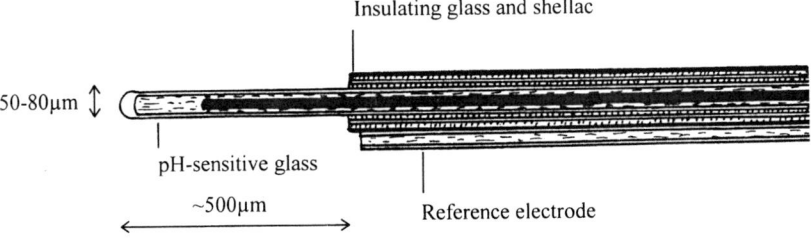

50-80μm

Insulating glass and shellac

pH-sensitive glass

~500μm

Reference electrode

Fig. 1. Glass pH-sensitive electrode used by Caldwell (1954) to measure pH_i in crab muscle fibres. The unexposed pH-sensitive (low-resistance) glass was insulated with shellac and insulating glass. In many cases, Caldwell used wax to attach a reference electrode. Modified from Caldwell (1954).

gasses, anions or buffers, and it is stable and can produce a relatively rapid response.

Only two particular microelectrode types are concentrated on in this section – the combination electrode, pH and reference, produced by Caldwell in 1954 (Fig. 1) and the recessed-tip design produced by Thomas in 1974 (Fig. 2). The Caldwell electrode was a design classic but is no longer in use because its relatively large size restricted its use to large cells such as crab muscle fibres. The electrode consisted of a portion of exposed pH-sensitive glass about 500 μm long. Such a long length of what was termed 'low-resistance glass' was necessary to obtain usable resistance of 1 GΩ. The unexposed length was insulated with pH-insensitive glass and shellac. This design was modified by

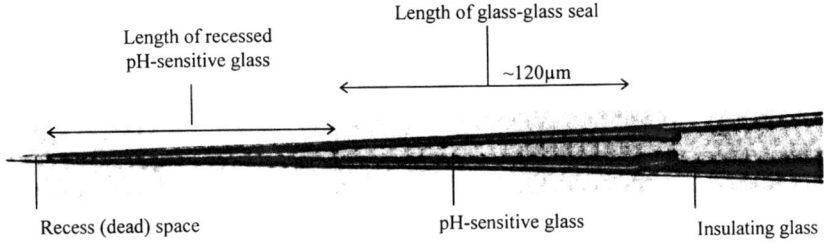

Length of recessed pH-sensitive glass

Length of glass-glass seal

~120μm

Recess (dead) space

pH-sensitive glass

Insulating glass

Fig. 2. Recessed-tip pH-sensitive microelectrode as used by Thomas (1974) to measure pH_i in snail neurons. A microforge was necessary for the construction of these electrodes. A heated filament was used to soften the low-melting-point, inner, pH-sensitive glass. Simultaneous application of high pressure down the pH-sensitive glass caused a high-resistance glass-to-glass seal to form.

Hinke (1967), who elegantly replaced the shellac with a glass-to-glass seal, but in the process had to abandon the integral reference electrode. Improvements in both the pH-sensitive glass and the input impedance of modern electrometers allowed the exposed length of pH-sensitive glass to be reduced to about 100 μm, but this still restricted the use of these electrodes to giant cells. However, in cells such as the squid giant axon and barnacle muscle fibres these electrodes are still the method of choice for measuring pH_i.

The first true pH-sensitive *micro*electrode was produced in 1974 by Thomas. Thomas introduced a design whereby the exposed length of pH-sensitive glass (still at least 100 μm in length) was recessed within the insulating glass such that the pH of the recessed space was measured. The result was that only the 1–2 μm tip had to be placed within the cell. Such microelectrodes require considerable skill to manufacture. The recessed-tip pH-sensitive microelectrode has been used to measure pH_i in numerous cell types (snail neurons, skeletal muscle and cardiac muscle) and, despite its relatively slow response, for those who can produce them, it remains the method of choice for measuring pH_i in cells around 100 μm in diameter. Even with such large cells, the requirement to place two microelectrodes into one cell can be difficult to fulfil, especially where the cell boundaries are obscured. Although double-barrelled recess-tip electrodes partly overcame this problem, their construction was too difficult for all but a few (de Hemptinne, 1979).

The problems of slow response, large tips and difficult construction of recessed-tip pH-sensitive microelectrodes were partially solved by the development of liquid ion-exchange microelectrodes.

Liquid ion-exchange pH-sensitive electrodes

The first liquid ion-exchange resin used to measure pH_i was a bicarbonate-sensitive exchanger (Khuri, Bogharian & Agulian, 1974). The microelectrodes had tip sizes of less than 1 μm and had a rapid response time. However, they were also sensitive to carbon dioxide. Nonetheless, they were used to measure pH_i in a number of cell types. Further progress was made in the early 1980s, resulting in neutral carrier microelectrodes based on tri-n-dodecylamine (see Ammann, 1986). There is now a wide range of hydrogen ionophores available with varying characteristics; however, they all share some basic properties.

Using pH-sensitive microelectrodes

The voltage measured from a pH-sensitive microelectrode when placed inside a cell is the sum of both the membrane potential and a potential

sensitive to pH. In order to measure pH$_i$ it is therefore also necessary
to measure the membrane potential. This requires either two separate
electrodes (Fig. 3A & B) or one double-barrelled electrode (Fig. 3C).
The measurement of membrane potential is a major strength of the
technique because it provides good information about cell health and
electrical activity. Double-barrelled electrodes are more difficult to construct and are generally only used on cells that are difficult to impale
with two electrodes. Fig. 3D shows a range of different glass forms that
have been used to construct double-barrelled microelectrodes.

Before the organic ligand can be introduced into the micropipette,
the glass surface must be made hydrophobic (Fig. 4). This is achieved
by baking the micropipette with a silane. There is a wide range of
different silanes and protocols used, all with variable success (see
Deyhimi & Coles, 1982, for more details). Silanisation both aids the

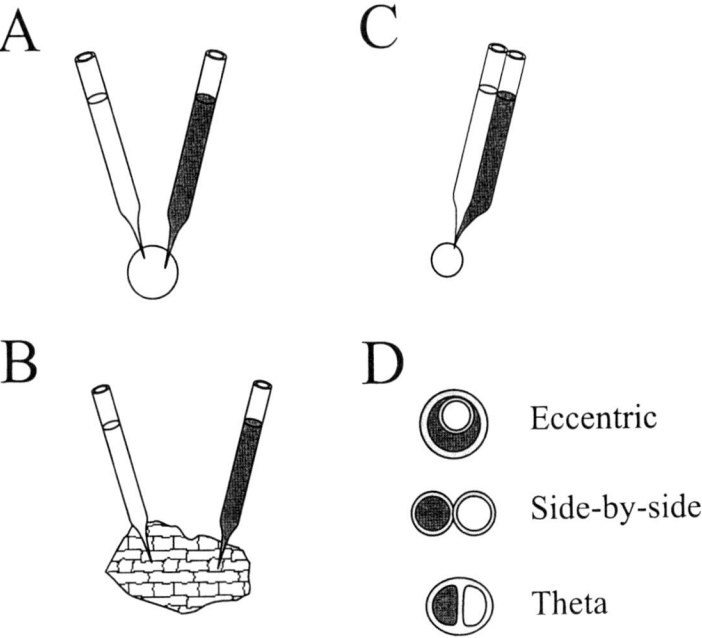

Fig. 3. Microelectrode measurement of pH$_i$ requires the measurement
of the membrane potential. This can be achieved in three ways. **A.**
Separate electrodes to measure membrane potential and pH placed in
the same cell. **B.** Membrane and pH electrodes inserted in different
but electrically coupled cells. **C.** A combination pH and membrane
potential electrode placed in a cell. **D.** Profiles of the three major
double-barrelled electrode configurations.

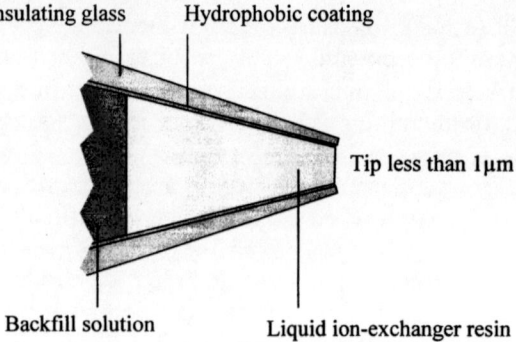

Fig. 4. Liquid ion-exchanger microelectrodes consist of a glass micro-pipette which is made hydrophobic by treatment with a silane and then filled with a small amount of exchanger resin.

retention of the ligand within the microelectrode tip and reduces the electrical shunt along the glass surface which would otherwise short-circuit the high-resistance ligand. This process is relatively straightfor-ward for single-barrelled electrodes where all of the glass surface can be treated. However, combination electrodes require the selective silanisation of only one barrel of the electrode (see Fig. 3C). Once the electrode is filled with a short column of ligand and a back-filling solu-tion, the electrode has a limited life (usually hours rather than days). Loss of the ligand, blockage of the tip and electrical shunts all have the potential to cause the electrode signal to drift and respond in a sub-Nernstian fashion.

Placing pH microelectrodes inside cells undoubtedly causes some damage, but the membrane potential is measured and the damage can be assessed. The damage usually consists of a 'leak' around the elec-trodes. This leak rarely causes a large change in pH_i because intracellu-lar buffering power is high and the pH gradient across the cell mem-brane is low. The extent to which the membrane potential reflects the amount of damage is, of course, dependent upon the input resistance of the cell. The damage to large cells appears less than that to small ones. However, in both small and large cells the damage, if sufficient, will lead to a large influx of other ions such as calcium and sodium. This physical damage to the integrity of the cell membrane has limited the used of pH-sensitive microelectrodes to large and robust cells and rep-resents a constant source of anxiety to those using ion-sensitive micro-electrodes. The leak around the pH-sensitive microelectrode can also cause another major problem. If for some reason the pH-sensitive microelectrode fails to record the full membrane potential, a relatively common phenomenon, then the pH signal following subtraction of the full membrane potential, as measured with the second electrode, will not be an accurate measure of pH_i. The only way to test for this subtraction error is to hyperpolarise the cell, a manoeuvre which in most cells (but not glial) does not change pH_i.

pH-sensitive dyes

Early dyes

Following De Vries' (1871) observations of colour changes in beetroot cells, others extracted and chemically identified various coloured pH-sensitive dyes. Attempts to 'load' such dyes into cells date back to early work on protozoa (Metchnikoff, 1893) using particulate litmus. By 1906, dyes like neutral red were being used to stain living cells and then 'measure' pH_i. Dye injection and early forms of 'scrape-loading' were also in their infancy around 1920. However, by the 1930s the use of pH_i indicators was in decline. The problems associated with their use were serious: salt, metachromatic and protein errors made absolute measurements difficult. However, many used absorbance dyes such as phenol red to measure relative pH changes (for example Ahmed & Connor, 1980), with some success. The introduction of fluorescent dyes in the early 1980s began a new era for pH_i measurement.

Fluorescent dyes

Fluorescent pH-sensitive dyes appeared to overcome many of the previous problems of optical pH measurement. The ratiometric approach – that is, dual excitation or dual emission – allowed for the simple correction for changes in path length, indicator concentration, leakage or photobleaching. There are now several different classes of pH-sensitive fluorescent dyes: fluoresceins, benzoxanthenes, rhodols and pyrenes. Each class of dye has a range of different properties (pK, excitation and emission wavelengths, lipid solubility, photostability, esterified forms and dynamic range), making them suitable for varying applications. However, since its introduction by Tsien in the early 1980s, 2′,7′-bis(carboxyethyl)-5(and 6)-carboxyfluoresceine (BCECF) has been by far the most popular. The fluorescent dye could be incorporated into cells by simply bathing them in the lipophilic acetoxymethyl (AM) ester form of the dye (BCECF-AM). The dye enters the cell where native esterases hydrolyse it, releasing the charged membrane impermeant BCECF. This allowed for pH_i measurements in even the smallest of cells. It was recognised fairly early on that the probe could not be calibrated satisfactorily *in vitro* (for example, Chaillet & Boron, 1985). The antibiotic nigericin, however, proved to be a simple solution. Nigericin is lipid soluble and is easily incorporated into the cell membrane, where it exchanges protons for potassium. By altering the extracellular potassium concentration and pH, whilst knowing the intracellular potassium concentration, the pH_i can be set and an *in vivo* calibration obtained. The AM loading of BCECF and the simple nigericin calibration resulted in an explosion of work on pH_i. However, serious problems remained with the technique that are only now starting to be appreciated. The remainder of this chapter is focused on three particular problems: cell health, inhibition of transporters, and contamination with exogenous transporters.

Recent problems with dyes

Assessing cell health

Electrode measurement of pH_i necessitates measuring the cell membrane potential – this allows for an obvious and simple way of judging cell health. A loss of membrane integrity usually results in a profound depolarisation. The measurement of pH_i with fluorescent dyes provides no such information; cell health must instead be assessed visually. Most of the fluorescence microscopes available in the 1980s and early 1990s did not allow for the continuous visualisation of cells whilst recording pH_i. Now, however, cheap, charged-coupled devices are making long wavelength recording of images increasingly common.

Since pH_i and extracellular pH (pH_e) are similar (approximately 7.2 and 7.4 respectively) it is not possible to use pH_i as a discriminator of cell health. A near-complete collapse of the membrane potential as a result of a loss in membrane integrity, and therefore a Nernstian distribution of H^+, can result in little change in pH_i. However, one can use the dye signal to provide some information about membrane integrity. Schwiening and Boron (1992) and later Bevensee, Schwiening and Boron (1995) showed that the rate constant for dye loss, as assessed by the pH-insensitive fluorescence of BCECF, could be used as an indicator of cell health. Figure 5 shows records of pH_i, intracellular BCECF concentration and the rate of dye loss. A simple interpretation made by Schwiening and Boron (1992) was that the loss of dye in Figure 5, and therefore the profound alkalinisation, resulted from a loss in membrane integrity.

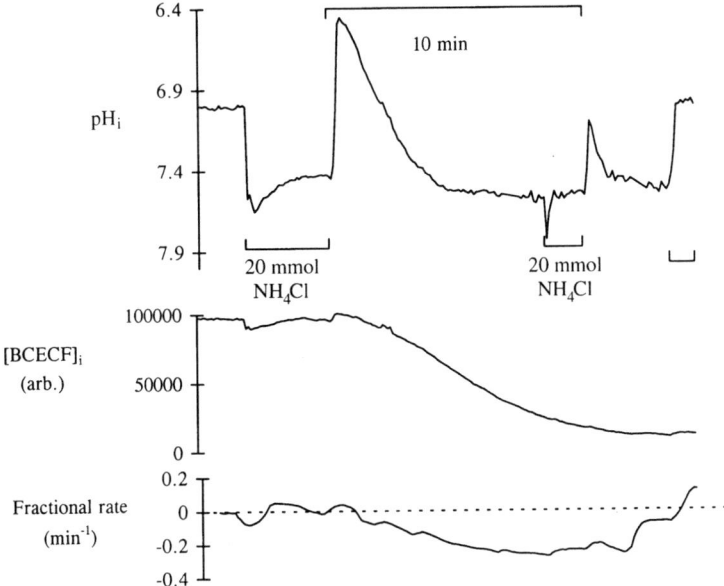

Fig. 5. pH_i, BCECF concentration (440 nm excited fluorescence) and the rate of dye loss measured in a rat hippocampal neuron. NH_4Cl application initially produced normal pH_i changes. However, within five minutes, pH_i approached the pH_e (7.4). A second NH_4Cl application produced only very small, transient pH_i changes. A nigericin/ high K^+ calibration to pH 7.00 is shown at the end of the experiment. Data modified from Schwiening and Boron (1992).

They proposed the use of the instantaneous rate of dye loss as an indicator of membrane integrity. The technique is not entirely straightforward because some BCECF is tightly bound to cellular proteins and is not lost during a modest loss of cell integrity.

It is remarkable that so few choose routinely to show such data. For instance, Figure 6 is a recording of pH_i made by Hays and Alpern (1991) in the kidney, showing a striking spontaneous change in pH_i. This result was interpreted as the delayed activation of proton pumps. It is noticeable that the signal-to-noise ratio in the experiment increased during the alkalinisation – a common result of a loss of dye. It is surprising that the delayed activation of H^+ pumps in Figure 6 and the loss of cell integrity in Figure 5 can result in such superficially similar pH_i changes.

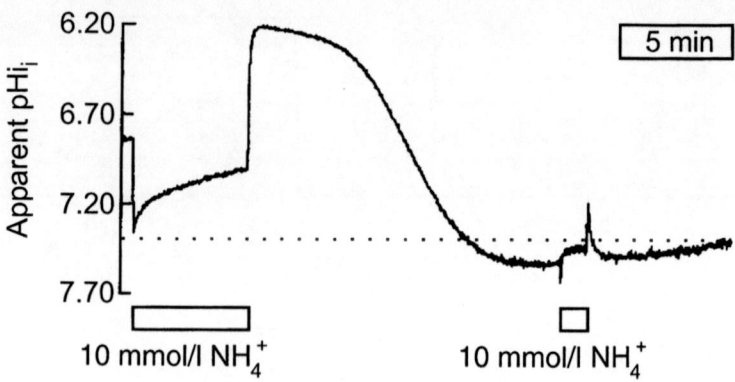

Fig. 6. pH$_i$ measured with BCECF in renal outer medullary collecting duct cells. NH$_4$Cl application initially produced normal pH$_i$ changes. After about ten minutes, pH$_i$ spontaneously approached the pH$_e$, indicated by the dotted line. A second NH$_4$Cl application produced only very small, transient pH$_i$ changes. Modified from Hays and Alpern (1991).

Inhibition of transporters

Free BCECF is highly charged, and hence retained within the cytosol. Unfortunately, this high charge causes the binding of BCECF to cellular proteins resulting in a large spectral shift of the dye. For this reason, the *in vitro* calibration cannot reliably be applied to *in vivo* measurement. Furthermore, one of the proteins to which BCECF binds is the calcium-ATPase. The inhibition of the calcium-ATPase by BCECF and other pH-sensitive dyes (see Table 1) was first demonstrated by Gatto and Milanick (1993) in red blood cells. This may at first sight seem of little consequence to those measuring pH$_i$. However, in 1993 Schwiening, Kennedy and Thomas showed that the only calcium extrusion mechanism in snail neurons, the calcium-ATPase, counter-transported H$^+$. It is thus possible that the calcium-ATPase or, more accurately, the calcium–hydrogen pump, is a major acid-loading mechanism. Measuring pH$_i$ with BCECF is clearly fraught with possible distortions. Gatto and Milanick (1993) reported different IC$_{50}$ values for BCECF obtained from different suppliers. Nevertheless, many of the pH$_i$ measurements made with BCECF may have been done in systems lacking what could be a major acid-loading mechanism, especially during calcium loading that follows electrical activity.

Fortunately, there are pH-sensitive dyes available that do not inhibit

Table 1. *Various inhibitors, including pH-sensitive dyes, and their IC$_{50}$ concentrations for the calcium pump in red blood cells*

Compound	IC$_{50}$ (μM)
Fluorescein	1000
DM-NERF	500
BCECF	100
5,6-Carboxy-SNAFL-1	65
Eosin B	0.05

Data from Gatto and Milanick (1993).

the calcium pump, for example 8-hydroxypyrene-1,3,6-trisulphonic acid (HPTS). Figure 7 shows acid pH$_i$ transients caused by the counter-transport of H$^+$ on the calcium pump following depolarisation. HPTS was loaded into the neuron through the patch-pipette. Figure 7 also shows the inhibition of the calcium pump by eosin (see Table 1). HPTS provides a good signal-to-noise ratio when compared to BCECF, especially at acidic pH (less than about 7.2). The major drawback of HPTS is that, unlike BCECF, it is not available in a membrane-permeant form.

Contamination with ionophores

The nigericin technique is almost universally used to calibrate pH-sensitive dyes. However, it has a potentially very serious drawback. Richmond and Vaughan-Jones (1993) showed that when nigericin, a potassium–hydrogen exchanger, was used for pH$_i$ calibration it could subsequently contaminate cells used in the same apparatus, even several days later. This contamination occurred even though the bath and tubing had been thoroughly rinsed. The result of the contamination was a marked acid loading and profound pH$_i$ changes on altering external potassium. Wilding, Cheng and Roos (1992) had shown exactly these effects and concluded that rat carotid body possessed a native potassium–hydrogen exchanger. Richmond and Vaughan-Jones (1993), working on the same cell type, showed that washing the experimental apparatus for several days with 20% Decon removed the nigericin contamination. They concluded that carotid body type-1 cells showed no native potassium–hydrogen exchanger.

It is likely that many of the researchers using nigericin to calibrate pH-sensitive dyes contaminated their cells with some molecules of

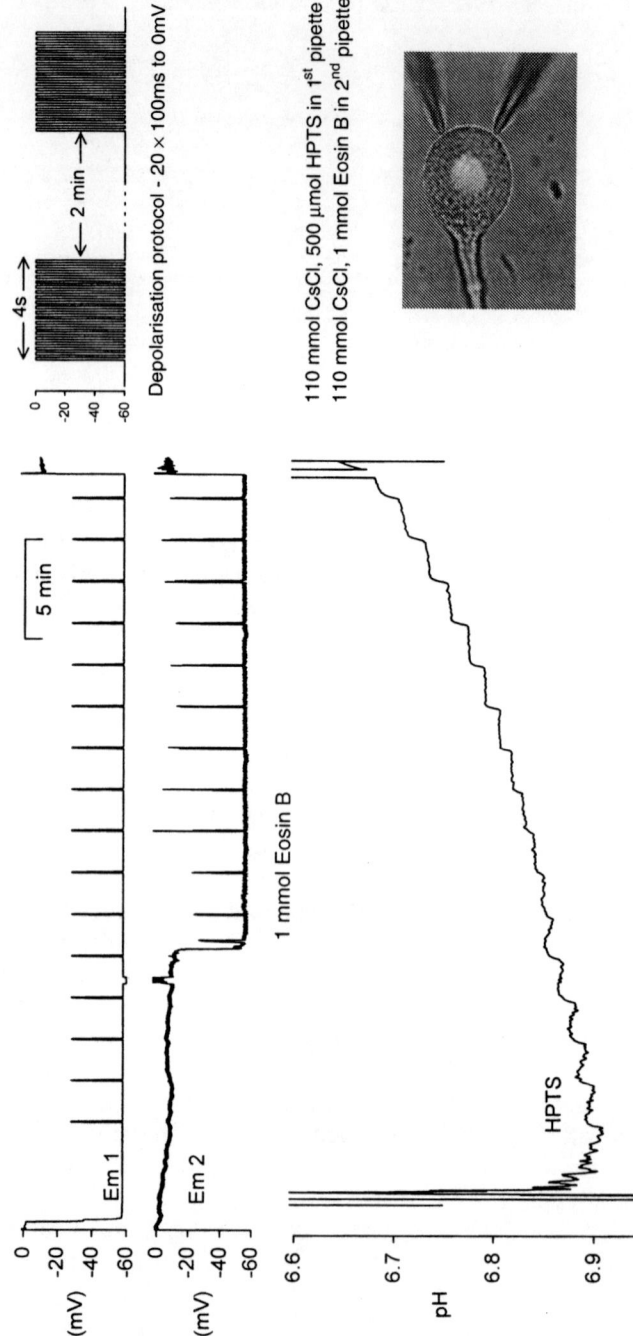

Fig. 7. pH_i measured in an isolated snail neuron ($\varnothing \sim 50\ \mu m$) using HPTS during depolarisation. The first patch-pipette (Em 1) was used to introduce HPTS. The second patch-pipette (Em 2) was used to introduce eosin B. Modified from Schwiening (1997).

exogenous potassium–hydrogen exchanger. There are alternative methods for calibrating pH_i. In cells too small to be impaled with a pH-sensitive microelectrode, the weak acid–base method (Szatkowski & Thomas, 1986; Eisner *et al.*, 1989) can be used. This technique involves the application of a weak acid and base in proportions that result in no pH_i change. The pH_i can be calculated from the proportions of the weak acid and base. This technique is cumbersome, and only results in a one-point calibration; however, it avoids any possibility of nigericin contamination. pH-sensitive dyes may also be calibrated *in vitro* but solutions may need to mimic the intracellular environment (e.g. high potassium, low calcium and include ATP and proteins).

Highs and lows of dyes and electrodes

Having presented an account of some serious problems associated with fluorescent pH-sensitive dyes it would be wrong to leave the reader with such negative feelings about them. Table 2 shows a comparison between dyes and electrodes. It can be seen that dyes are easy to load into cells, causing little damage, require minimal manual dexterity and can be used on commonly available inverted microscopes. Dyes can be used to monitor pH_i in several cells at once, even moving ones, and records are not complicated by electrical artefacts. If one can assess the problems (uncertain calibration, possible nigericin contamination, lack of membrane potential measurement, limited pH range and inhibition of transporters), then dyes remain the method of choice.

The disadvantages of pH-sensitive electrodes are clear in Table 2, the main one being damage. Most workers measuring pH_i in mammalian non-muscle cells would find it very difficult to make pH-sensitive microelectrodes small enough to use, and it is likely that dyes will remain the method of choice for them. However, pH-sensitive micro-electrodes are easy to calibrate, stable and can be used on electrophysiological rigs with minimal extra expense.

Recommendations for further reading

Although now slightly dated, Thomas (1978) can be recommended for more details on using microelectrodes to measure pH. Ammann (1986) contains much information on pH-sensitive ligands, whilst Purves (1981) discusses some of the wider issues relating to microelectrode use. *Microelectrode techniques: The Plymouth Workshop Handbook* (Ogden, 1994) also contains useful methodological nuggets of wisdom. Finally, this author always enjoys a trawl through the latest *Molecular*

Table 2. *Comparison of pH-sensitive microelectrodes and fluorophores*

	pH-sensitive microelectrodes	Fluorescent dyes
What is measured?	Nernstian for H^+ across a membrane	Buffering of H^+ causing a change in fluorescence
Where is pH_i measured?	Point source defined by electrode tip location	Diffuse volume dependent on optics and system
Can pH_i be measured at multiple locations simultaneously?	No	Yes, depending upon system
Cost of equipment	Low	High but many off-the-shelf products
Probe fabrication	Either difficult to make or temperamental	Available off the shelf
Reproducibility of probe characteristics	Low	High
Ease of use	Requires manual dexterity	Mostly trivial
Insertion of probe	Requires two barrels to be inserted into cell	By diffusion (AM loading across membrane or free acid through patch-pipette)
Damage caused by probe	Physical to membrane	Inhibition of transporters and possible toxicity of dye or by-products
Factors which disrupt signal	Electrical activity, fluctuations in temperature or CO_2	Various drugs, for example harmaline and caffeine
Response speed	Slow (~1 s)	Fast
Linearity of response	Very good over wide range	Limited linear range
Calibration technique	In vitro	Nigericin/high K^+ or weak acid and bases
Calibration accuracy	High	Relatively uncertain
Calibration problems	None	Nigericin contamination
Resolution	Good	Poor
Size of cells	$>\sim50$ μm	Any size
Need to visualise individual cells?	Yes, unless working on electrically tightly coupled cells	No
Can pH_i be measured during cell movement?	No, absolute requirement for stability	Yes, depending upon system

probes handbook (http://www.probes.com/) and the *Fluka selectophore* catalogue.

Acknowledgements

The author is grateful to Roger Thomas and Debbie Willoughby for their helpful comments. This work was supported by funding from The Wellcome Trust.

References

Ahmed, Z. & Connor, J.A. (1980). Intracellular pH changes induced by calcium influx during electrical activity in molluscan neurones. *Journal of General Physiology* **75**, 403–26.

Ammann, D. (1986). *Ion-selective microelectrodes*. Berlin: Springer-Verlag.

Ammann, D., Lanter, F., Steiner, R.A., Schulthess, P., Shijo, Y. & Simon, W. (1981). Neutral carrier based hydrogen ion selective microelectrode for extra- and intracellular studies. *Annals of Chemistry* **53**, 2267–9.

Bevensee, M.O., Schwiening, C.J. & Boron, W.F. (1995). Use of BCECF and propidium iodide to assess membrane integrity of acutely isolated CA1 neurons from rat hippocampus. *Journal of Neuroscience Methods* **58**, 61–75.

Buytendijk, F.J.J. & Woerdeman, M.W. (1927). Die physico-chemischen Erscheinungen waehrend der Entwicklung. I. Die Mesung der Wasserstoffionenkonzentration. *Wilhelm Roux Archiv fur Entwicklungsmechanik der Organismen* **112**, 387–410.

Caldwell, P.C. (1954). An investigation of the intracellular pH of crab muscle fibres by means of micro-glass and micro-tungsten electrodes. *Journal of Physiology – London* **128**, 169–80.

Caldwell, P.C. (1956). *Intracellular pH*, vol. V. New York: Academic Press.

Chaillet, J.R. & Boron, W.F. (1985). Intracellular calibration of a pH-sensitive dye in isolated, perfused salamander proximal tubules. *Journal of General Physiology* **86**(6), 765–94.

Crozier, W.J. (1918). On indicators in animal tissues. *Journal of Biological Chemistry* **35**, 455–60.

de Hemptinne, A. (1979). A double barrelled pH micro-electrode for intracellular use. *Journal of Physiology – London* **295**, 5P–6P.

De Vries, H. (1871). Sur la permeabilite du protoplasma des Betteraves rouges. *Archives Neerlandaises des Sciences Exactes et Naturelles* **6**, 118–26.

Deyhimi, F. & Coles, J.A. (1982). Rapid silylation of a glass surface: choice of reagent and effect of experimental parameters on hydrophobicity. *Helvetica Chimica Acta* **65**(6), 1752–9.

Eisner, D.A., Kenning, N.A., O'Neill, S.C., Pocock, G., Richards, C. D. & Valdeolmillos, M. (1989). A novel method for absolute calibration of intracellular pH indicators. *Pflugers Archiv – European Journal of Physiology* **413**, 553–8.

Fridericia, L.S. (1920). Exchange of chloride ions and of carbon dioxide between blood corpuscles and blood plasma. *Journal of Biological Chemistry* **42**, 245–57.

Gatto, C. & Milanick, M.A. (1993). Inhibition of the red blood cell calcium pump by eosin and other fluorescein analogs. *American Journal of Physiology* **264(33)**, c1577–86.

Hays, S.R. & Alpern, R. J. (1991). Inhibition of Na^+-independent H^+ pump by Na^+-induced changes in cell Ca^{2+}. *Journal of General Physiology* **98(4)**, 791–814.

Hinke, J.A.M. (1967). Cation selective microelectrodes for intracellular use. In: *Glass electrodes for hydrogen and other cations. Principles and practice*, ed. G. Eisenman, p. 464. New York: Marcell Dekker.

Khuri, R.N., Bogharian, K.K. & Agulian, S.K. (1974). Intracellular bicarbonate in single skeletal muscle fibres. *Pflugers Archiv – European Journal of Physiology* **349**, 285–99.

Matsumura, Y.S., Aoki, K., Jajino, K. & Fujimoto, M. (1980). The double-barrelled microelectrode for the measurement of intracellular pH, using liquid ion-exchanger, and its biological application. *Proceedings of the International Congress of Physiological Sciences* **14**, 572.

Metchnikoff, E. (1893). *Lectures on the comparative pathology of inflammation*. London: Trubner & Co.

Michaelis, L. & Davidoff, W. (1912). Methodisches und Sachliches zur elektrometrischen Bestimmung der Blut-Alkalescenz. *Biochemische Zeitschrift* **46**, 131–50.

Ogden, D. (1994). Microelectrode techniques: The Plymouth Workshop handbook, 2nd edn. Cambridge: The Company of Biologists Ltd.

Purves, R.D. (1981). *Microelectrode methods for intracellular recording and iontophoresis*. London: Academic Press.

Richmond, P. & Vaughan-Jones, R.D. (1993). K^+–H^+ exchange in isolated carotid body type-1 cells of the neonatal rat is caused by nigericin contamination. *Journal of Physiology – London* **467**, 277P.

Roos, A. & Boron, W.F. (1981). Intracellular pH. *Physiology Review* **61**, 296–434.

Schwiening, C.J. (1997). The effects of intracellular eosin B on depolarization-induced pH changes in isolated snail neurones. *Journal of Physiology – London* Dublin Meeting, 159P.

Schwiening, C.J. & Boron, W.F. (1992). Fractional rate of dye signal decrease as an aid in assessing membrane-permeability. *Journal of Physiology – London* **452**, 187.

Schwiening, C.J., Kennedy, H.J. & Thomas, R.C. (1993). Calcium–hydrogen exchange by the plasma membrane Ca-ATPase of voltage-clamped snail neurons. *Proceedings of the Royal Society London B* **253**, 285–9.

Szatkowski, M. S. & Thomas, R. C. (1986). New method for calculating pH from accurately measured changes in pH induced by a weak acid–base. *Pflugers Archiv – European Journal of Physiology* **407**, 59–63.

Taylor, C.V. & Whitaker, D.M. (1927). Potentiometric determinations in the protoplasm and cell-sap of nitella. *Protoplasm* **3**, 1–6.

Thomas, R.C. (1974). Intracellular pH of snail neurones measured with a new pH-sensitive glass micro-electrode. *Journal of Physiology – London* **238**, 159–80.

Thomas, R.C. (1978). *Ion-sensitive intracellular microelectrodes: how to make and use them.* London: Academic Press.

Waddell, W.J. & Butler, T.C. (1959). Calculation of intracellular pH from the distribution of 5,5-dimethyl-2,4-oxazolidinedione (DMO). Application to skeletal muscle of the dog. *Journal of Clinical Investigation* **38**, 720–9.

Wilding, T.J., Cheng, B. & Roos, A. (1992). pH regulation in adult rat carotid body glomus cells. Importance of extracellular pH, sodium, and potassium. *Journal of General Physiology* **100(4)**, 593–608.

H.H. FELLE

pH-sensitive microelectrodes: how to use them in plant cells

Introduction

pH-sensitive microelectrodes of various designs have been used in plant physiology for over two decades. Davis (1976), using an antimony electrode, presented cytosolic and vacuolar data following light/dark changes in *Phaeoceros laevis*, and Bowling (1976), using the same electrode type, reported vacuolar pH profiles across the root cortex of *Helianthus annuus*. Due to some fundamental problems with this electrode, it did not prove very successful and is not in use nowadays. The recessed-tip Thomas-type glass–glass electrode proved to be a far more reliable tool. Using this electrode, Sanders and Slayman (1982) carried out an important study on pH regulation in *Neurospora crassa*, in which they demonstrated the predominant role of oxidative phosphorylation for cytosolic pH control. Unfortunately, however, the manufacture of this electrode is difficult and it has subsequently been rarely used in plant physiology.

In contrast, neutral carrier-based liquid membrane electrodes (Ammann, 1986) have gained wider acceptance and they have been used successfully for various problems of cell physiology. Although in principle these electrodes can be used to measure the free concentrations of a whole variety of ions, the number of laboratories that use this technique has remained comparatively small. The main reason for this may be that intracellular applications are technically difficult, requiring some electrophysiological background, training and patience (Felle, 1992).

This chapter outlines some principles of fabrication and possible applications of the pH microelectrode in order to clarify the technique for those who have little or no experience with it but who may want to measure pH (or free concentrations of other ions) in plant cells. Demonstration of the possibilities this technique offers may help the reader to decide which of the currently available techniques – dimethyloxazolidine-dione (DMO), nuclear magnetic resonance (NMR), dyes or microelectrodes – is the most appropriate for solving a specific problem. For reasons of clarity, this chapter puts little emphasis on profound physical properties of ion-selective electrodes, for which the reader is referred to Ammann (1986). In order to demonstrate the usefulness of this tool, a variety of applications to different physiological problems will be summarised. This chapter is not a review, and therefore the author offers apologies to all those whose valuable work is not mentioned.

Technical considerations and experiences

How to build and prepare a pH microelectrode

There are probably as many different ways to build ion-selective micro-electrodes as there are laboratories that use them. One successful way is the following (Felle & Bertl, 1986a):

1. pull the electrode from borosilicate glass tubing (a solid filament inside the tubing helps filling);
2. silanise (make hydrophobic) the inner surface of the glass tubing to hold resin;
3. backfill with resin;
4. fill the remainder of the electrode with reference solution;
5. connect the electrode to a high-impedance amplifier and equilibrate in test solution until drift is minimal;
6. calibrate and measure.

The silanisation is a critical step! Some laboratories do this by placing the electrodes in a silane vapour under heat which will coat the entire electrode. As the internal as well as the external surfaces are now hydrophobic, there is a possibility that some resin will creep out of the tip, move along the surface, and thus may cause an electrical leak while measuring intracellularly. Therefore, wet internal silanisation is preferable: after heating the electrode in an oven at about 200 °C, the blunt end of the electrode is dipped into the silane solution (0.02% silane in chloroform). Due to the solid filament inside the electrode, the tip rapidly fills and the chloroform evaporates leaving the silane on the inner glass surface only. To ensure proper salinisation, this procedure is repeated after approximately one hour. In order to form a covalent bond between the silane and the glass, the electrodes are heated for about two hours at approximately 200 °C. The silane concentration is critical: too little will result in displacement of the resin by the buffer; too much will plug the tip. There is a choice of several commercially available silane solutions which differ with respect to the type and amount of their hydrophobic groups. For example, dimethyldichlorosilane (DDS) will make the glass surface less hydrophobic than tributylchlorosilane (TBS).

After cooling, the silanised electrodes are back-filled with the resin mixture (see below) under microscopic control using a long (glass) pipette. While the resin moves along the filament to fill the tip, air bubbles form; in general, these will vanish spontaneously within 20 to 30 minutes. At this stage the electrodes can be stored for later use, preferably somewhere that is cool, dark and dry. Before use, the reference solution (pH buffer, for example 100 mM Mes/Tris plus 0.5 mM KCl) is filled in. Prior to calibration, it is advisable to equilibrate the electrode under constant medium flow for at least one hour. Because tip damage during impalement may change the calibration properties, an intracellular functional test should be carried out, for example by testing the effect of a weak acid (see below).

There are some basic properties of ion-selective microelectrodes the potential user has to get used to.

> pH is measured as voltage, which is calibrated and transformed to pH units.
>
> The pH electrode measures not only pH, but the sum of all voltages occurring within the circuit. Under regular measuring conditions, the two main voltages to be considered will be the membrane potential and the voltage arising from the pH difference across the liquid ion-exchanger membrane. This problem is solved by inserting a second (voltage-) electrode into the same cellular compartment (cytoplasm or vacuole). Both electrodes are connected to a differential amplifier. The channel to which the pH electrode is connected has to have a high-impedance input ($R_{in} = 10^{15}$ Ω) to avoid drifts. The two signals coming from the cell are amplified and the membrane potential is simultaneously subtracted from the pH electrode signal to give the net pH. With single cells, this can be achieved by inserting either two separate electrodes or one double-barrelled electrode. In tissues, however, only double-barrelled electrodes apply (Felle, 1986).
>
> pH microelectrodes, as all ion-selective microelectrodes, respond to voltage changes considerably more slowly (about 90 per cent in 2–10 seconds) than conventional voltage electrodes. Therefore, a fast voltage change across the membrane may cause an artefactual deflection on the difference trace, which, however, is not a pH change. During this time, no

information on pH is available, a problem which is either accepted (information on pH may not be essential) or is circumvented by choosing other experimental conditions. In any case, this electrode property can be used to test whether the electrode is still operating in a proper manner: a fast-responding pH electrode is just picking up voltage, and not the pH! Under optimal conditions (good shielding and grounding), pH changes of 0.01 to 0.001 pH units can be recorded reliably.

Intracellular applications of the pH microelectrode

The electrode just described is ready for intracellular use in most animal cells. What makes the difference between a pH measurement in the cytosol of, say, a mammalian cell and a plant cell? The typical plant cell has a large vacuole, a tough cell wall, and is turgescent. These properties can make pH measurements within a plant cell rather demanding: the turgor because it presses the resin back into the shank of the electrode; the vacuole because the tip of the electrode may penetrate the tonoplast or measure in a tonoplast pocket; the cell wall because it may break the tip of the electrode and change its precalibration properties.

The problem of the location of the sensitive tip (vacuole or cytosol) can be dealt with by simply observing the electrode signal. Since little or no voltage difference exists across the tonoplast (Bethmann *et al.*, 1995), and the vacuole is an acidic compartment, the signal coming from a pH electrode placed within the vacuole will be roughly 59 mV per pH unit less negative than the signal coming from an electrode measuring in the cytosol. Therefore, the actual localisation of the sensitive tip will be revealed upon a clean impalement right away. In contrast to the widespread opinion that microelectrodes always measure in the vacuole, experience teaches that the sensitive tip mostly measures in the cytosol (Felle & Bertl, 1986a). It is possible, however, that laboratories that use other cell types may have different experiences.

The toughness of the cell wall has to be dealt with through experimenting with a simple voltage electrode to find a non-bending and non-breaking tip or by bevelling to produce a wider orifice and a sharper tip.

The largest problem by far is the cell turgor, which under typical experimental conditions will be 5 to 10 bar. Upon impalement, this pressure will inevitably push the resin into the shank of the electrode and terminate the measurement sooner or later. There are several possibilities for dealing with this problem:

partly plug the tip with polyvinylchloride (Felle & Bertl, 1986a);

decrease the cell turgor (Steigner *et al.*, 1988);

mix the resin with polyvinylchloride (Tsien & Rink, 1985) or another polymer (Reid & Smith, 1988) to produce a firm gel in the tip that can withstand the pressure difference;

apply a pressure from the rear of the electrode (Felle *et al.*, 1996).

All of these methods have been applied successfully at different laboratories.

Finally, three advantages of the pH electrode that are not associated with other techniques should be mentioned.

1. All recordings are continuous, which means there is information on pH from the first to the last moment of the experiment. This is important for all tests concerning regulatory or signal transduction processes.
2. Membrane potential control: due to the requirement that a voltage reference be placed within the same compartment of a cell, a voltage control runs at all times.
3. Recorded data need no further interpretation as they are intrinsically final signals from the cell in time and value.

Application of the pH microelectrode to physiological problems

This section presents applications of the pH microelectrode to some physiological problems. Naturally, not every physiological problem can be investigated in the same plant, tissue or cell. For that reason the technique has to be adapted to the material under investigation, which can mean one has to pull a tougher or sharper tip, manufacture a double-barrelled instead of single-barrelled electrode or use a different resin. Because there is only limited space, the treatment of the separate problems will have to be short and concentrate on the most essential aspect. For a more profound discussion, the reader is referred to the original literature cited.

Light/dark responses of green plant cells

A truly non-invasive pH test is the switch from light to dark. As Figure 1(a) shows, the green thallus cells of the aquatic liverwort *Riccia fluitans* respond to 'light off' with an immediate pH decrease of approximately 0.3 units, a change that is reversed rapidly. 'Light on' yields the opposite response (Felle & Bertl, 1986b). Similar changes were meas-

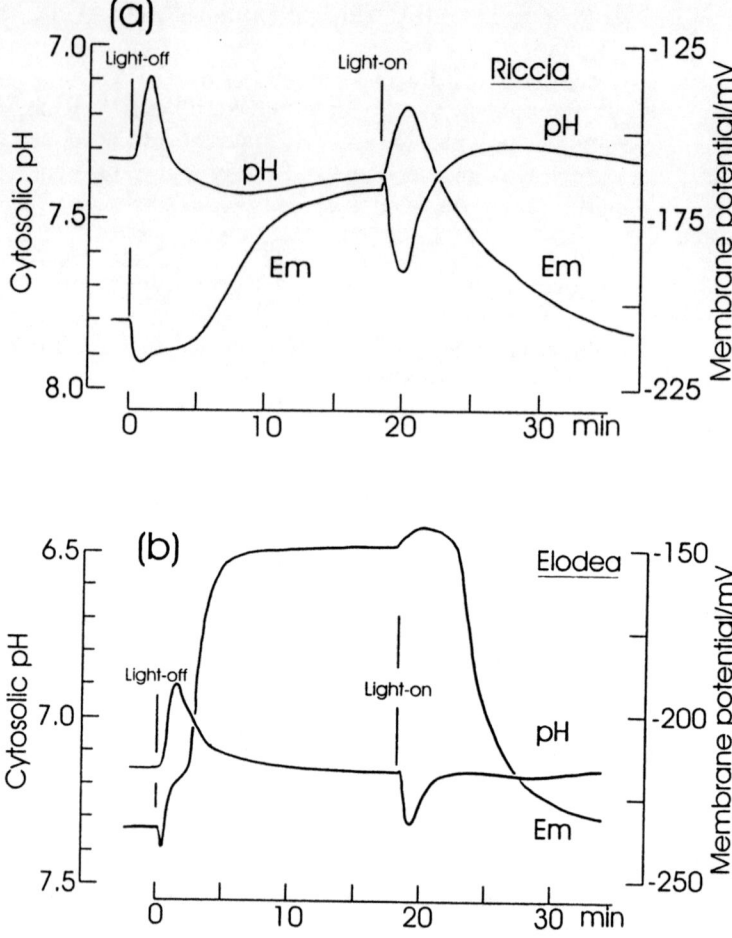

Fig. 1. Light/dark effects. Cytosolic pH (pH) and membrane potential (Em) of (**a**) green *Riccia fluitans* thallus cells and (**b**) *Elodea densa* leaf cells, responding to 'light off' with a transient acidification and to 'light on' with a transient alkalinisation. Measured with double-barrelled microelectrodes. Copied from original recordings.

ured in the green alga *Eremosphera viridis* (Thaler, Simonis & Schönknecht, 1989) and in higher plants like *Kalanchoe, Lemna* (Felle & Bertl, 1986a) and *Elodea* (Fig. 1b). Because the pH responses to light/dark only occur in green cells, are abolished by 5,5-dimethyloxazolidine-2,4,-dione and apparently do not depend on

CO_2, it was concluded that these pH changes originate from proton fluxes across the thylakoid membranes (Hansen *et al.*, 1993). The cells respond with an initial hyperpolarisation to the 'light-off'-induced acidification and with a depolarisation to the 'light-on'-induced alkalisation, both of which can be explained in terms of the response of the plasma membrane H^+ ATPase to the concentration changes of its transport substrate, the proton (see below). However, the regulatory background of the pH recovery is not yet solved.

Acid–base interactions

A rather convenient way to manipulate cytosolic pH is by the application of weak acids or bases (Roos & Boron, 1981; Sanders & Slayman, 1982; Frachisse, Johannes & Felle, 1988). This approach can be used to determine the cytosolic buffer capacity, which is the ratio of weak acid anions accumulated in the cytosol ($= \Delta[H^+]$) to the measured cytosolic pH change. The amount of weak acid anions accumulated can be *calculated* with the Henderson–Hasselbalch equation according to the transmembrane ΔpH and pK_a of the weak acid, whereas the cytosolic pH and the acid-induced pH changes are *measured* (Sanders & Slayman, 1982). Typical cytosolic buffer capacities are in the range of 30–60 mmol H^+/pH unit (Sanders & Slayman, 1982; Felle, 1986; Guern *et al.*, 1991). For example in Figure 2(a), the apparent buffer capacity, revealed by acetic acid and procaine, would be around 40–45 mmol H^+/pH.

The cytosolic pH kinetics measured following the external addition of weak acids (or bases) reveal an interesting aspect of intracellular pH (pH_i) regulation. According to the external pH (pH_e), the original cytosolic pH, and the pK_a of the weak acid added, a monophasic cytosolic acidification without recovery should be observed. Such a 'pH clamp' would be imposed to the cells because any removal of protons by membrane transport would be countered by import of more protonated acid and its immediate dissociation. Experience teaches, however, that the weak acid-induced pH change is in fact partly reversed (Fig. 2a; Guern *et al.*, 1991), which clearly indicates metabolic control, and very strong mechanisms compensating for the proton load keeping the distribution of the unprotonated form of acid across the plasma membrane out of equilibrium.

A major drawback of the use of weak acids to modify the cytosolic pH is due to their lipophilic solubility, which can alter the properties of the plasma membrane as well as of other intracellular membranes. In fact, it has been demonstrated that fatty acids become more and more

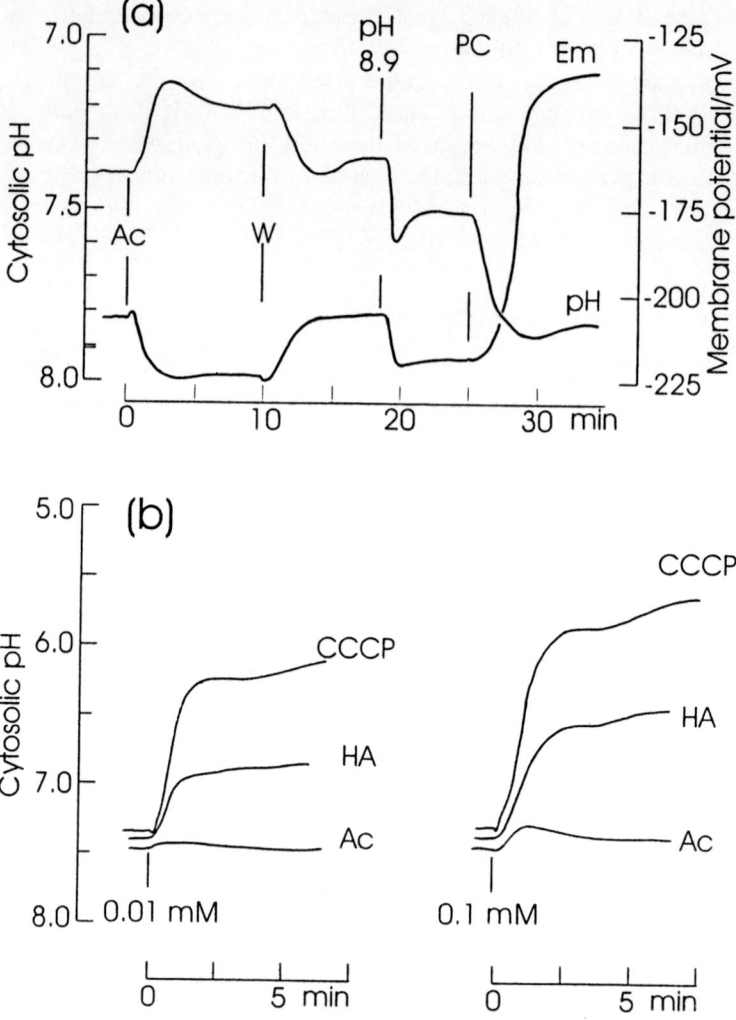

Fig. 2. Weak acid–base effects. (**a**) Action of acetic acid and procaine on the cytosolic pH (pH) and the membrane potential (Em) of *Riccia fluitans* rhizoid cells. 1 mM acetic acid (Ac) was added at pH_e of 5.6. W = removal of acetic acid. 1 mM procaine (PC) was added after changing the pH_e to 8.9. (**b**) Comparison of the effect of weak acids with different membrane solubilities on cytosolic pH of *Riccia fluitans* rhizoid cells at the indicated concentrations. Ac = acetic acid, HA = heptanoic acid, CCCP = carbonyl cyanide-m-chlorophenyl-hydrazone. pH_e = 5.6. Copied from original recordings.

effective in changing cytosolic pH and membrane potential as the anion gets increasingly membrane permeant (Frachisse *et al.*, 1988). Uncouplers of oxidative phosphorylation – 2,4-dinitrophenol (DNP), carbonyl cyanide m-chlorophenyl hydrazone (CCCP) etc. – are also weak acids, which, due to their ability to cross the membranes in both their protonated and unprotonated forms, should be much more effective in perturbing cytosolic pH than those weak acids that cross a membrane mainly in their protonated form. In principle, this is demonstrated in Figure 2b by comparing the effects of equal amounts of acetic acid, heptanoic acid and CCCP on cytosolic pH: 0.1 mM acetic acid hardly affects cytosolic pH, whereas 0.01 mM CCCP almost completely equalises the cytosolic pH with the external pH of 5.6 within minutes.

Apart from being an important tool for the investigation of pH_i regulation, the action of weak acids on cellular compartments has important physiological implications. Organic acids, produced and consumed by the cellular compartments, contribute considerably to a stable pH in the cytosol or in organelles in that they distribute across the membranes of the organelles according to their dissociation properties (different pK_a). Changing their concentration will lead to an altered buffer capacity. According to the 'biochemical pH stat' of Davies (1986), balancing the production and consumption of malic acid is a key process in pH regulation in plants, whereby, due to its pH sensitivity in the range 6.8 to 8.0, the phosphoenol pyruvate (PEP) carboxylase is the critical enzyme. This view, however, has remained controversial.

Proton cotransport

Protons, driving the transport of their cosubstrate, should change the pH on either side of the membrane, i.e. alkalise the external medium and acidify the cytoplasm. The finding that sugars indeed alkalise the external medium of *Neurospora* (Slayman & Slayman, 1974) or *Chlorella* (Komor & Tanner, 1976) apparently proves the above assumption. The cytosolic pH, however, does not necessarily follow this prediction in all cases. Thus, transport of substrates like sugars or amino acids leads to a clear-cut alkalinisation of roughly 0.2 pH units (Fig. 3(a); Johannes & Felle, 1987). Because the external pH also alkalises, plasma membrane proton transport cannot account for this effect, and therefore metabolic pH shifts must be taken into consideration. One could argue that the pH_e changes have been measured in a weakly buffered solution, which means the amount of protons leaving the medium would have barely shifted the well-buffered cytosol. But, basically, this finding remains an unsolved problem, because 3-oxymethylglucose or amino-

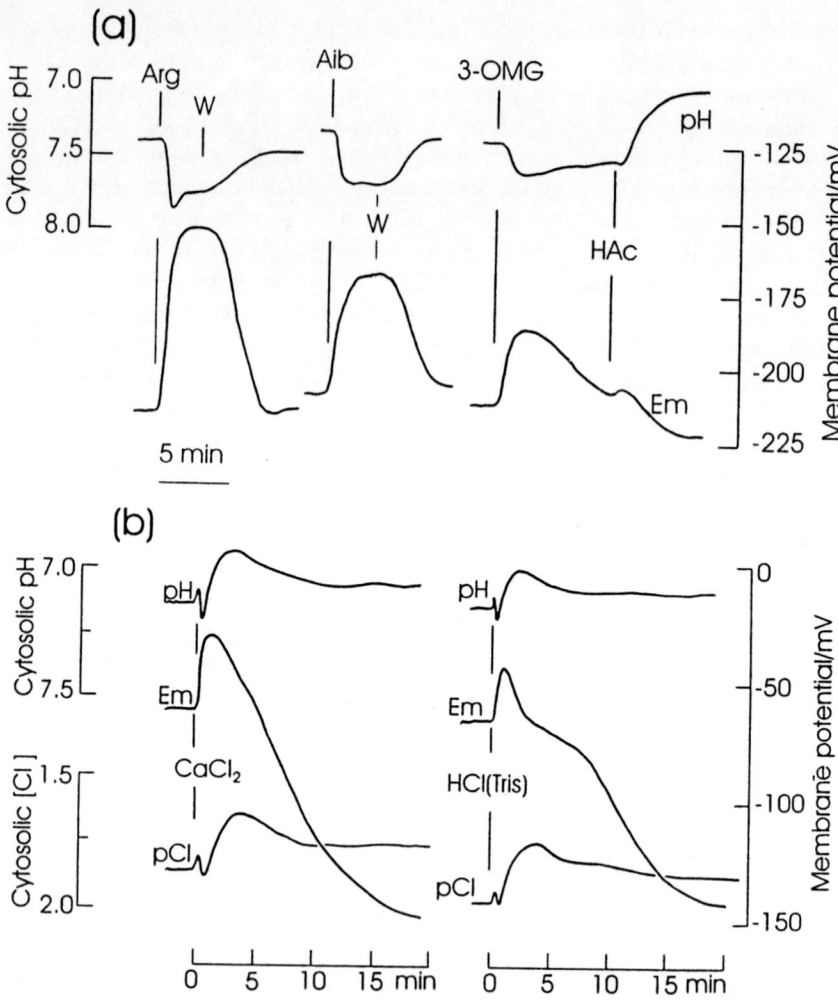

Fig. 3. Proton cotransport. (a) Cytosolic pH (pH) and membrane potential (Em) of *Riccia fluitans* rhizoid cells in response to external additions of arginine (Arg) or amino-isobutyric acid (Aib), and 3-oxy-methylglucose (3-OMG), representative of amino acids and hexoses, respectively. W = removal of the substrate. At the end of the experiment, acetic acid (HAc) was tested as internal functional control. (b) Cytosolic pH (pH), cytosolic [Cl⁻] (pCl) and membrane potential (Em) of *Sinapis alba* root hair cells before and after the addition of 10 mM CaCl₂ or 10 mM HCl (+Tris) to the external medium. pH$_e$ = 4.5. Copied from original recordings.

isobutyric acid (supposedly non-metabolisable substrates of these transporters) also has the same alkalising effect. One has to assume that, although these compounds are not metabolised as their fellow substrates, they nevertheless enter parts of the metabolic cycles and thus shift equilibria. Ullrich and Novacky (1990) demonstrated that import of NO_3^- also leads to cytosolic alkalinisation (interpreted with rapid reduction), but that import of Cl^- acidified the cytosol of *Limnobium* root hairs, a finding which was confirmed in *Sinapis* root hairs (Fig. 3(b); Felle, 1994).

The proton motive force electrode

A pH microelectrode measures the sum of both membrane potential and pH difference. Provided that the pH_e is kept constant throughout the test, a pH microelectrode can be used to measure the proton motive force as a whole across a membrane directly and continuously (Frachisse *et al.*, 1988; Felle, 1994). This is a rather elegant way to record in one go the development of a thermodynamic quantity, which is important for the transport of proton-driven substrates (cotransport). Remis, Bulychev and Kurella (1987) have used this approach to determine the proton motive force in chloroplasts.

Signal transduction

Although the maintenance of a constant pH is essential for the metabolic network, protons must be regarded as an important *cellular messenger* (Felle, 1989; Kurkdjian & Guern, 1989; Blatt, 1992). Because protons cannot be excluded from their intracellular milieu, their activity must be regulated. Besides being both the substrate and the product of metabolic pathways, protons communicate information about the cellular energy balance to enzymes and structures that may share no other common effector. Because even small shifts in pH_i may cause large effects on the activity of enzymes, and on membrane transporters, localised pH changes are of great importance. As such, the perception of a signal may not necessarily be followed by a general cytosolic pH change, but by local variations which may only be picked up by fluorescent dyes with a proper spatial resolution. In spite of this restriction, using pH-sensitive microelectrodes, spatial differences in pH in rhizoid cells of *Pelvetia* (Gibbon & Kropf, 1994) and signal-induced pH changes have been measured (Herrmann & Felle, 1995), thus boosting the idea of protons acting as cellular messengers in plants.

Because phytohormones may alter the growth pattern of a plant organ, changes in metabolism that go along with cytosolic pH shifts are likely. Indeed, maize coleoptiles respond to auxins with a cytosolic acidification accompanied by a hyperpolarisation and an external acidification (Felle *et al.*, 1986; Peters & Felle, 1991). Several different approaches have provided functional evidence that it is the plasma membrane H^+ ATPase that causes this hyperpolarisation (Barbier-Brygoo *et al.*, 1989; Rück *et al.*, 1993). Although the origin of the acidification is unknown, the pH oscillations with similar periods indicate that the acidification could be one factor in stimulating the pump, which in turn then acidifies the external space (Fig. 4(a); Felle, 1988a; Peters & Felle, 1991). In this context, it was interesting to observe that cytosolic Ca^{2+} also oscillated, which led to the assumption of an interaction of Ca^{2+} and pH (Felle, 1988b).

The responses of *Sinapis alba* root hair cells to auxins are different. Indolyl-3-acetic acid (IAA) causes a sharp depolarisation which is accompanied by a cytosolic alkalinisation of about 0.3 pH (Fig. 4(b); Tretyn, Wagner & Felle, 1991). Although the changes in membrane potential and pH appear closely related, the alkalinisation is not likely to be the cause of the depolarisation. Presumably, this is caused by a deactivation of the pump or by the activation of an anion channel which releases Cl^- (Hedrich, 1994). Due to their steep outwardly directed electrochemical gradient, Cl^--charge rapidly would leave the cell and thus depolarise the membrane.

Changes in cytosolic pH in response to external signals are not uncommon. While oligopeptide elicitors (inducing plant defence reactions) apparently acidify the cytoplasm in parsley cells (Nürnberger *et al.*, 1994), chitolipo-oligosaccharides (Nod factors), initiating symbiotic responses in legumes, alkalise the cytoplasm of their target cell (Felle *et al.*, 1996). Moreover, K^+ channels are apparently mediated by cytosolic pH. Steigner *et al.* (1988) report that in the green alga *Eremosphera viridis* the acidification of the cytosol induces an 'action potential-like response', corresponding to an increase in the conductivity of a K^+ channel. Blatt and Armstrong (1993) report that micromolar concentrations of abscisic acid (ABA) alkalise the cytosolic pH, while the outward-rectifying K^+ channel current rises. The data are supported by the observation that acid loads, imposed with external butyrate, abolish the ABA-evoked rise in this K^+ current.

pH regulation: the role of the H^+ ATPase

Since pH electrodes measure continuously, they are a useful tool for investigating various problems of pH regulation. One of the most controversially discussed issues is the role of the plasma membrane

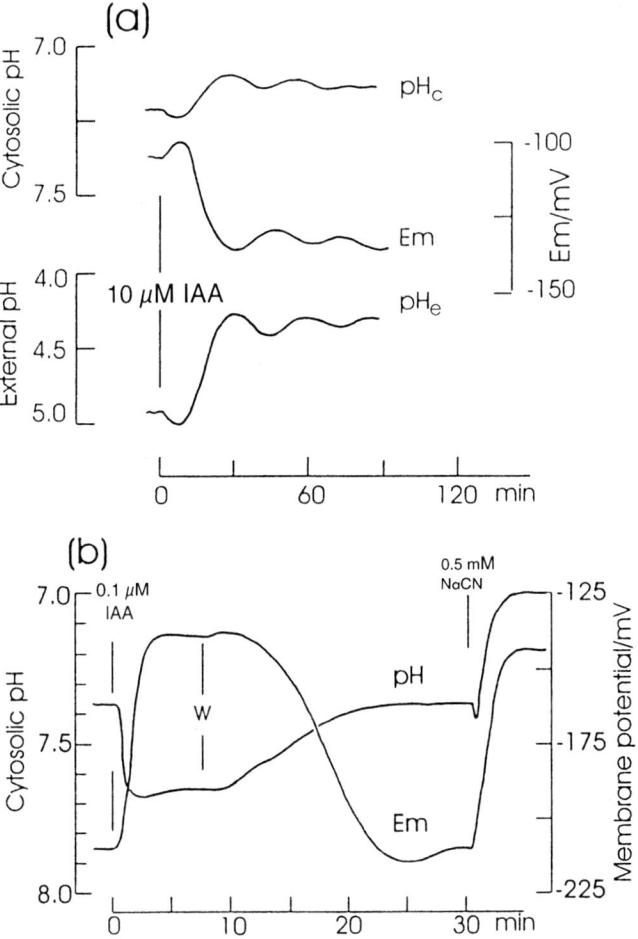

Fig. 4. Auxin effects. (**a**) Effect of 10 μM indole acetic acid (IAA) on cytosolic pH (pHc), membrane potential (Em), and external pH (pH$_e$) of *Zea mays* coleoptiles. (**b**) Effect of 0.1 μM IAA on cytosolic pH (pH) and membrane potential (Em) of *Sinapis alba* root hair cells. At the end of the experiment, a cyanide (NaCN) test shows that a depolarisation does not necessarily cause an alkalinisation. pH$_e$ = 6.0. W = removal of IAA. Copied from original recordings.

proton pump in cytosolic pH regulation (Smith & Raven, 1979; Sanders & Slayman, 1982; Felle, 1988c; 1991). This has led to a polarisation amongst the laboratories, some of which favour the idea of the so-called 'biochemical pH stat' (Davies, 1986), while others prefer cytosolic pH to be regulated through the 'biophysical pH stat'. Whereas membrane transport is the basis for the latter, the biochemical pH stat highlights metabolic processes.

All protons produced in excess by metabolism have to be exported from the cytosol sooner or later. While mammalian cells, with no H^+ ATPase, use the H^+/Na^+-antiporter, which is energised by the $Na^+(K^+)$ATPase, plants and fungi use the H^+ ATPase to get rid of excess H^+. There is no doubt as to the long-term role of the H^+ ATPase for pH regulation (Raven, 1985), but considerable confusion exists in the literature regarding its short-term role. The observation that the H^+ ATPase responds to cytosolic pH changes and thus reverses pH perturbations is trivial, because H^+ is a transport substrate, forcing the enzyme to respond to an acidification with proton extrusion which is usually associated with a hyperpolarisation (Figs. 1, 2a, 3b, 4), unless it is controlled to do otherwise. Therefore, the important question is not whether the H^+ pump can restore a perturbed cytosolic pH, but whether the pH can be shifted by changes in pump activity. That is, will a stimulation of the H^+ pump (not by protons) lead to an increase and a deactivation to a decrease in cytosolic pH? The answer to this can only be obtained through proper experiments.

> *Stimulation of the proton pump.* The fungal toxin fusicoccin has been proven to be a powerful tool to stimulate the plasma membrane H^+ ATPase of plant cells (Marrè, 1979), an effect that is usually recorded as hyperpolarisation or as increase in membrane current (proton extrusion). Using the DMO technique, which is based on the pH-dependent transmembrane distribution of weak acids, it has been demonstrated in *Elodea densa* leaves that the cytosolic pH indeed increased following treatment with fusicoccin, an observation that would support the notion that the pump does in fact regulate cytosolic pH (Marrè, Beffagna & Romani, 1988; Kurkdjian & Guern, 1989). However, experiments in other laboratories shed a different light on this problem. Using nuclear magnetic resonance (NMR) spectroscopy on cells of pea internodes, Talbott, Ray and Roberts (1988) did not find significant pH changes in response to fusicoccin. Using pH-sensitive microelectrodes, Ullrich and Novacky (1990) demonstrated in root

hairs of *Limnobium stoloniferum* that fusicoccin may increase or decrease cytosolic pH or may not have any effect at all, depending on the respective experimental conditions; i.e. on the fluxes of the so-called stronger ions (Stewart, 1983). In Figure 5a it is demonstrated that 2 µM fusicoccin clearly hyperpolarises root hairs of *Medicago sativa*, but does not alkalise the cytosol; occasionally even slight acidifications are observed.

Inhibition or deactivation of the proton pump. In a microelectrode study, Sanders and Slayman (1982) reported that vanadate (an inhibitor of P-type ATPases) inhibited the pump (depolarisation), but did not acidify the cytosol. On the other hand, cyanide, which blocks oxidative phosphorylation and hence deactivates the pump, acidified the cytosol by roughly half a pH unit, an effect which was interpreted as metabolic. This effect has been confirmed since by other laboratories (see also Fig. 4b), and in fact appears to be typical for the switch from aerobic to anaerobic metabolism (Fig. 5b), as similarly demonstrated with the NMR technique (Fox, McCallan & Ratcliffe, 1995; Ratcliffe, 1997). Figure 5b shows that low concentrations of oligomycin (or antimycin A, not shown) separate the pH shift from the inhibition of the oxidative phosphorylation, leaving the activity of the H^+ ATPase almost unaffected. When under these conditions the H^+ ATPase is deactivated, for example with cyanide or with vanadate (Bowman & Slayman, 1979), there is no further acidification (Felle, 1995).

Thus, both sets of experiments in which the activity of the H^+ ATPase was altered indicate that the proton pump is apparently *not* able to shift the control pH. It must remain open, however, whether the difference in results is based on the use of different techniques or conditions (which would be less desirable) or due to the use of different plants or cells which live in different biospheres and thus may follow different strategies to control their cytosolic pH. (For an extended treatment of this problem see Ratcliffe, this volume.)

Vacuolar pH

Vacuolar pH is not measured routinely. As stated above, one reason is that location of the electrode tip within the vacuole is a rare event and a vacuolar measurement cannot usually be chosen at will. Also, since test substances must first pass through the plasma membrane and the

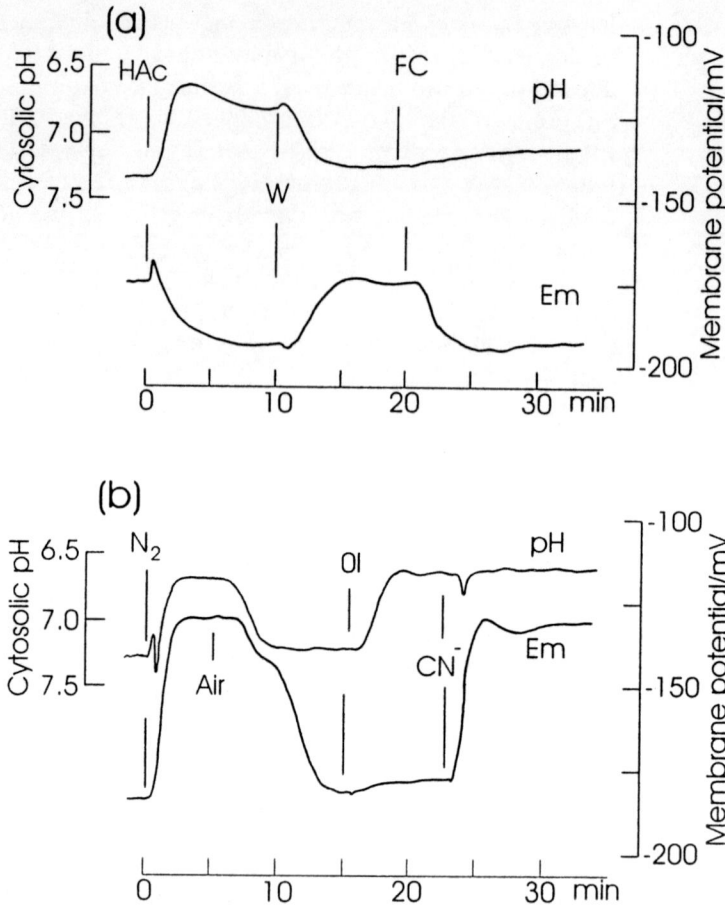

Fig. 5. Proton pump activity and cytosolic pH. (a) The effect of the fungal toxin fusicoccin (FC) on cytosolic pH (pH) and membrane potential (Em) of *Sinapis alba* root hairs. Prior to the FC experiment, acetic acid (HAc) was given as an internal functional test of the electrode. $pH_e = 6.0$. (b) Anoxia: cytosolic pH (pH) and membrane potential (Em) of *Medicago sativa* root hair cells respond to withdrawal of O_2 through flushing with N_2 with acidification and depolarisation. 2 µM oligomycin (Ol) acidifies the cytosol, but only slightly depolarises the cells. 1 mM NaCN (CN^-), added in the presence of oligomycin, depolarises but does not change cytosolic pH. Copied from original recordings.

cytosol before they reach the tonoplast, interpretations from such *in vivo* tests often leave many questions unanswered. However, there have been some successful impalements which at least give some information as to the absolute vacuolar pH and impressively demonstrate the variability of plant vacuoles with respect to their acidity: rhizoid cells of *Riccia fluitans*, pH 4.8 (Bertl, Felle & Bentrup, 1984); root hairs of *Sinapis alba*, pH 5.6 (Felle, 1986); cotyledonal mesophyll cells of radish, pH 6.0 (Strack, Sharma & Felle, 1987); lime fruits, pH 2.0 (Echeverria, Burns & Felle, 1992); *Eremosphera*, pH 5.0 (Bethmann *et al.*, 1995). Attempts to insert pH electrodes into isolated vacuoles have been of limited success, however (Kurkdjian & Barbier-Brygoo, 1983).

Extracellular applications of the pH microelectrode

The pH of the apoplast

Because various transporters translocate protons in and out of the cell, the extracellular environment, i.e. the ionic milieu of the apoplast, has to be known in order to understand the cytosolic processes, especially pH regulation. Whereas with single cells this can be controlled to some extent, the apoplast of tissues and organs is an area that is not so easily accessible. However, the use of pH microelectrodes offers an opportunity for measuring the apoplast pH continuously. Because the apoplast is a rather small space, certain procedures are required to get the sensitive tip into place (Fig. 6).

1. A double-barrelled microelectrode (pH barrel and voltage reference) is pushed into the tissue. Provided the impalement is carried out with care, the first measurement will be an intracellular one.
2. Pushing the electrode further will result in loss of this recording. In most cases, the electrode tip will slip into the next cell and show another intracellular recording.
3. In case the electrode tip is located between cells (apoplast), the signal from the electrode differs from intracellular recordings and indicates the pH of the apoplast. On average, two out of ten attempts will result in a successful recording.

The pH of the unstirred layer

One can show that the pH on the surface of a root is quite well regulated. A blunt pH microelectrode is placed at a given distance (10 μm) from the root surface and is moved along the root. As shown in Fig. 7,

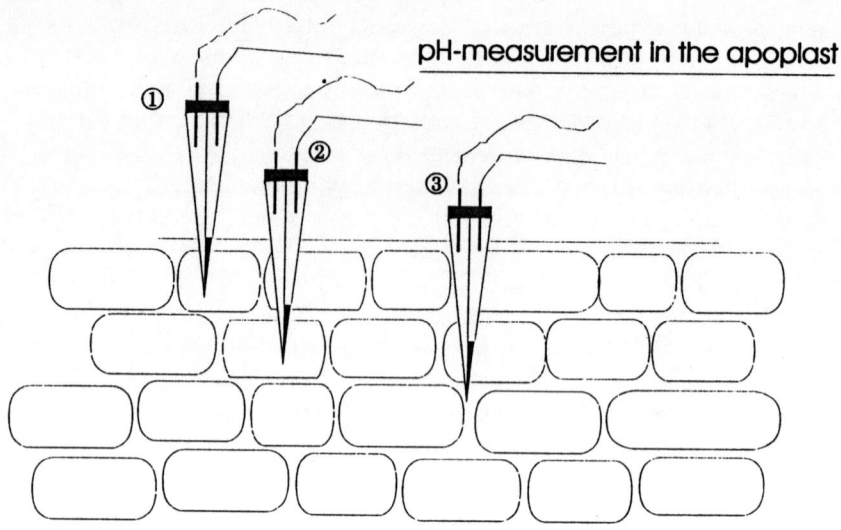

Fig. 6. pH measurement in the apoplast. Principle of approach. (1) A double-barrelled microelectrode, inserted into the (root cortex) tissue, measures intracellularly first. (2) Driving the electrode further results in another intracellular measurement, or in an apoplastic measurement (3), when the tip hits a radial cell wall.

this surface pH is considerably different from the bulk pH. Comparison with the direct apoplast measurements reveals that in the case of the outer layers of the root cortex, the surface pH behaves in a similar way to the apoplast. Some laboratories have used this approach: Lucas and Kochian (1986) and Newman *et al.* (1987) have characterised fluxes of H^+ and K^+ in maize roots; Kochian, Shaff and Lucas (1989) have investigated high-affinity K^+ uptake in maize roots; whereas, Monshausen, Zieschang and Sievers (1996) investigated gravitropism – just to name a few. An extended form of this extracellular approach is the ion-selective vibrating probe system (Kühtreiber & Jaffe, 1990). This ingenious technique allows quantitative analysis of ion fluxes from and to a single cell by moving the electrode at a given angle to the cell surface forward and backward at a low frequency.

Signal transduction

Extracellularly placed pH microelectrodes can be applied successfully to problems of signal transduction. Since root hairs are microantennas

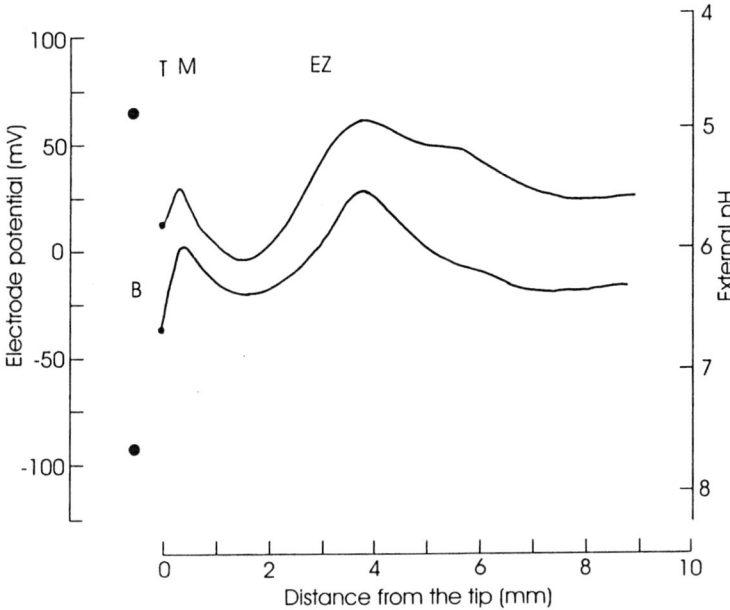

Fig. 7. pH profiles, measured with a blunt pH electrode placed at a 10 μm distance from the surface of a *Zea mays* root. Tests were carried out at pH 4.9 and 7.8 (B = bulk of medium). First measuring point was at the root tip (T). M = meristematic zone; EZ = elongation zone. Buffer concentration was 0.5 mM Mes/Tris, mixed to the respective pH. Measuring points were connected to obtain continuous lines.

reaching into the rhizosphere, they are in contact with micro-organisms. These fungi or bacteria excrete substances to which the root hairs may respond with either a defence or a symbiotic reaction. For example, the first response of legume root hairs to rhizobial lipochito-oligosaccharides (Nod factors) is a depolarisation (Erhardt, Atkinson & Long, 1992; Felle *et al.*, 1995), accompanied by a host-specific, concentration-dependent alkalinisation of the root hair space (Fig. 8). The latter can be measured most conveniently with a pH microelectrode placed within the root hair space. Typically, these responses to Nod factors are bell shaped and are elicited by concentrations as low as 0.1 nM. Their magnitude depends on the pH_e and on other environmental conditions, and the lag phase of about 15 seconds indicates the involvement of a specific receptor.

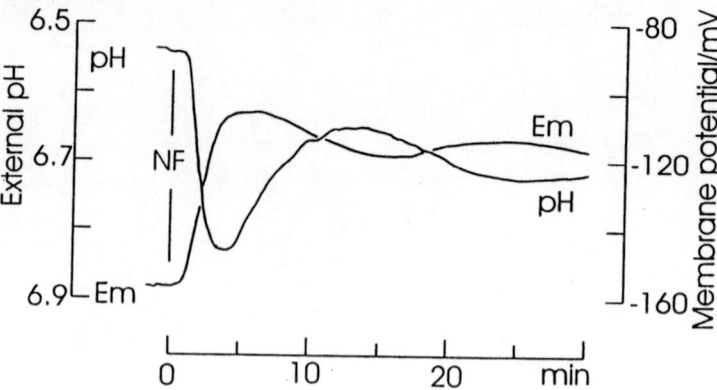

Fig. 8. Nod factors (NF). pH$_e$ was measured with a blunt pH electrode in the root hair zone of *Medicago sativa*. Perfusion of chamber was 1–2 ml/min. Simultaneously, the membrane potential (Em) of a root hair was measured. Following the addition of 0.1 μM Nod factor (a sulphated lipochito-oligosaccharide), a transient alkalinisation and a depolarisation are observed. Bulk pH was 7.0. Buffer concentration was 0.5 mM Mes/Tris. Copied from original recordings.

Conclusions

As just demonstrated, a wide field of applications is open for the user of pH-sensitive microelectrodes. The possibility of using the same technique to measure the free concentrations of other ions (currently, resin cocktails of NH_4^+, CA^{2+}, Cl^-, H^+, Li^+, Mg^{2+}, NO_3^-, K^+ and Na^+ are commercially available) by simply filling the electrodes with the respective resin makes this technique extremely attractive. The invasiveness of the technique has been criticised, and this aspect cannot be disregarded. However, this apparent disadvantage is more than compensated for by the voltage reference. Because the membrane potential is measured simultaneously throughout the test, the physiological state of the system can be well controlled.

What is the future? There is no doubt that the microelectrode technique is difficult, but once learned and adapted to the system under investigation, it offers a way to measure dynamic processes of regulation and signal transduction inside the cell, at the membrane and outside the cell, simultaneously if required.

References

Ammann, D. (1986). *Ion-selective microelectrodes. Principles, design and application.* Berlin, Heidelberg: Springer.

Barbier-Brygoo, H., Ephritikhine, G., Klämbt, D., Ghislain, M. & Guern, J. (1989). Functional evidence for an auxin receptor at the plasmalemma of tobacco mesophyll protoplasts. *Proceedings of the National Academy of Sciences USA* **86**, 891–6.

Bethmann, B., Thaler, M., Simonis, W. & Schönknecht, G. (1995). Electrochemical potential gradients of H^+, K^+, Ca^{2+}, and Cl^- across the tonoplast of the green alga *Eremosphera viridis*. *Plant Physiology* **109**, 1317–26.

Bertl, A., Felle, H. & Bentrup, F.-W. (1984). Amine transport in *Riccia fluitans*. *Plant Physiology* **76**, 75–8.

Blatt, M.R. (1992). K^+ channels of stomatal guard cells: characteristics of the inward rectifier and its control by pH. *Journal of General Physiology* **99**, 615–44.

Blatt, M.R. & Armstrong, F. (1993). K^+ channels of stomatal guard cells: abscisic-acid-evoked control of the outward rectifier mediated by cytoplasmic pH. *Planta* **191**, 330–41.

Bowling, D.J.F. (1976). Measurement of intracellular pH in roots using a H^+-sensitive microelectrode. In *Membrane transport in plants*, ed. U. Zimmermann & J. Dainty, pp. 386–90. Berlin, Heidelberg, New York: Springer.

Bowman, B. & Slayman, C.W. (1979). The effect of vanadate on the plasma membrane ATPase of *Neurospora crassa*. *Journal of Biological Chemistry* **254**, 2954–9.

Davies, D.D. (1986). The fine control of cytosolic pH. *Physiologia Plantarum* **67**, 702–6.

Davis, R.F. (1976). Photoinduced changes in electrical potentials and H^+-activities of the chloroplast, cytoplasm and vacuole of *Phaeoceros laevis*. In *Membrane transport in plants*, ed. U. Zimmermann & J. Dainty, pp. 197–201. Berlin, Heidelberg, New York: Springer.

Echeverria, E., Burns, J. & Felle, H. (1992). Compartmentation and cellular conditions controlling sucrose breakdown in mature acid lime fruits. *Phytochemistry* **31**, 4091–5.

Erhardt, D.W., Atkinson, E.M. & Long, S.R. (1992). Depolarization of alfalfa root hair membrane potential by *Rhizobium meliloti* Nod factors. *Science* **256**, 998–1000.

Felle, H. (1986). Proton transport and pH control in *Sinapis alba* root hairs: a study carried out with double-barreled pH microelectrodes. *Journal of Experimental Botany* **38**, 340–54.

Felle, H. (1988a). Auxin causes oscillations of cytosolic free calcium and pH in *Zea mays* coleoptiles. *Planta* **174**, 495–9.

Felle, H. (1998b). Cytoplasmic free calcium in *Riccia fluitans* L. and *Zea mays* L.: interaction of Ca^{2+} and pH? *Planta* **176**, 248–55.

Felle, H. (1988c). Short-term pH regulation in plants. *Physiologia Plantarum* **74**, 583–91.

Felle, H. (1989). pH as a second messenger in plants. In *Second mess-*

engers in plant growth and development, ed. W.F. Boss & D.J. Morré, pp. 145–66. New York: Alan R. Liss.

Felle, H.H. (1991). The role of the plasma membrane proton pump in short-term pH regulation in the aquatic liverwort *Riccia fluitans* L. (1991). *Journal of Experimental Botany* **42**, 645–52.

Felle, H.H. (1992). Ion-selective microelectrodes: their use and importance in modern plant cell biology. *Botanica Acta* **106**, 5–12.

Felle, H.H. (1994). The H^+/Cl^- symporter in root-hair cells of *Sinapis alba*. *Plant Physiology* **106**, 1131–6.

Felle, H.H. (1995). Control of cytoplasmic pH under anoxic conditions and its implications for plasma membrane proton transport in *Medicago sativa* root hairs. *Journal of Experimental Botany* **47**, 967–73.

Felle, H. & Bertl, A. (1986a). The fabrication of H^+-selective liquid-membrane microelectrodes for use in plant cells. *Journal of Experimental Botany* **37**, 1416–28.

Felle, H. & Bertl, A. (1986b). Light-induced cytoplasmic pH changes and their interrelation to the activity of the electrogenic proton pump in *Riccia fluitans*. *Biochimica et Biophysica Acta* **848**, 176–82.

Felle, H., Brummer, B., Bertl, A. & Parish, R.W. (1986). Indole-3-acetic acid and fusicoccin cause cytosolic acidification of corn coleoptile cells. *Proceedings of the National Academy of Sciences USA* **83**, 8992–5.

Felle, H.H., Kondorosi, E., Kondorosi, A. & Schultze, M. (1995). Nod signal-induced plasma membrane potential changes in alfalfa root hairs are differentially sensitive to structural modifications of the lipochitooligosaccharide. *The Plant Journal* **7**, 939–47.

Felle, H.H., Kondorosi, E., Kondorosi, A. & Schultze, M. (1996). Rapid alkalinization in alfalfa root hairs in response to rhizobial lipochitooligosaccharide signals. *The Plant Journal* **10**, 295–301.

Fox, G.G., McCallan, N.R. & Ratcliffe, R.G. (1995). Manipulating cytoplasmic pH under anoxia: a critical test of the role of pH in the switch from aerobic to anaerobic metabolism. *Planta* **195**, 323–30.

Frachisse, J.-M., Johannes, E. & Felle, H. (1988). The use of weak acids as physiological tools: a study of the effects of fatty acids on intracellular pH and electrical plasmalemma properties of *Riccia fluitans* rhizoid cells. *Biochimica et Biophysica Acta* **938**, 199–210.

Gibbon, B.C. & Kropf, D.L. (1994). Cytosolic pH gradients associated with tip growth. *Science* **263**, 1419–21.

Guern, J., Felle, H., Mathieu, Y. & Kurkdjian, A. (1991). Regulation of intracellular pH in plant cells. *International Review of Cytology* **127**, 111–73.

Hansen, U.-P., Moldaenke, C., Tabrizi, H. & Ramm, D. (1993). The effect of transthylakoid proton uptake on cytosolic pH and the imbalance of ATP and $NADPH/H^+$ production as measured by CO_2-

and light-induced depolarization of the plasmalemma. *Plant Cell Physiology* **34**, 681–95.

Hedrich, R. (1994). Voltage-dependent chloride channels in plant cells: identification, characterization, and regulation of a guard cell anion channel. *Current Topics in Membranes* **42**, 1–33.

Herrmann, A. & Felle, H.H. (1995). Tip growth in root hair cells of *Sinapis alba* L.: significance of internal and external Ca^{2+} and pH. *New Phytologist* **129**, 523–33.

Johannes, E. & Felle, H. (1987). Implications for cytoplasmic pH, protonmotive force, and amino-acid transport across the plasmalemma of *Riccia fluitans*. *Planta* **172**, 53–9.

Kochian, L.V., Shaff, J.E. & Lucas, W.J. (1989). High affinity K^+ uptake in maize roots. A lack of coupling with H^+ efflux. *Plant Physiology* **91**, 1202–11.

Komor, E. & Tanner, W. (1976). Proton movement associated with hexose transport in *Chlorella vulgaris*. In *Membrane transport in plants*, ed. U. Zimmermann & J. Dainty, pp. 209–15. Berlin, Heidelberg, New York: Springer.

Kühtreiber, W.M. & Jaffe, L.F. (1990). Detection of extracellular calcium gradients with a calcium-specific vibrating electrode. *Journal of Cellular Biology* **110**, 271–86.

Kurkdjian, A. & Barbier-Brygoo, H. (1983). A hydrogen ion-selective liquid-membrane microelectrode for measurement of the vacuolar pH of plant cells in suspension culture. *Analytical Chemistry* **123**, 96–104.

Kurkdjian, A. & Guern, J. (1989). Intracellular pH: measurement and importance in cell activity. *Annual Review of Plant Physiology* **40**, 271–303.

Lucas, W.J. & Kochian, L.V. (1986). Ion transport processes in corn roots: An approach utilizing microelectrode techniques. In *Advanced agricultural instrumentation*, ed. W.G. Gensler, pp. 402–25. Dordrecht, Boston, Lancaster: Martinus Nijhoff.

Marrè, E. (1979). Fusicoccin: a tool in plant physiology. *Annual Review of Plant Physiology* **30**, 273–88.

Marrè, E., Beffagna, N. & Romani, G. (1988). Potassium transport and regulation of intracellular pH in *Elodea densa* leaves. *Botanica Acta* **101**, 24–31.

Monshausen, B.G., Zieschang, H.E. & Sievers, A. (1996). Differential proton secretion in the apical elongation zone caused by gravistimulation is induced by a signal from the root cap. *Plant, Cell and Environment* **19**, 1408–14.

Newman, I.A., Kochian, L.V., Grusak, M.A. & Lucas, W.J. (1987). Fluxes of H^+ and K^+ in corn roots. *Plant Physiology* **84**, 1177–84.

Nürnberger, T., Nennstiel, D., Jabs, T., Sacks, W.R., Hahlbrock, K. & Scheel, D. (1994). High affinity binding of a fungal oligopeptide elicitor to parsley plasma membranes triggers multiple defense responses. *Cell* **78**, 449–60.

Peters, W.S. & Felle, H. (1991). Control of apoplast pH in corn coleoptile segments. II. The effects of various auxins and auxin analogues. *Journal of Plant Physiology* **137**, 691–6.

Ratcliffe, R.G. (1997). *In vivo* NMR studies of the metabolic response of plant tissue to anoxia. *Annals of Botany* **79**, 39–48.

Raven, J.A. (1985). pH regulation in plants. *Science Progress (Oxford)* **69**, 495–509.

Reid, R.J. & Smith, F.A. (1988). Measurements of the cytoplasmic pH of *Chara corallina* using double-barrelled pH microelectrodes. *Journal of Experimental Botany* **39**, 1421–32.

Remis, D., Bulychev, A. & Kurella, G.A. (1987). The electrical and chemical components of the protonmotive force in chloroplasts as measured with capillary and pH-sensitive microelectrodes. *Biochimica et Biophysica Acta* **852**, 68–73.

Roos, A. & Boron, W.F. (1981). Intracellular pH. *Physiological Reviews* **61**, 296–434.

Rück, A., Palme, K., Venis, M.A., Napier, R.M. & Felle, H.H. (1993). Patch-clamp analysis establishes a role for an auxin binding protein in the auxin stimulation of plasma membrane current in *Zea mays* protoplasts. *The Plant Journal* **4**, 41–6.

Sanders, D. & Slayman, C.L. (1982). Control of intracellular pH. Predominant role of oxidative metabolism, not proton transport, in the eucaryotic microorganism *Neurospora*. *Journal of General Physiology* **80**, 377–402.

Slayman, C.L. & Slayman, C.W. (1974). Depolarization of the plasmalemma of *Neurospora* during active transport of glucose: evidence for a proton-dependent cotransport system. *Proceedings of the National Academy of Sciences USA* **71**, 1935–9.

Smith, F.A. & Raven, J.A. (1979). Intracellular pH and its regulation. *Annual Review of Plant Physiology* **30**, 289–311.

Steigner, W., Köhler, K., Simonis, W. & Urbach, W. (1988). Transient cytoplasmic pH-changes in correlation with opening of potassium channels. *Plant Physiology* **99**, 103–10.

Stewart, P.A. (1983). Modern quantitative acid–base chemistry. *Canadian Journal of Physiological Pharmacology* **62**, 1444–61.

Strack, D., Sharma, V. & Felle, H. (1987). Vacuolar pH in radish cotyledonal mesophyll cells. *Planta* **172**, 563–5.

Talbott, L.D., Ray, P. & Roberts, J.K.M. (1988). Effect of indoleacetic acid- and fusicoccin-stimulated proton extrusion on internal pH of pea internode cells. *Plant Physiology* **87**, 211–16.

Thaler, M., Simonis, W. & Schönknecht, G. (1989). Light-dependent changes of the cytoplasmic H^+ and Cl^- activity in the green alga *Eremosphera viridis*. *Journal of Experimental Botany* **40**, 1195–203.

Tretyn, A., Wagner, G. & Felle, H.H. (1991). Signal transduction in *Sinapis alba* root hairs: auxins as external messengers. *Journal of Plant Physiology* **139**, 187–93.

Tsien, R.J. & Rink, T.J. (1985). Neutral carrier ion-selective microelectrodes for measurement of intracellular free calcium. *Biochimica et Biophysica Acta* **599**, 623–38.

Ullrich, C.I. & Novacky, A.J. (1990). Extra- and intracellular pH and membrane potential changes induced by K^+, Cl^-, $H_2PO_4^-$, and NO_3^- uptake and fusicoccin in root hairs of *Limnobium stoloniferum*. *Plant Physiology* **94**, 1561–7.

STEPHEN T. KINSEY and
TIMOTHY S. MOERLAND

The use of nuclear magnetic resonance for examining pH in living systems

Introduction

In what has become a classic paper, Moon and Richards (1973) demonstrated that phosphate compounds which occur naturally in cells could be used to measure intracellular pH (pH_i) non-invasively using ^{31}P nuclear magnetic resonance (NMR). The ability to probe a quantity as integral to cellular function as pH in living, unperturbed cells offered the promise of vast new insights into cellular metabolism. Indeed, hundreds of papers have been published as a direct result of Moon and Richards' original contribution, and the indirect impact of their work on the development of other applications of NMR to biology may be even greater. Today, NMR methods of pH measurement are common in clinical as well as in academic settings. In many cases, ^{31}P-spectra are collected routinely as a means of assessing the energetic and acid–base status of a plant or animal sample during an experimental procedure. Many new NMR pH indicator compounds have been characterized, both endogenous and exogenous, which probe both pH_i and extracellular pH (pH_e). In addition, a wide variety of experimental protocols have been developed to examine acid–base balance non-invasively, including spectroscopy of isolated tissues and organs, *in vivo* localized spectroscopy, and imaging methods.

This brief review discusses the principles of pH measurement using NMR and highlights many of the pH indicator compounds that are currently employed. The strengths and weaknesses of NMR measurement of pH are discussed, as are sources of error. Finally, applications to comparative animal and plant physiology are described to provide practical examples of the kinds of experiments that are possible and information that can be obtained using this methodology. This is in no way a comprehensive review of the literature, and is intended to provide an assessment of the current standing of NMR as an experimental tool in acid–base studies.

Practical considerations

The NMR signal is derived from the tiny current that is induced in a transceiver coil by atomic nuclei precessing about a strong magnetic field. The precessional, or resonance, frequency is determined by the magnetogyric ratio of the nucleus (highest for ^1H), and the field strength of the magnet. This frequency is also modified slightly by the electron shielding a nucleus experiences, such that nuclei in different chemical environments have peaks that appear at different positions in the NMR frequency spectrum. This chemical modification of resonance frequency is known as chemical shift. Chemical shift has units of Hz, but is usually normalized to parts per million (ppm).

Because the detected current is weak, NMR is a very insensitive technique. This shortcoming means that spectra cannot usually be observed after a single excitation. Instead, a number of scans must be acquired and averaged together, which means that it may take several seconds to tens of minutes to acquire a final spectrum. For the purposes of measuring pH, the ability to acquire a well-resolved spectrum over a short time is highly desirable, particularly when the pH is changing rapidly. Since higher resonance frequencies equate to greater sensitivity, the use of nuclei with large magnetogyric ratios and of high field magnets improves time resolution for a given sample. However, sensitivity is also a function of the number of nuclei within the transceiver coil, which is directly related to the volume of the sample and the concentration of the metabolite being observed. Although it is difficult to put an absolute limit on the minimum volume and concentration which will yield a spectrum with an adequate signal-to-noise ratio (SNR), in general, sample volumes for biological applications must be of the order of a few cubic millimetres and concentrations must be greater than 0.1 mM.

Detailed descriptions of the fundamentals of NMR and its applications in biology have been covered in several recent reviews which focus on comparative animal physiology (Ellington & Wiseman, 1989; Wasser, Lawler & Jackson, 1996) and plants (Ratcliffe, 1994), as well as the more complete text by Gadian (1995).

Principles of NMR pH measurement

NMR pH measurement is possible because some resonances arise from chemical species that are in fast-exchange equilibrium between acid and base forms. Differences in electron shielding of the protonated and unprotonated nucleus result in the acid and base forms having slightly different chemical shifts. In the case of the commonly used pH indicator inorganic phosphate (P_i), the equilibrium is as follows:

$$H_2PO_4^- \leftrightarrow H^+ + HPO_4^{2-} \tag{1}$$

The chemical shift of the acid form, δ_{acid}, is approximately 2.6 ppm less than that of the base form, δ_{base}. Because there is fast-exchange relative to the NMR time scale between the two forms, a single peak is observed for P_i which has a chemical shift, δ_{obs}, that is a function of the concentrations of the acid and base forms (Fig. 1A). The equilibrium concentrations of each form are determined by the pK_a (approximately 6.6 for P_i), and the pH is calculated from the Henderson–Hasselbalch equation:

$$pK_a = -\log_{10}\left(\frac{[H^+][HPO_4^{2-}]}{H_2PO_4^-}\right) \tag{2}$$

$$pH = pK_a + \log_{10}\left(\frac{[HPO_4^{2-}]}{[H_2PO_4^-]}\right) \tag{3}$$

For the NMR experiment, the Henderson–Hasselbalch equation is expressed in terms of chemical shift:

$$pH = pK_a + \log_{10}\left(\frac{\delta_{obs} - \delta_{acid}}{\delta_{base} - \delta_{obs}}\right) \tag{4}$$

The pH can then be calculated using NMR by measuring δ_{obs} in a spectrum (relative to the chemical shift of a pH-independent reference peak) and knowing the pK_a, δ_{acid}, and δ_{base} for the pH indicator species. In practice, the three unknowns in equation 4 are obtained empirically by measuring the chemical shift of the indicator species over a range of pH values in a model solution and fitting equation 4 to the experimental data (Fig. 1B). If two protons are dissociable near physiological pH ($H_2A \leftrightarrow HA^- + H^+ \leftrightarrow A^{2-} + 2H^+$), the observed chemical shift is a function of the chemical shift of the fully protonated form, δ_1, the mono-protonated form, δ_2, and the unprotonated form, δ_3, and the two pK_a values, pK_1 and pK_2 (Robitaille *et al.*, 1991):

$$\delta_{obs} = \frac{\delta_1}{1 + 10^{pH-pK_1} + 10^{2pH-pK_1-pK_2}} + \frac{\delta_2}{1 + 10^{pK_1-pH} + 10^{pH-pK_2}} + \tag{5}$$
$$\frac{\delta_3}{1 + 10^{pK_2-pH} + 10^{pK_1+pK_2-2pH}}$$

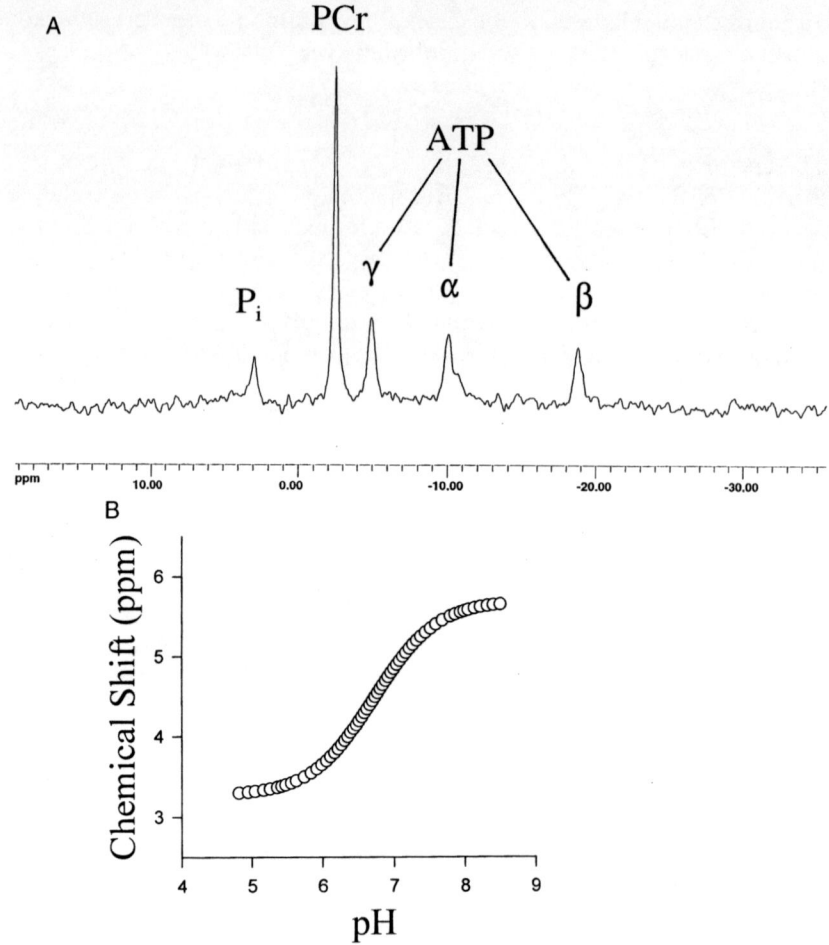

Fig. 1. **A**. Typical ^{31}P-NMR spectrum from skeletal muscle. The chemical shift of the P_i peak changes according to the pH_i (as does the chemical shift of the ATP peaks). The chemical shift of the phosphocreatine (PCr) peak is pH independent over the physiological pH range and is often used as an internal chemical shift reference. **B**. Relationship between pH and ppm for P_i calculated from equation 4 (see p. 47) using $\delta_{acid} = 3.27$, $\delta_{base} = 5.69$, and $pK_a = 6.72$ (Van Ginneken et al., 1995).

For most pH indicators used in biological applications, however, the complication of multiple pK_as can usually be ignored.

NMR pH indicators

The minimal criteria for a useful NMR pH indicator are: (1) adequate SNR, (2) pK_a in the physiological pH range, and (3) large dynamic range of chemical shift ($\delta_{acid} - \delta_{base}$). The SNR of the indicator must be high enough so that a well-resolved chemical shift can be measured without noise spikes contributing significantly to the peak shape (Madden *et al.*, 1991). Also, the pH range to be examined should be within approximately ± 1 pH unit from the pK_a because otherwise the pH-dependent changes in chemical shift become unacceptably small. Finally, accurate measurement of the indicator's chemical shift requires a suitable pH-independent chemical shift reference peak.

The most important and widely used endogenous indicator of pH_i is the resonance arising from P_i as observed in ^{31}P-spectra (Moon & Richards, 1973; reviewed in Rudin & Sauter (1992) and van den Thillart & van Waarde (1996); see Fig. 1). This naturally occurring pH_i probe has a pK_a of approximately 6.6 and it occurs in sufficient concentration to be observable in many tissue types. The phosphagen peak, which is phosphocreatine (PCr) in vertebrates, is generally used as a pH-independent chemical shift reference. A disadvantage of using P_i is the dependence of its chemical shift on the concentration of free Mg^{2+} (see below). Besides pH_i information, the ^{31}P-spectra of biological samples allows assessment of the energetic condition of the tissue of interest by measuring the relative peak areas (= relative concentrations) of ATP, PCr, and P_i, and by calculating the free $[Mg^{2+}]$ from the chemical shifts of the ATP peaks (Gupta, Benovic & Rose, 1978; Gupta & Moore, 1980; Kushmerick *et al.*, 1986; Douman & Ellington, 1992; Combs & Ellington, 1995). Adverse physiological conditions almost always affect the relative concentrations of the high-energy phosphate compounds, either directly or indirectly, because ATP-dependent compensatory mechanisms are usually activated (for an extensive review of ^{31}P-NMR energetic studies see van den Thillart & van Waarde, 1996). This ability to assess the physiological 'health' of a tissue is a tremendous advantage of ^{31}P-NMR, particularly when perfused organs or tissues are being used.

The P_i-NMR signal is usually well resolved in vertebrate and plant tissues, but it has been shown to decrease to immeasurable levels following aerobic exercise in skeletal muscle (Pan *et al.*, 1988). However, most physiological perturbations actually lead to an increase in the intensity of the P_i resonance due to net phosphagen hydrolysis. In invertebrates which utilize phosphagens other than PCr (for example phosphoarginine), the equilibrium constant of the phosphotransferase reaction is substantially reduced relative to that of the creatine kinase reaction (Ellington, 1989), yielding a much smaller P_i peak at high ATP/ADP ratios which may be insufficient for pH measurement. Robitaille et al. (1991) conducted a thorough analysis of the pH characteristics of many alternative endogenous pH-sensitive phosphorus-containing compounds that can be observed with ^{31}P-NMR. However, the visibility of other phosphorus metabolites in ^{31}P-spectra is largely tissue specific.

In cases in which a suitable endogenous ^{31}P indicator resonance is not available, a useful alternative is the exogenous pH probe, 2-deoxy-glucose-6-phosphate (2DG6P), as first suggested by Navon et al. (1977), which has a slightly lower pK_a (approximately 6.2) than P_i. From the perfusion medium, the substrate 2-deoxyglucose enters the cell via the glucose transporter and is phosphorylated by hexokinase to form 2DG6P, which is not further metabolized and accumulates in the cell. 2DG6P is an exclusively cytosolic pH indicator, unlike P_i, which may have a mitochondrial component (Garlick, Soboll & Bullock, 1992). However, pH measurements using both indicators have been in very good agreement since usually most of the P_i is cytosolic (Hamm & Yue, 1987; Wiseman & Ellington, 1989; Soto et al., 1996). A disadvantage of using 2DG6P is that it traps phosphate and causes a reciprocal reduction in the phosphagen concentration, and it inhibits glycolysis at high concentrations (Allen et al., 1985).

Measurement of pH using 2DG6P and the vast majority of phosphorus-containing compounds is hindered by their fairly low pK_as, which are not suitable for measuring pH in the alkaline physiological range (Robitaille et al., 1991). Notable exceptions include several phosphonate compounds. Szwergold, Brown and Freed (1989) demonstrated the utility of several 2-amino-phosphono-carboxylic acid compounds which have pK_as ranging from 6.9 to 7.6. Ideally, these compounds should be fit to equation 5 because of the presence of two titratable protons, but in practice a titration curve which does not extend to the alkaline extremes can be fit to equation 4 for a single pK_a. In addition, phosphonates which do not cross cell membranes can be used as pH_e probes, such as phenylphosphonate (Meyer, Brown & Kushmerick, 1985; Graham, Taylor & Brown, 1994), methylphosphonate (which is permeable in some cells; DeFronzo & Gillies, 1987) and 3-aminopropyl phosphonate (Gillies, Liu & Bhujwalla, 1994), all of which are non-toxic and have pK_as ranging from 6.9 to 7.6.

Pan et al. (1988) demonstrated that the imidazole protons on C-2 and

C-4 of carnosine (β-Ala-His) could be observed in ^{1}H-spectra as indicators of pH in vertebrate skeletal muscle. The high NMR sensitivity of protons makes this method advantageous when spectra need to be collected over a short time period, such as under conditions when pH is changing rapidly. Further, the chemical shifts of the imidazole resonances are not sensitive to the concentration of intracellular Mg^{2+}, giving them a distinct advantage over P_i (Pan *et al.*, 1988). However, ^{1}H-spectroscopy requires the suppression of the intense water resonance, which will otherwise fill up the NMR digitizer and limit the dynamic range of the resonance of interest (and, hence, SNR). While soluble carnosine is present in fairly high concentrations in vertebrate muscle (Okuma & Abe, 1992), it is generally absent or present at very low concentrations in invertebrates (Morris & Baldwin, 1984; Wiseman & Ellington, 1989).

Recently, Aime *et al.* (1996) demonstrated that ^{1}H-NMR could be used to observe the pH sensitivity of paramagnetic complexes of the lanthanide, ytterbium, with the macrocyclic ligand DOTP (1,4,7,10-tetra-azacyclododecane-N,N′,N″,N‴-tetrakis methylenephosphonic acid). Nuclei that are bound to paramagnetic centers can undergo extremely rapid relaxation, or dramatic changes in chemical shifts that far outweigh the typical diamagnetic shifts observed in most spectra. The pH-sensitive Yb(DOTP) complex has multiple pK_as, but the relationship between chemical shift and pH is linear over a pH range of 5.0 to 7.5. Aime *et al.* (1996) proposed this compound as an exquisitely sensitive *in vivo* pH_e indicator because of the large changes in chemical shift that are induced by the paramagnetic center for only slight changes in pH. The potential for enhanced sensitivity to pH changes makes this class of compounds intriguing, but their potential toxicity may ultimately limit their application in biology (Gadian, 1995).

^{13}C-NMR is being used increasingly to examine metabolism because specific carbons in a metabolite can be selectively enriched with the ^{13}C isotope by applying the proper substrate, and the fate of the label can be followed simultaneously through multiple metabolic pathways in living tissue (reviewed in Sherry & Malloy, 1996). Using ^{13}C-enriched substrates is usually necessary because of the extreme insensitivity of the ^{13}C nucleus (0.018 percent as sensitive as ^{1}H), which results from the low natural abundance of this isotope (1.1 percent of total carbon) and its small magnetogyric ratio. Chacko and Weiss (1993) showed that perfusion of a rat heart with glucose which was ^{13}C enriched at the C-1 position led to an enhanced peak in the ^{13}C spectrum for the C-3 of *sn*-glycerol-3-phosphate. This peak was pH sensitive, with a pK_a of 6.2. Measurement of pH_i using this peak allows the simultaneous measurement of intermediary metabolism and pH_i (using the alanine methyl protons as a chemical shift reference). Further, because sn-glycerol-3-phosphate is a glycolytic product, it is exclusively cytosolic, and offers some advantage over the P_i resonance which may be observed in the cytosol and in the mitochondria.

^{19}F-NMR offers sensitivity nearly equal to that of protons (83 percent of ^{1}H), and has the advantage that no naturally occurring resonances exist which could cause peak overlap in the ^{19}F spectrum. The obvious disadvantage of using ^{19}F-labeled pH indicator compounds is that they must be loaded and trapped in the cell. Taylor and Deutsch (1983; 1988; 1989) have developed a number of pH-dependent fluorinated compounds, the most useful of which are the fluorinated α-methylamino acids. These compounds do not readily enter many cell types but can be loaded as methyl ester precursors, which are taken up by cells and hydrolysed by intracellular methyl esterases to yield the amino acid pH indicator (Taylor & Deutsch, 1983). Cells lacking methyl esterase activity have been shown to cleave the p-chlorophenyl ester precursor of one of these pH indicators, difluoromethylalanine (Taylor & Deutsch, 1988). This pH indicator is one of the most useful because its pK_a is 7.3, it has two peaks with a pH-dependent spacing (so no chemical shift reference is needed), and it shows a large dynamic range of chemical shift, making it very sensitive to pH changes. This class of compounds is also non-toxic to cells (Taylor & Deutsch, 1983).

Sources of error

The most important source of error in NMR pH measurement results from inaccurate calibration curves which are generated from titration solutions that do not reflect the conditions in the compartment of interest. This is particularly problematic when measuring pH_i because of the uncertainty associated with the composition of the cytosol in living cells. Inappropriate model solutions lead to errors in the empirically determined parameters, pK_a, δ_{acid}, and δ_{base}, which ultimately results in systemic errors in the calculation of pH_i. The principal factor affecting the relationship between chemical shift and pH_i for all NMR pH indicators is ionic strength, which in the cytosol is primarily a result of differences in the $[K^+]$ (Roberts, Wade-Jardetzky & Jardetzky, 1980; Taylor & Deutsch, 1983; Robitaille et al., 1991). For P_i, the free $[Mg^{2+}]$ also induces changes in the behavior of the chemical shift with pH (Roberts et al., 1980; Robitaille et al., 1991). Great care should therefore be taken in preparing a titration solution, particularly when experimental conditions might cause the ion concentration to change during pH$_i$ measurement, such as occurs for free Mg^{2+} during acidosis (Combs & Ellington, 1995). Because of the interaction between pH$_i$ and free [Mg^{2+}], Williams, Mosher and Smith (1993) developed a protocol for determining both quantities simultaneously from the chemical shift differences between the three ATP peaks. Temperature also has a strong effect on the pH–chemical shift relationship (Kost, 1990), but this is easily controlled by generating calibration curves at the experimental temperature.

Ackerman *et al.* (1996) created a general model for pH dependence of the chemical shift of the indicator resonance that also accounts for the pH dependence of the reference peak. The traditional model (equation 4) assumes no pH dependence of the reference species, and is invalid when the pH approaches the pK$_a$ of the reference peak (which may occur during extreme pH excursions). The usual simplification is acceptable for cases in which the pH does not approach the reference pK$_a$, when truly non-titratable reference probes are used, or when external references are used (for example a reference within the NMR coil in a separate capillary tube).

Graham, Taylor and Brown (1994) demonstrated the error inherent in the non-linear relationship between pH and chemical shift. This problem is exacerbated as the spread in the true pH distribution increases and as the offset of the pH from the pK$_a$ of the indicator increases. This effect is manifested as a skewing of the NMR peak as the true mean pH changes, or as a falsely indicated change in the NMR-derived pH as the standard deviation of the true pH distribution increases (without a change in the true mean pH). Graham *et al.* (1994) offered a remedy for this problem which consists of: (1) converting the ppm axis to pH using the Henderson–Hasselbalch equation, and (2) dividing the intensity at each point in the spectrum by the derivative of the Henderson–Hasselbalch equation with respect to chemical shift:

$$\frac{d\text{pH}}{d\delta_{obs}} = \frac{d}{d\delta_{obs}}\left[\text{pK}_a + \log\left(\frac{\delta_{acid} - \delta_{obs}}{\delta_{obs} - \delta_{base}}\right)\right] = \frac{\delta_{base} - \delta_{acid}}{(\delta_{acid} - \delta_{obs})(\delta_{obs} - \delta_{base})} \quad (6)$$

This latter treatment accounts for the non-linearity between ppm and pH and yields spectral peaks that accurately reflect the distribution of pH in a tissue.

It should be emphasized that, despite the uncertainties discussed above, evidence derived from comparisons with other pH measurement methods strongly suggests that NMR measurements of pH$_i$ and pH$_e$ usually have an absolute accuracy ranging from 0.05 to 0.1 pH units, and changes in pH can be measured with a precision of ± 0.03 pH units (Madden *et al.*, 1991; Gadian, 1995).

Applications in comparative physiology

NMR measurement of pH has been widely employed in comparative physiology, almost exclusively by using ^{31}P-spectroscopy and either the P_i or 2DG6P resonance as the pH indicator (reviewed in Ellington & Wiseman, 1989; van den Thillart & van Waarde, 1996; Wasser *et al.*, 1996). Several experimental modes have evolved which depend on the type of spectrometer available and on the type of experiment: (1) spectroscopy of perfused or superfused tissues or organs, (2) spectroscopy of very small, homogeneous tissue samples, (3) whole-animal spectroscopy, including localized spectroscopy, and (4) spectroscopic imaging methods.

^{31}P-NMR has been used extensively to monitor pH_i transients in perfused or superfused molluscan muscle preparations during contractile activity (Chih & Ellington, 1985), environmental anoxia (Barrow, Jamieson & Norton, 1980; Ellington, 1983; Graham & Ellington, 1985), and experimentally induced acidifications (Ellington, 1985; Combs & Ellington, 1995). Changes in the intracellular free $[Mg^{2+}]$ as a function of changes in pH_i have been examined in superfused muscles from horseshoe crab (Douman & Ellington, 1992) and from whelk (Combs & Ellington, 1995). Among vertebrates, pH_i has been examined in perfused turtle hearts during anoxia and lactic acidosis (Wasser *et al.*, 1990; Jackson *et al.*, 1991). These methodologies are useful for attaining global pH within a physiologically relevant whole tissue or organ, but cellular heterogeneity and the diffusive time lags associated with larger tissues elicit some disadvantages for certain types of experiments.

Dubyak and Scarpa (1983) developed a probe with a small microsolenoid transceiver coil (2.2 mm ID) which could be used to obtain phosphorus spectra from a single, cannulated barnacle (*Balanus nubilis*) depressor muscle fiber. This design removed potential variation resulting from cellular heterogeneity of multicellular tissue preparations. Hamm and Yue (1987) altered this design to incorporate a flow-through superfusion system to examine individual cells of the same muscle, and variations of this basic design have been used for examining pH in small tissue preparations or single cells (Wiseman & Ellington, 1989; Wiseman, Moerland & Kushmerick, 1993; Kinsey & Ellington, 1995; 1996; Combs & Ellington, 1995). The reduced diffusive distances across these small tissue preparations make them easy to keep well oxygenated, and more suitable for certain experimental manipulations that require rapid equilibration of the medium across the tissue. This latter advantage of small preparations has been exploited to measure buffering capacity in molluscs by measuring pH_i transients during acidification with the weak acid DMO (5,5-dimethyl-oxazolidine-2,4-dione; Wiseman & Ellington, 1989; Kinsey & Ellington, 1995) and CO_2 (Zange, Grieshaber & Jans, 1990), as well as by employing NH_4Cl prepulses (Hamm & Yue, 1987; Zange *et al.*, 1990). These experiments can also be used to examine the rate of ion exchange as measured by the slope of the recovery of pH_i following acidification (Hamm & Yue, 1987; Zange *et al.*, 1990; Kinsey & Ellington, 1995).

Figure 2 shows examples of such experiments using DMO and NH_4Cl prepulse methods, and illustrates the need to acquire spectra

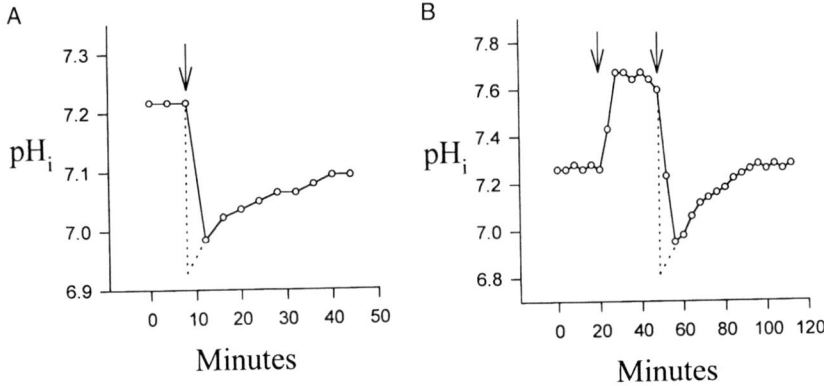

Fig. 2. **A.** DMO acid pulse experiment using a small strip of cardiac muscle (2 mm diameter) from the whelk *Fasciolaria tulipa*. The arrow indicates addition of 60 mM DMO. **B.** NH_4Cl prepulse experiment in cardiac muscle from the whelk *Busycon contrarium*. The first arrow indicates the addition of 20 mM NH_4Cl, which causes an alkaliniz-ation of the pH_i. The second arrow indicates clearance of extracellular NH_4Cl, which causes an intracellular acidification due to dissociation of NH_4^+ into NH_3 (which rapidly diffuses out of the cell) and H^+ (which stays in the cell). The rapid acidification in both experiments can be used to calculate buffering capacity, and the recovery of pH_i following acidification can be used to measure rates of 'proton pum-ping'. In both experiments, diffusive time lags are minimized by using small tissue preparations. However, the inability of NMR to collect data at a single point in time causes an underestimate of the actual acidification (dashed lines).

rapidly during periods of fast pH change, as well as to have rapid equilibration of the medium across the tissue. Even when using these very small preparations, there is an underestimate of the pH_i drop during the rapid acidification (dashed line) which results from the inability of NMR to sample a 'point in time'. As discussed earlier, the NMR signal (and hence the time resolution) is improved if either the sample size or the magnetic field strength is increased. Kinsey and Ellington (1996) made use of a very high-field spectrometer (14 Tesla or 600 MHz for ^{1}H) to examine lactate transport and its effect on pH_i in individual muscle fibers from lobster. ^{1}H-spectra (for lactate measurement) and ^{31}P-spectra (for pH_i measurement – unpublished results) could be acquired every 15 seconds. Increased availability of high-field magnets, as well as continued improvements in microcoil design (Wiseman *et al.*, 1993; Webb & Grant, 1996) should provide greater opportunities for examining small tissue preparations with excellent time resolution.

^{31}P-spectroscopy was used by Busa and Crowe (1983) to evaluate the relationship between cellular acidification and the onset of dormancy in suspensions of *Artemia* embryos. However, this interesting application is not suitable for most whole-animal studies because pH information is usually desired from a particular region or tissue in the body. One approach is to position the appropriate portion of the body within the sensitive volume of the probe's transceiver coil. Using this methodology, pH_i excursions were monitored in lugworm body wall muscles during environmental anoxia (Kamp & Juretschke, 1989) and changes in the pH of the medium (Juretschke & Kamp, 1990), in shrimp abdominal muscle during contraction (Thebault, Raffin & Le Gall, 1987), in fish muscle during movement (Chiba *et al.*, 1990), and in salamander tail muscle during changes in ambient temperature (Johnson *et al.*, 1993). One of the more interesting applications of this approach were the studies of Johnson *et al.* (1993) and Hitzig *et al.* (1994) which examined the fractional dissociation of imidazole (α-imidazole) in newt and lungless salamander tail muscle. These authors were testing the hypothesis that α-imidazole does not change with temperature as suggested by the alphastat hypothesis. Direct measurement of α-imidazole was achieved using ^{1}H-spectroscopy of the dipeptide carnosine (Pan *et al.*, 1988) over a temperature range of 10–30 °C. Although pH_i changed with temperature, α-imidazole was found to be constant in the newt, as predicted by the alphastat hypothesis. This was not the case for the lungless salamander, which showed temperature-dependent changes in α-imidazole.

A more precise method of attaining spectra from a localized region is to place a surface coil against the region of the body from which the spectrum is desired. The major shortcoming of surface coils is that they do not excite nuclei homogeneously across the entire region sampled in the spectrum, meaning that the boundaries of the localized regions are 'fuzzy' and do not contribute equivalently to the final spectrum. However, surface coils have been used with great success. Rees *et al.* (1991) used this approach to demonstrate the acidification of tissues in the land snail *Oreohelix strigosa* which accompanies estivation. Fan, Higashi and Macdonald (1991) examined pH_i in the phasic adductor muscle of the bivalve *Mytilus edulis* during emergence-induced anoxia by using a surface coil placed on one of the animal's valves. Wegener, Bolas and Thomas (1991) used the NMR transceiver coil as a harness to hold a locust in place while they monitored pH_i *in vivo* from active flight muscles. Fishes make excellent NMR specimens if they can be properly immobilized, because they have large regions of muscle of homogeneous fiber type. Van den Thillart *et al.* (1989) developed a flow-through NMR probe for maintaining living fishes which contained an inflatable bladder inside the sample chamber to immobilize the fish by pressing it against the chamber wall. A surface coil positioned on the outside wall of the chamber was used to collect ^{31}P-spectra from homogeneous regions of epaxial muscle. This methodology has been used extensively to examine pH_i (as well as energetics) in fish muscle *in vivo* under a variety of experimental treatments (reviewed in van den Thillart & van Waarde, 1996).

More exact methods of obtaining localized pH measurements utilize magnetic field gradients and require imaging capability of the spectrometer. These methods allow the sample to be placed within a normal transceiver coil and have a spectrum collected from a small volume of the total sample (Ordidge *et al.*, 1985; Ordidge, Connelly & Lohman, 1986; Frahm, Merboldt & Hanicke, 1987). Similarly, chemical shift imaging allows the acquisition of spatially resolved spectra over the entire sample, and in the case of ^{31}P, pH maps can be constructed (Morikawa *et al.*, 1993). The shortcoming of these methods is typically a low SNR, which becomes more problematic as the localized region becomes smaller and the concentrations of the metabolites in the spectrum become lower. Applications of these methods to comparative acid–base questions are limited. Skibbe *et al.* (1995; 1996) have used chemical shift imaging to acquire pH maps and observe pH gradients across the midgut of lepidopteran larvae. Although their results are the highest resolution to date for pH mapping, the voxel size of 2.0 mm^3 is still large relative to the size of the animal and spatial resolution is quite poor. Although localization using imaging gradients is appealing due to its elegance and ability to control precisely the localized region, the

SNR limitations currently make surface coils more useful for most applications involving small animals.

Applications in plant physiology

Although the literature on the NMR measurement of pH in plants is more modest than that for animals, a number of groups have used ^{31}P-NMR with great success (reviewed in Ratcliffe, 1991; 1994). The experimental considerations are much the same as described above, and methodologies have been developed for maintaining higher plants, excised tissues, and cell suspensions in good physiological condition (Ratcliffe, 1994). For experiments under illumination, adequate light must be supplied to the entire sample, which can be difficult for preparations such as tightly packed leaves or high-density photosynthetic cell suspensions. In the latter case, light is usually supplied with a fiber optic light guide, which may be inserted into the cell suspension. Of particular interest in NMR studies of plants are the separate peaks that occur in most ^{31}P-spectra for cytosolic and vacuolar P_i, which allows unequivocal multicompartmental analyses of pH regulation (Ratcliffe, 1994). As in most animal tissues, however, the cytosolic component of the P_i signal probably also contains contributions from organelles, which in plants include mitochondria, chloroplasts and amyloplasts. Hentrich *et al.* (1993) have demonstrated that P_i from chloroplasts can be observed in the alga *Chlamydomonas reinhardtii*, but in most cases a single peak is observed for cytosolic P_i.

The responses of pH_i to anoxia in plant cells have received considerable attention and ^{31}P-NMR studies have been recently reviewed (Ratcliffe, 1995; 1997). Attempts have been made to correlate metabolic proton production with changes in cytoplasmic and vacuolar pH during anoxia. Roberts *et al.* (1984; 1992) found that lactate production was closely correlated with intracellular acidification in anoxic maize root tips, while Saint-Ges *et al.* (1991) did not observe a tight coupling between lactate and protons in this tissue. Lactate efflux is known to be an important means of pH_i regulation in plants during anoxia, and Xia and Roberts (1994) demonstrated with ^{31}P-NMR that exposure of maize root tips to 3% O_2 prior to anoxia (so-called acclimation) led to an enhanced efflux of lactate and an increased ability to regulate pH_i. Cytoplasmic pH has also been measured in the context of the pH-stat model, which states that the switch from lactate to ethanol production following the imposition of anoxia is triggered by a drop in cytoplasmic pH (Davies, Grego & Kenworthy, 1974). NMR was first used to test this model by Roberts *et al.* (1984), who showed that experimental reductions in pH$_i$ in maize root tips prior to the onset of anoxia accelerated the switch to ethanol production. Fox, McCallan and Ratcliffe (1995) offered further evidence for the pH-stat model in the same tissue by observing that ethanol production was halted when anoxia was imposed under conditions which facilitated the recovery of pH$_i$ to normal, aerobic values. These results showed that pH$_i$ can reversibly trigger the switch from lactate to ethanol production. Similar NMR results have shown that the production of γ-aminobutyrate during hypoxia is also pH triggered in plants (Carroll *et al.*, 1994; Ford, Ratcliffe & Robins, 1996).

^{31}P-NMR has also been used to examine cytoplasmic and vacuolar pH in plants during changes in the external ionic environment. Hyperosmotic conditions caused a substantial increase in vacuolar pH, while cytoplasmic pH continued to be tightly regulated in barley roots (Martinez & Läuchli, 1993) and in mung bean root tips (Nakamura *et al.*, 1992). In both studies, high external $[Ca^{2+}]$ was found to diminish vacuolar alkalinization. Spickett, Smirnoff and Ratcliffe (1992) monitored cytoplasmic and vacuolar pH, as well as phosphorus metabolite levels in ^{31}P-spectra of maize root tips during hyperosmotic shock. These authors observed treatment-induced increases in pH of both compartments, increases in the levels of phosphocholine and vacuolar phosphate, and transient increases followed by recovery to initial levels of cytoplasmic phosphate.

The high surface area to volume of cell suspensions makes them ideal for examining membrane transport phenomena. Sakano, Yazaki and Mimura (1992) monitored the cytoplasmic and vacuolar acidification associated with inorganic phosphate uptake in cultured cell suspensions of *Catharanthus roseus*. Gout, Bligny and Douce (1992) examined the effects of changes in pH$_e$ on transmembrane proton movement in sycamore cell suspensions. In this study, ^{13}C-NMR was used to monitor the pH-sensitive resonances of the CH_2-linked carboxyl groups of citric acid and the ^{13}C-enriched carbon of bicarbonate in order to measure pH at the acid and basic extremes, respectively, while ^{31}P-NMR was used to measure pH at intermediate values.

Conclusions

NMR is well established as an accurate, and in many ways a straightforward, method of measuring pH in living systems. The tremendous advantages offered by the non-invasive nature of NMR promise to maintain it as a powerful tool, while its insensitivity secures it as one of many methods for examining acid–base balance. The advances in hardware made in each new generation of spectrometers, and the increasing availability of high-field magnets to physiologists, are reassuring signs that NMR methods of studying pH in living systems will continue to improve. In fact, the development of magnetic susceptibility-matched or superconducting transceiver coils may be the next major advancement for biological studies. These improvements are also sure to widen the applicability of imaging and localized spectroscopy methods to plant and animal acid–base studies. Although multicompartmental pH analyses are routine in NMR studies of plants, this important area has yet to be satisfactorily addressed in animals, but the further characterization of pH indicators may make this possible in the future.

Acknowledgements

The authors thank Professor W. Ross Ellington for his review of this chapter and for many helpful discussions. The research was supported by a National Institutes of Health postdoctoral fellowship to S.T.K. (F32 DK09571-01) while at the National High Magnetic Field Laboratory, and an American Heart Association Florida Affiliate Grant-in-Aid to T.S.M.

References

Ackerman, J.J.H., Soto, G.E., Spees, W.M., Zhu, Z. & Evelhoch, J.L. (1996). The NMR chemical shift pH measurement revisited: analysis of error and modeling of a pH dependent reference. *Magnetic Resonance in Medicine* **36**, 674–83.

Aime, S., Botta, M., Milone, L. & Terreno, E. (1996). Paramagnetic complexes as novel NMR pH indicators. *Chemical Communications* **11**, 1265–6.

Allen, D.G., Morris, P.G., Orchard, C.H. & Pirolo, J.S. (1985). A nuclear magnetic resonance study of metabolism in the ferret heart during hypoxia and inhibition of glycolysis. *Journal of Physiology, London* **361**, 185–204.

Barrow, K.D., Jamieson, D.D. & Norton, R.S. (1980). [31]P Nuclear-magnetic-resonance studies of energy metabolism in tissue from the marine invertebrate *Tapes watlingi. European Journal of Biochemistry* **103**, 289–97.

Busa, W.B. & Crowe, J.H. (1983). Intracellular pH regulates transitions between dormancy and development of brine shrimp (*Artemia salina*) embryos. *Science* **221**, 366–8.

Carroll, A.D., Fox, G.G., Laurie, S, Phillips, R., Ratcliffe, R.G. & Stewart, G.R. (1994). Ammonium assimilation and the role of γ-aminobutyric acid in pH homeostasis in carrot cell suspensions. *Plant Physiology* **106**, 513–20.

Chacko, V.P. & Weiss, R.G. (1993). Intracellular pH determination by ^{13}C-NMR spectroscopy. *American Journal of Physiology* **264**, C755–60.

Chiba, A., Hamaguchi, M., Kosaka, M., Tokuno, T., Asai, T. & Chichibu, S. (1990). Energy metabolism in unrestrained fish with *in vivo* ^{31}P-NMR. *Comparative Biochemistry and Physiology* **96A(2)**, 253–355.

Chih, C.P. & Ellington, W.R. (1985). Metabolic correlates of intracellular pH change during rapid contractile activity in a molluscan muscle. *Journal of Experimental Zoology* **236**, 27–34.

Combs, C.A. & Ellington, W.R. (1995). Graded intracellular acidosis produces extensive and reversible reductions in the effective free energy change of ATP hydrolysis in a molluscan muscle. *Journal of Comparative Physiology B* **165**, 203–12.

Davies, D.D., Grego, S. & Kenworthy, P. (1974). The control of the production of lactate and ethanol by higher plants. *Planta* **118**, 297–310.

DeFronzo, M. & Gillies, R.J. (1987). Characterization of methylphosphonate as a ^{31}P NMR pH indicator. *Journal of Biological Chemistry* **262**, 11032–7.

Douman, C. & Ellington, W.R. (1992). Intracellular free magnesium in the muscle of an osmoconforming marine invertebrate: measurement and effect of metabolic and acid–base perturbations. *Journal of Experimental Biology* **261**, 394–405.

Dubyak, G.R. & Scarpa, A. (1983). Phosphorus-31 nuclear magnetic resonance studies of single muscle cells isolated from barnacle depressor muscle. *Biochemistry* **22**, 3531–6.

Ellington, W.R. (1983). The extent of intracellular acidification during anoxia in the catch muscles of two bivalve molluscs. *Journal of Experimental Zoology* **227**, 313–17.

Ellington, W.R. (1985). Metabolic impact of experimental reductions of intracellular pH in molluscan cardiac muscle. *Molecular Physiology* **7**, 155–64.

Ellington, W.R. (1989). Phosphocreatine represents a thermodynamic and functional improvement over other muscle phosphagens. *Journal of Experimental Biology* **143**, 177–94.

Ellington, W.R. & Wiseman, R.W. (1989). Nuclear magnetic resonance spectroscopic techniques for the study of cellular function. *Advances in Comparative and Environmental Physiology* **5**, 77–113.

Fan, T.W.-M., Higashi, R.M. & Macdonald, J.M. (1991). Emergence and recovery response of phosphate metabolites and intracellular pH in intact *Mytilus edulis* as examined *in situ* by *in vivo* ^{31}P-NMR. *Biochimica et Biophysica Acta* **1092**, 39–47.

Ford, Y.Y., Ratcliffe, R.G. & Robins, R.J. (1996). Phytohormone-induced GABA production in transformed root cultures of *Datura stramonium*: an *in vivo* ^{15}N NMR study. *Journal of Experimental Botany* **47**, 811–18.

Fox, G.G., McCallan, N.R. & Ratcliffe, R.G. (1995). Manipulating cytoplasmic pH under anoxia: a critical test of the role of pH in the switch from aerobic to anaerobic metabolism. *Planta* **195**, 324–30.

Frahm, J., Merboldt, K.D. & Hanicke, W. (1987). Localized proton spectroscopy using stimulated echoes. *Journal of Magnetic Resonance* **72**, 502–8.

Gadian, D.G. (1995). *NMR and its applications to living systems.* Oxford: Oxford University Press.

Garlick, P.B., Soboll, S. & Bullock, G.R. (1992). Evidence that mitochondrial phosphate is visible in ^{31}P NMR spectra of isolated, perfused rat hearts. *NMR in Biomedicine* **5**, 29–36.

Gillies, R.J., Liu, Z. & Bhujwalla, Z. (1994). ^{31}P MRS measurements of extracellular pH of tumors using 3-aminopropyl phosphonate. *American Journal of Physiology* **267**, C195.

Gout, E., Bligny, R. & Douce, R. (1992). Regulation of intracellular pH values in higher plant cells: carbon-13 and phosphorus-31 nuclear magnetic resonance studies. *Journal of Biological Chemistry* **267(20)**, 13903–9.

Graham, R.A. & Ellington, W.R. (1985). Phosphorus nuclear magnetic resonance studies of energy metabolism in molluscan tissues: intracellular pH change and the qualitative nature of anaerobic end products. *Physiological Zoology* **58(4)**, 478–90.

Graham, R.A., Taylor, A.H. & Brown, T.R. (1994). A method for calculating the distribution of pH in tissues and a new source of pH error from the ^{31}P-NMR spectrum. *American Journal of Physiology* **266**, R638–45.

Gupta, R.K., Benovic, J.L. & Rose, Z.B. (1978). Magnetic resonance studies of the binding of ATP and cations to human hemoglobin. *Journal of Biological Chemistry* **253**, 6165–71.

Gupta, R.K. & Moore, R.D. (1980). ^{31}P NMR studies of intracellular free Mg^{2+} in intact frog skeletal muscle. *Journal of Biological Chemistry* **255**, 3987–93.

Hamm, J.R. & Yue, G.M. (1987). ^{31}P nuclear magnetic resonance measurements of intracellular pH in giant barnacle muscle. *American Journal of Physiology* **252**, C30–7.

Hentrich, S., Hebeler, M., Grimme, L.H., Leibfritz, D. & Mayer, A. (1993). P-31 NMR saturation transfer experiments in *Chlamydomonas reinhardtii*: evidence for the NMR visibility of chloroplastidic P_i. *European Biophysics Journal* **22**, 31–9.

Hitzig, B.M., Perng, W.-C., Burt, T., Okunieff, P. & Johnson, D.C. (1994). ^{1}H-NMR measurement of fractional dissociation of imidazole in intact animals. *American Journal of Physiology* **266**, R1008–15.

Jackson, D.C., Arendt, E.A., Inman, K.C., Lawler, R.G., Panol, G. &

Wasser, J.S. (1991). ^{31}P-NMR study of normoxic and anoxic perfused turtle heart during graded CO_2 and lactic acidosis. *American Journal of Physiology* **260**, R1130–6.

Johnson, D.C., Burt, C.T., Perng, W.-C., & Hitzig, B.M. (1993). Effects of temperature on muscle pH_i and phosphate metabolites in newts and lungless salamanders. *American Journal of Physiology* **265**, R1162–7.

Juretschke, H.P. & Kamp, G. (1990). Influence of intracellular pH on reduction of energy metabolism during hypoxia in the lugworm *Arenicola marina*. *The Journal of Experimental Zoology* **256**, 255–63.

Kamp, G. & Juretschke, H.P. (1989). Hypercapnic and hypocapnic hypoxia in the lugworm *Arenicola marina*: a ^{31}P NMR study. *Journal of Experimental Zoology* **252**, 219–27.

Kinsey, S.T. & Ellington, W.R. (1995). Interspecific comparisons of capacity for regulation of intracellular pH in molluscan muscle. *Physiological Zoology* **68(1)**, 26–42.

Kinsey, S.T. & Ellington, W.R. (1996). ^{1}H- and ^{31}P-nuclear magnetic resonance studies of L-lactate transport in isolated muscle fibers from the spiny lobster *Panulirus argus*. *Journal of Experimental Biology* **199**, 2225–34.

Kost, G.J. (1990). pH standardization for phosphorus-31 magnetic resonance heart spectroscopy at different temperatures. *Magnetic Resonance in Medicine* **14**, 496–506.

Kushmerick, M.J., Dillon, P.F., Meyer, R.A., Brown, T.R., Krisanda, J.M. & Sweeney, H.L. (1986). ^{31}P NMR spectroscopy, chemical analysis, and free Mg^{2+} of rabbit bladder and uterine smooth muscle. *Journal of Biological Chemistry* **261**, 14420–9.

Madden, A., Leach, M.O., Sharp, J.C., Collins, D.J. & Easton, D. (1991). A quantitative analysis of the accuracy of *in vivo* pH measurements with ^{31}P NMR spectroscopy: assessment of pH measurement methodology. *NMR in Biomedicine* **4(1)**, 1–10.

Martinez, V. & Läuchli, A. (1993). Effects of Ca^{2+} on the salt-stress response of barley roots as observed by *in vivo* ^{31}P-nuclear magnetic resonance and *in vitro* analysis. *Planta* **190**, 519–24.

Meyer, R.A., Brown, T.R. & Kushmerick, M.J. (1985). Phosphorus nuclear magnetic resonance of fast- and slow-twitch muscle. *American Journal of Physiology* **248**, C279–87.

Moon, R.B. & Richards, J.H. (1973). Determination of intracellular pH by ^{31}P magnetic resonance. *Journal of Biological Chemistry* **248(20)**, 7276–8.

Morikawa, S., Inubushi, T., Kito, K. & Kido, C. (1993). pH mapping in living tissues: an application of *in vivo* ^{31}P NMR chemical shift imaging. *Magnetic Resonance in Medicine* **29**, 249–51.

Morris, G.M. & Baldwin, J. (1984). pH buffering capacity of invertebrate muscle: correlations with anaerobic work. *Molecular Physiology* **5**, 61–70.

Nakamura, Y., Ogawa, T., Kasamo, K., Sakata, M. & Ohta, E. (1992). Changes in cytoplasmic and vacuolar pH in intact cells of mung bean root-tips under high-NaCl stress at different external concentrations of Ca^{2+} ions. *Plant Cell Physiology* **33(7)**, 849–58.

Navon, G., Ogawa, S., Shulman, R.G. & Yamane, T. (1977). ^{31}P nuclear magnetic resonance studies of Ehrlich ascites tumor cells. *Proceedings of the National Academy of Science USA* **74**, 87–91.

Okuma, E. & Abe, H. (1992). Major buffering constituents in animal muscle. *Comparative Biochemistry and Physiology* **102A**, 37–41.

Ordidge, R.J., Bendal, M.R., Gordon, R.E. & Connelly, A. (1985). Volume selection for *in vivo* biological spectroscopy. In *Magnetic resonance in biology and medicine*, ed. G. Govil, C.L. Khetrapal & A. Saran, pp. 387–97. New Delhi: McGraw-Hill.

Ordidge, R.J., Connelly, A. & Lohman, J.A.B. (1986). Image-selected *in vivo* spectroscopy (ISIS). A new technique for spatially selective NMR spectroscopy. *Journal of Magnetic Resonance* **66**, 283–94.

Pan, J.W., Hamm, J.R., Rothman, D.L. & Shulman, R.G. (1988). Intracellular pH in human skeletal muscle by ^{1}H NMR. *Proceedings of the National Academy of Science USA* **85**, 7836–9.

Ratcliffe, R.G. (1991). Nuclear magnetic resonance in plant science research. *Botanical Society of Scotland* **46**, 107–20.

Ratcliffe, R.G. (1994). *In vivo* nuclear magnetic resonance studies of higher plants and algae. In *Advances in botanical research*, ed. J.A. Callow, Vol. 20, pp. 43–123. London: Academic Press.

Ratcliffe, R.G. (1995). Metabolic aspects of the anoxic response in plant tissue. In *Environment and plant metabolism: flexibility and acclimation*, ed. N. Smirnoff, pp. 111–27. Oxford: BIOS Scientific Publishers.

Ratcliffe, R.G. (1997). *In vivo* NMR studies of the metabolic response of plants tissues to anoxia. *Annals of Botany* **79 (Suppl. A)**, 39–48.

Rees, B.B., Malhotra, D., Shapiro, J.I. & Hand, S.C. (1991). Intracellular pH decreases during entry into estivation in the land snail *Oreohelix strigosa*. *Journal of Experimental Biology* **159**, 525–30.

Roberts, J.K.M., Callis, J., Wemmer, D., Walbot, V. & Jardetzky, O. (1984). Mechanism of cytoplasmic pH regulation in hypoxic maize root tips and its role in survival under hypoxia. *Proceedings of the National Academy of Sciences USA* **81**, 3379–83.

Roberts, J.K.M., Hooks, M.A., Miaullis, A.P., Edwards, S. & Webster, C. (1992). Contribution of malate amino acid metabolism to cytoplasmic pH regulation in hypoxic maize root tips studied using nuclear magnetic resonance spectroscopy. *Plant Physiology* **98**, 480–7.

Roberts, J.K.M., Wade-Jardetzky, N. & Jardetzky, O. (1980). Intracellular pH measurements by ^{31}P nuclear magnetic resonance.

Influence of factors other than pH on the ^{31}P chemical shifts. *Biochemistry* **20**, 5389–94.

Robitaille, P.-M.L., Robitaille, P.A., Brown, G.G. Jr & Brown, G.G. (1991). An analysis of the pH-dependent chemical shift behavior of phosphorus-containing compounds. *Journal of Magnetic Resonance* **92**, 73–84.

Rudin, M. & Sauter, A. (1992). *In vivo* phosphorus-31 NMR potential and limitations. In *In vivo magnetic resonance spectroscopy III: In vivo MR spectroscopy: potential and limitations*, ed. P. Diehl, E. Fluck, H. Gunther, R. Kosfield & J. Seelig, Vol. 28, pp. 161–88. Berlin: Springer-Verlag.

Saint-Ges, V., Roby, C., Bligny, R., Pradet, A. & Douce, R. (1991). Kinetic studies of the variations of cytoplasmic pH, nucleotide triphosphates (^{31}P NMR) and lactate during normoxic and anoxic transitions in maize root tips. *European Journal of Biochemistry* **200**, 477–82.

Sakano, K., Yazaki, Y. & Mimura, T. (1992). Cytoplasmic acidification induced by inorganic phosphate uptake in suspension cultured *Cataranthus roseus* cells: measurement with fluorescent pH indicator and ^{31}P-nuclear magnetic resonance. *Plant Physiology* **99**, 672–80.

Sherry, A.D. & Malloy, C.R. (1996). Isotopic methods for probing organization of cellular metabolism. *Cell Biochemistry and Function* **14**, 259–68.

Skibbe, U., Christeller, J.T., Callaghan, P.T., Eccles, C.D. & Laing, W.A. (1996). Visualization of pH gradients in the larval midgut of *Spodoptera litura* using ^{31}P-NMR microscopy. *Journal of Insect Physiology* **42(8)**, 777–90.

Skibbe, U., Christeller, J.T., Eccles, C.D., Laing, W.A. & Callaghan, P.T. (1995). Phosphorus imaging as a tool for studying the pH metabolism in living insects. *Journal of Magnetic Resonance, Series B* **108**, 262–8.

Soto, G.E., Zhu, Z., Evelhoch, J.L. & Ackerman, J.J.H. (1996). Tumor ^{31}P NMR pH measurements *in vivo*: A comparison of inorganic phosphate and intracellular 2-deoxyglucose-6-phosphate as pH_{nmr} indicators in murine radiation-induced fibrosarcoma-1. *Magnetic Resonance in Medicine* **36**, 698–704.

Spickett, C.M., Smirnoff, N. & Ratcliffe, R.G. (1992). Metabolic response of maize roots to hyperosmotic shock: an *in vivo* ^{31}P nuclear magnetic resonance study. *Plant Physiology* **99**, 856–63.

Szwergold, B.J., Brown, T.R. & Freed, J.J. (1989). Bicarbonate abolishes intracellular alkalinization in mitogen-stimulated 3T3 cells. *Journal of Cellular Physiology* **235**, 227–35.

Taylor, J.S. & Deutsch, C. (1983). Fluorinated α-methylamino acids as ^{19}F NMR indicators of intracellular pH. *Biophysical Journal* **43**, 261–7.

Taylor, J.S. & Deutsch, C. (1988). ^{19}F-Nuclear magnetic resonance: measurements of [O_2] and pH in biological systems. *Biophysical Journal* **53**, 227–33.

Taylor, J.S. & Deutsch, C. (1989). New class of ^{19}F pH indicators: fluoroanilines. *Biophysical Journal* **55**, 799–804.

Thebault, M.T., Raffin, J.P. & Le Gall, J.Y. (1987). *In vivo* ^{31}P NMR in crustacean muscles: fatigue and recovery in the tail musculature from the prawn *Palaemon elegans*. *Biochemical and Biophysical Research Communications* **145(1)**, 453–9.

Van den Thillart, G., Körner, F., van Waarde, A., Eckelens, C. & Lugtenburg, J. (1989). A flow through probe for *in vivo* ^{31}P NMR spectroscopy of unanesthetized aquatic vertebrates at 9.4 tesla. *Journal of Magnetic Resonance* **84**, 573–9.

Van den Thillart, G. & van Waarde, A. (1996). Nuclear magnetic resonance spectroscopy in living systems: applications to comparative physiology. *Physiological Reviews* **76(3)**, 799–837.

Van Ginneken, V., van den Thillart, G., Addink, A. & Erkelens, C. (1995). Fish muscle energy metabolism measured during hypoxia and recovery: an *in vivo* ^{31}P-NMR study. *American Journal of Physiology* **268**, 1178–87.

Wasser, J.S., Inman, K.C., Arendt, E.A., Lawler, R.G. & Jackson, D.C. (1990). ^{31}P-NMR measurements of pH_i and high-energy phosphates in isolated turtle hearts during anoxia and acidosis. *American Journal of Physiology* **259**, R521–30.

Wasser, J.S., Lawler, R.G. & Jackson, D.C. (1996). Nuclear magnetic resonance spectroscopy and its applications in comparative physiology. *Physiological Zoology* **69(1)**, 1–34.

Webb, A.G. & Grant, S.C. (1996). Signal-to-noise and magnetic susceptibility trade-offs in solenoidal microcoils for NMR. *Journal of Magnetic Resonance, Series B*, **113**, 83–7.

Wegener, G., Bolas, N.M. & Thomas, A.A.G. (1991). Locust flight muscle metabolism studied *in vivo* by ^{31}P NMR. *Journal of Comparative Physiology* **161B**, 247–56.

Williams, G.D., Mosher, T.J. & Smith, M.B. (1993). Simultaneous determination of intracellular magnesium and pH from the three ^{31}P NMR chemical shifts of ATP. *Analytical Biochemistry* **214**, 458–67.

Wiseman, R.W. & Ellington, W.R. (1989). Intracellular buffering capacity in molluscan muscle: superfused muscle *versus* homogenates. *Physiological Zoology* **62**, 541–58.

Wiseman, R.W., Moerland, T.S. & Kushmerick, M.J. (1993). Biological applications for small solenoids: NMR spectroscopy of microliter volumes at high-fields. *NMR in Biomedicine* **6**, 153–6.

Xia, J.-H. & Roberts, J.K.M. (1994). Improved cytoplasmic pH regulation, increased lactate efflux, and reduced cytoplasmic lactate levels are biochemical traits expressed in root tips of whole maize seedlings acclimated to a low-oxygen environment. *Plant Physiology* **105**, 651–7.

Zange, J., Grieshaber, M.K. & Jans, A.W.H. (1990). The regulation of intracellular pH estimated by ^{31}P-NMR spectroscopy in the anterior byssus retractor muscle of *Mytilus edulis* L. *Journal of Experimental Biology* **150**, 95–109.

H.O. PÖRTNER and F.J. SARTORIS

Invasive studies of intracellular acid–base parameters: quantitative analyses during environmental and functional stress

Introduction

In recent years, studies of the effects of environmental variables on the physiology and biochemistry of different animal species have increasingly included an analysis of acid–base status and regulation. pH values in different body compartments are widely accepted to play a key role in the maintenance of physiological function or its limitation under functional or environmental stress. pH affects protein function in metabolism and O_2 transport. Also, acid–base and metabolic regulation are interdependent processes such that changes in pH may affect metabolic rate, the mode of catabolism and energetic parameters. Ideally, these analyses should not only describe correlated changes in the different processes under investigation, but should also provide a quantitative picture of the changes involved and the processes responsible for them (Heisler, 1989b).

However, acid–base regulation not only means adjustment or defence of pH, which is traditionally seen as being the key acid–base parameter determining regulatory processes, but it may, under certain conditions and with the help of the respective membrane carriers (see below), also give priority to the regulation of the levels of base (carbonate, bicarbonate) or acid (carbonic acid, proportional to P_{CO_2}) in the respective body fluids. In that sense, pH would become a dependent variable. Also, for some treatments it is not pH which is of interest but rather the activity of protons (pH = $-\log a_{H^+}$), when protons contribute to some biochemical reactions in a concentration-dependent manner (see equation 1, p. 73, as an example). In general, biochemical treatments of acid–base regulation usually focus on intracellular pH (pH_i) as a key parameter related to protein function, whereas physiological, cellular and especially whole-animal studies have always considered the close interrelationships between pH and the CO_2/bicarbonate system in intracellular and extracellular fluids (Siggaard-Andersen, 1974). In support of the latter concept, the involvement of the species CO_2 and bicarbonate as substrates or products in enzymatic reactions has become apparent (Pörtner, 1989; Walsh & Milligan, 1989; Hardewig, Pörtner & Grieshaber, 1994), justifying the adoption of physiological concepts of acid–base regulation in metabolic biochemistry.

Acid–base regulation is an energy-dependent process because some of the acid–base equivalents are transported by H^+-ATPases or by secondary active processes, for example *via* the Na^+/H^+ exchanger, which depends upon the Na gradient established by Na^+/K^+-ATPase. It has recently been suggested that certain species are capable of modulating the cost of acid–base regulation as a means of adjusting the rate of energy turnover to environmental requirements (Reipschläger & Pörtner, 1996). In addition, the significance of metabolism for acid–base regulation has been increasingly discussed. Interest has focused not only on disturbances of the acid–base status by metabolism, but also on the contribution to acid–base homeostasis by metabolism. These investigations have been applied to both aerobic (Atkinson & Camien, 1982; Häussinger *et al.*, 1988; Atkinson & Bourke, 1995; Pörtner, 1989; 1995) and anaerobic metabolism (Hochachka & Mommsen, 1983; Pörtner, Heisler & Grieshaber, 1984; Pörtner, 1987a; 1989).

A full set of acid–base parameters in whole-animal research

Bearing these general ideas in mind, studying acid–base status involves the choice of an appropriate methodology for the analysis of acid–base parameters. Acid–base regulation occurs at systemic, cellular and subcellular levels. This chapter focuses on intracellular acid–base parameters. For research concentrating on environmental and functional issues, these parameters need to be largely investigated in whole animals, unconfined and as close as possible to their natural situation (for example dwelling in burrows), or in animals during and after exercise. Sometimes, acid–base balance needs to be studied after long-term exposures (between hours and months) to fluctuations in environmental variables such as O_2, CO_2, salinity, temperature, or, most recently, hydrogen peroxide (Abele-Oeschger, Sartoris & Pörtner, 1997), whereas short-term modifications (between seconds and minutes) occur with the use of muscular activity during attack and escape (e.g. Milligan, 1996).

Microelectrode and fluorescent probes and, to some extent, ^{31}P-NMR are most suitable for cellular and subcellular investigations of acid–base parameters (see Schwiening & Thomas and Kinsey & Moerland, this volume). At higher levels of complexity, in whole animals, microelectrode and fluorescent probes are no longer applicable. Historically, the first reliable method to be used in whole animals was the measurement of the pH-dependent distribution of weak acids and bases, in particular the weak acid dimethyloxazolidine-dione (DMO), between intracellular and extracellular spaces (Waddell & Butler, 1959; for review see Roos & Boron, 1981). In brief, DMO is infused into the animal *via* an indwelling catheter and pH_i is calculated from DMO distribution and the measured values of extracellular pH (pH_e; Table 1). By using radio-labelled DMO, it is possible to determine pH_i invasively in not just one but various tissues collected from the same individual animal. However, the measurement of rapidly occurring pH_i changes, for example during muscle activity, is limited by the velocity of DMO distribution. Further disadvantages arise from the fact that pH_i can only be mathematically estimated. To do this, the following parameters must be measured: pH_e, water content of the tissue, and concentrations of radiolabelled inulin and DMO in the tissue and plasma to allow evaluation of intracellular and extracellular DMO. Each of the necessary measurements has its own inherent errors. Since these errors may be additive, this leads to a relatively high variability in calculated pH_i values. More recently, pH_i in isolated tissues and whole animals has been investigated by the use of ^{31}P-nuclear magnetic resonance (^{31}P-NMR; Kinsey & Moerland, this volume; van den Thillart & van Waarde, 1996; Wasser, Lawler & Jackson, 1996) and by an improved and reliable version of the homogenisation technique (Pörtner *et al.*, 1990). ^{31}P-NMR requires an *in vivo* analysis of the immobilised animal. Therefore, the homogenate technique is most suitable to investigate acid–base parameters in tissue samples collected from the exercised animal or even from animals in the field.

For whole-animal approaches, a quantitative picture should include co-ordinated analyses of intracellular and extracellular acid–base parameters. Quantification of acid–base parameters in all compartments will reveal the net movement of acid–base equivalents across membranes or epithelial layers (cf. Heisler, 1989a). As a corollary, treatment of the acid–base status is complete with quantitative knowledge of changes in pH, $P\text{CO}_2$ and apparent bicarbonate levels in the compartments of interest as well as knowledge of non-bicarbonate buffers resisting such changes by proton binding or release.

Homogenate analyses of pH_i

The homogenate technique allows for a clear allocation of pH values to the experimental condition and for the elimination of time delays in pH_i assessment. This method follows the freeze-stop technique, which was established for determining the metabolite status of shock frozen tissues (Wollenberger, Ristau & Schoffa, 1960). Previous versions of the homogenate technique were applied to biopsy samples of human muscle (e.g. Sahlin, Harris & Hultman, 1975; Sahlin et al., 1976). Samples need to be taken in such a way that the resting and experimental states of the tissues are maintained under control and experimental conditions. Control and experimental animals are therefore anaesthetised prior to tissue sampling and decapitation, if required, to eliminate the influence of a potential stress response (Pörtner et al., 1990; 1991b; Pörtner, MacLatchy & Toews, 1991a; Tang & Boutilier, 1991). The frozen tissue is ground under liquid nitrogen using mortar and pestle, and the tissue powder is then thawed in a medium (in a volume about five times the wet weight of the tissue) containing potassium fluoride (KF) and nitrilotriacetic acid (NTA), thereby removing Mg^{2+} and Ca^{2+} and preventing ATP-dependent metabolism, which occurs through the action of Mg^{2+}-dependent and Ca^{2+}-dependent ATPases and kinases. ATP-dependent anaerobic metabolism is responsible for the pH changes that occur in tissue homogenates after subcellular structures have been destroyed. According to model calculations (Pörtner et al., 1990), any distortion in the measured pH values due to the pH of the medium, dilution by the medium or mixing with intracellular or extracellular fluids can be disregarded. The special merits of the homogenate technique are the simple methodological procedure, low costs, low variability and the small sample volume required (in reaction tubes with a small enough volume, analysis is possible with sample sizes down to 20 mg fresh weight).

On the practical side, pitfalls in the use of the homogenate technique can be avoided when it is considered that CO_2 condensation needs to be minimised during the process of homogenate preparation. The use of clean liquid nitrogen and dewars, mortars and pestle free of rime during a short but efficient grinding procedure under a nitrogen atmosphere excludes mixing with condensating CO_2. (It is usually sufficient to grind on the bottom of a box, e.g. Styrofoam, thereby allowing the evaporating nitrogen to fill up the volume above mortar and pestle.) pH is best measured in a thermostatted capillary pH electrode, but pH microelectrodes have also been used (Krause & Wegener, 1996) after preparation and centrifugation of the homogenate in a closed (usually 0.5 ml) Eppendorf cap. If required, complete tissue extraction can be ensured by treatment of the closed Eppendorf cap with ultrasound (Sommer, Klein & Pörtner, 1997). The pH electrode and supernatant (inside the capillary electrode) are thermostatted to the experimental temperature of the animal in order to avoid temperature artefacts. NTA supports rapid binding of Ca^{2+} and Mg^{2+}. However, the concentration of NTA needs to be minimised because at too high levels this substance releases protons, a process minimised or excluded by the formation and precipitation of magnesium or calcium fluorides.

The method has been used so far in studies of invertebrate (annelid, mollusc, sipunculid) and vertebrate (amphibian, fish, reptilian, mammalian) tissues (e.g. Pörtner, *et al.*, 1990; 1991a; 1991b; 1998; Hardewig *et al.*, 1991a; 1991b; Tang & Boutilier, 1991; Schulte, Moyes & Hochachka, 1992; Boutilier *et al.*, 1993; Branco, Pörtner & Wood, 1993; Ferguson, Kiefer & Tufts, 1993; Claiborne, Walton & Compton-McCullough, 1994; Kiefer, Currie & Tufts, 1994; Wang, Heigenhauser & Wood, 1994; 1996a; 1996b; Day & Butler, 1996; Pörtner, Finke & Lee, 1996; Schmidt *et al.*, 1996; Zielinski & Pörtner, 1996; Sommer *et al.*, 1997; Larsen, Pörtner & Jensen, 1997; see also Wood & Wang, this volume). Recently, the homogenate technique has been used to quantify passive and active contributions to temperature-induced changes in pH_i and to study the relevance of Mg^{2+} in acid–base regulation (van Dijk, Hardewig & Pörtner, 1997; Sartoris & Pörtner, 1997a; 1997b; see below).

The accuracy of the method for each individual pH measurement is confirmed by the demonstration of a strong correlation of measured pH_i changes with changes in metabolic parameters (Pörtner *et al.*, 1991b; 1996). Phosphagen breakdown and anaerobic glycolysis are predominantly responsible for metabolic changes in the acid–base status of working muscle starting beyond the anaerobic threshold (Pörtner *et al.*, 1996). These relationships have been studied in invertebrate (molluscan, arthropod, sipunculid) muscles in which the phosphagen is phospho-L-arginine (PLA) instead of phosphocreatine and where octopine, other opines or lactate are formed (Grieshaber *et al.*, 1994). A speciality compared to glycolysis and phosphocreatine depletion in vertebrate muscle is that L-arginine released during PLA depletion may be absorbed in octopine formation. Because ATP content is initially maintained at the expense of phosphagen breakdown, intracellular alkalosis is characteristic for the initial stages of anaerobic muscle activity (e.g. Chih & Ellington, 1985) based on equation (1):

$$\text{Phospho-L-arginine (PLA}^-) + \text{MgADP}^- + \text{H}^+ \Leftrightarrow \qquad (1)$$

$$\text{L-arginine (L-Arg}^+) + \text{MgATP}^{2-}$$

$$\text{L-arginine (L-Arg}^+) + 0.5 \text{ glucose} \Leftrightarrow \text{octopine} + \text{H}^+ \qquad (2)$$

After initial alkalisation, glycolytic ATP and H^+ production become predominant during maintained *anaerobic* muscle activity, causing pH_i

to fall below the control level (see equation 2; for details, Pörtner, 1987a).

Analyses in the mantle musculature of various squid species have revealed a highly significant linear correlation between changes in pH_i and glycolytic end-product (octopine) accumulation (Fig. 1), the formation of octopine being equivalent in its acidifying effect to lactate generation. The linearity of the relationship between octopine concentration and pH_i suggests a well-balanced response of intracellular physicochemical and metabolic buffering processes as well as proton equivalent ion exchange between tissue and extracellular space to glycolytic proton generation *in vivo*. An *in vivo* buffer value largely independent of pH results, which is higher than the non-bicarbonate buffer value (β_{NB}) determined for control conditions (resting muscle). The latter only reflects physicochemical buffering as measured *in vitro* (see Non-bicarbonate buffer values, p. 82). The increase in slope of the apparent buffer line compared to β_{NB} indicates how metabolic proton consumption can reduce the degree of glycolytic acidification (see Fig. 1). Phosphagen degradation is the main process associated with proton consumption, whereas ATP hydrolysis to ADP and Pi contributes a small amount of surplus protons. Metabolic processes that further reduce acidosis during activity include the deamination of AMP or adenosine further along the ATP degradation pathway, or the metabolism of dicarboxylic acids (malate, aspartate) in the early stages of anaerobic mitochondrial metabolism. The latter occurs when excessive oxygen requirements cannot be met by increased oxygen supply (Pörtner, 1987a).

These compensatory or additive processes in metabolism may change the slope of the octopine/pH relationship with no disruption of linearity in such a way that they either take place at a constant rate or are triggered by falling pH_i. The net channelling of proton equivalents into the extracellular space is minimal in squid (Pörtner, 1994; 1997). A fall in phospho-L-arginine levels with pH_i due to pH dependence of arginine kinase (equation 1) and a linear rise of H^+ binding by inorganic phosphate with falling pH were seen in squid mantle (see Pörtner *et al.*, 1996). These observations support the conclusion that release of inorganic phosphate during phosphagen hydrolysis is the main process affecting the degree of glycolytic acidification. The extensive use of phosphagen in *Illex illecebroses* (Pörtner, 1993) may also explain at least part of the difference seen between the squid species in Figure 1. The formation of octopine (removal of L-arginine), development of acidosis, and accumulation of free ADP co-operate in such a way that transphosphorylation of the phosphagen supports higher performance levels. Moreover, the available data strongly suggest that the develop-

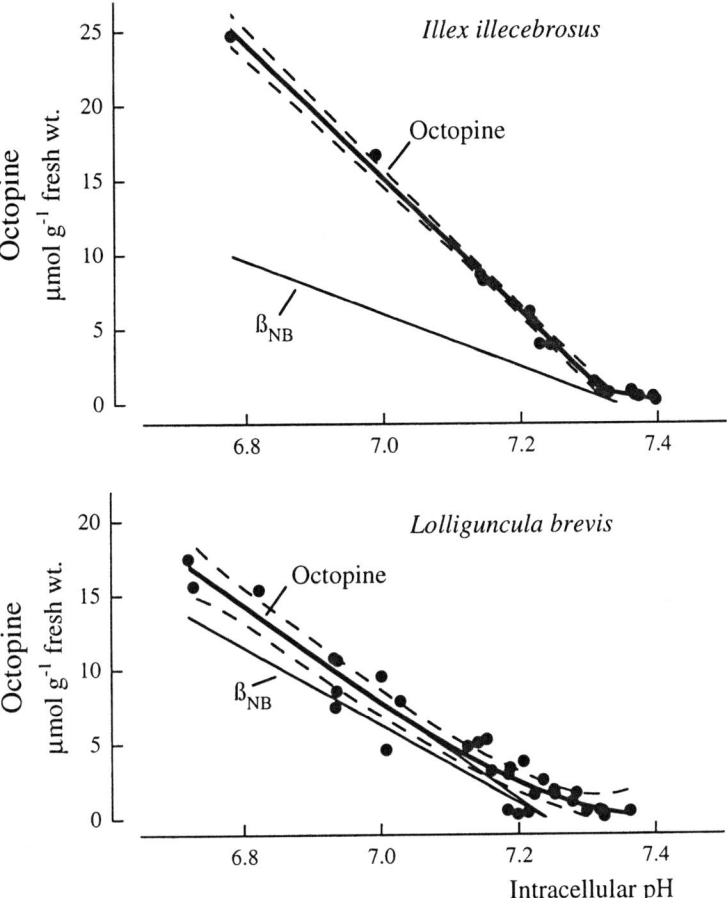

Fig. 1. Comparison of the relationship between octopine levels and pH$_i$ in mantle tissues of squids *Lolliguncula brevis* and *Illex illecebrosus*. This relationship reflects the extent to which glycolytic protons are buffered *in vivo*. Non-bicarbonate buffer lines (β_{NB}) determined *in vitro* reflect how physicochemical buffering alone would resist glycolytic acidification. The additional buffering on top of the non-bicarbonate buffer value is due to the transphosphorylation of the phosphagen triggered by the decrease in pH$_i$ and the rise in free ADP levels. Modified after Pörtner *et al.* (1996). Base release from the tissue into the blood may occur at the onset of muscular activity and explains the initial drop in pH$_i$ at rather constant octopine levels (Pörtner, 1994). Solid lines were determined by regression analysis and delineate significant relationships ($p<0.05$); dashed lines represent 95 per cent confidence intervals.

ment of acidosis may protect the adenylates from being largely degraded during fatigue (Pörtner *et al.*, 1993; 1996).

Intracellular P_{CO_2} and bicarbonate levels

The analysis of pH_i and total CO_2 levels (C_{CO_2}) in the homogenate allows the quantification of cellular bicarbonate and P_{CO_2} levels by calculation. In a first step, total CO_2 levels measured in the homogenate have to be corrected for extracellular water and C_{CO_2} content by use of an adequate marker for extracellular space (for example radiolabelled inulin) to obtain intracellular C_{CO_2} levels (Pörtner *et al.*, 1990). The calculation requires the use of pH_i and the apparent dissociation constant of the CO_2/apparent bicarbonate system, pK''', as well as the physical solubility of CO_2, α, which can be derived according to Heisler (1986a):

$$P_iCO_2 = C_{CO_2} / (10^{pH_i - pK'''} \cdot \alpha + \alpha) \tag{3}$$

Note that apparent bicarbonate includes all bicarbonate and carbonate species according to their effect in cellular buffering. For a full set of equations, see Pörtner *et al.* (1990).

Overall, the calculation of intracellular C_{CO_2} requires analysis or considerations of parameters analogous to those needed for the determination of pH_i from DMO distribution. In consequence, sources of error similar to those involved in DMO measurements arise, and values of intracellular P_{CO_2} calculated from the intracellular C_{CO_2} concentration (Pörtner *et al.*, 1990; 1991a; 1991b; 1996; 1998; Boutilier *et al.*, 1993; Reipschläger & Pörtner, 1996) exhibit a relatively large degree of variation. Variability is reduced by highly accurate estimates of pH_i, as with the homogenate technique which determines pH with much less signal-to-noise ratio than the DMO or ^{31}P-NMR methods. This means that each individual pH value has the same significance as that given to just the average value when using the DMO method. Estimations of intracellular P_{CO_2} would not be possible if based on individual DMO-derived pH_i values, due to their inaccuracy.

With adequate knowledge of the relationships between pH_i and pH_e, P_{CO_2} and bicarbonate levels, intracellular acid–base parameters in isolated muscle tissues could be varied and clamped by setting the adequate values in the extracellular medium (Reipschläger & Pörtner, 1996). That way it could unequivocally be demonstrated that among all acid–base parameters, only a decrease in pH_e (rather than pH_i) was suitable to cause metabolic depression during environmental stress in *Sipunculus nudus* (Fig. 2). The mechanisms responding to the decrease in pH_e are currently under investigation. A model suggests that a switch

Table 1. *The influence of pH heterogeneity in cellular compartments on compartmental levels of total CO₂ and calculated values of intracellular P_{CO_2}*

Calculations of DMO and CO_2 distribution follow a modified Henderson-Hasselbalch equation:

$$pH_i = pK' + \log[c_i/c_e \ (10^{pH_e - pK'} + 1) -1] \tag{a}$$

$$c_i = (F_c \cdot c_c + F_m \cdot c_m)/(F_c + F_m) \tag{b}$$

$$[HDMO]_m = [HDMO]_c \text{ or } (\alpha \cdot P_{CO_2})_m = (\alpha \cdot P_{CO_2})_c \tag{c}$$

$$[DMO^-]_m/[DMO^-]_c = 10^{\Delta pH} \text{ or } '[HCO_3^-]'_m \, /'[HCO_3^-]'_c = 10^{\Delta pH} \tag{d}$$

where pK' is the apparent dissociation constant of HDMO or pK''' of the CO_2/HCO_3^- system (see text), c is the concentration of ([HDMO] + [DMO⁻]) or the C_{CO_2} in mean intracellular (i), extracellular (e), mitochondrial (m) or cytosolic (c) compartment, and F is the fractional volume of that compartment.

Calculation of factors determining homogenate pH_i considers the contribution of mitochondrial and cytosolic volumes, total buffer values and pH differences (ΔpH).

$$pH_i = (F_c \cdot \beta_{c,tot} \cdot pH_c + F_m \cdot \beta_{m,tot} \cdot pH_m)/(F_c + F_m) \tag{e}$$

$$\beta_{tot} = \beta_{NB} + \beta_{CO_2} \text{ (closed system, at } pH_i) \tag{f}$$

$P_i_{CO_2}$ values (1 mmHg = 0.1333 kPa; 1 kPa = 7.502 mmHg) are calculated (see text, equation 3) and compared, based on homogenate or weak acid-derived pH_i values and mean intracellular C_{CO_2} evaluated in a toad, *Bufo marinus* (see Pörtner *et al.*, 1990; 1991b). The pH difference between mitochondrial matrix (m) and cytosol (c) and the fraction of mitochrondrial matrix fluid (F_m) in the cell is assumed to be similar in toad ventricle to that described for rat ventricle. With a mitochondrial content of around 35 per cent (Smith & Page, 1976; Hoppeler *et al.*, 1984), 19 per cent of the cell fluid is attributed to the mitochondrial matrix ($F_m = 0.19$, $F_c = 0.81$), with a pH gradient of 0.63 units between cytosol and matrix fluids (Kauppinen, Hiltunen & Hassinen, 1980; Kauppinen, 1983; see Table 2).

Ventricle: $C_{CO_2,i} = 8.45$ mmol l^{-1} cell water
$\Delta pH_{m-c} = 0.63$
$pH_{i,hom} = 7.13$ → $P_i_{CO_2} = 17.2$ mmHg
$pH_{i,weak \ acid} = 7.35$ → $P_i_{CO_2} = 10.8$ mmHg

Precondition: mitochondrial and cytosolic P_{CO_2} values are more or less equal. Test case: the adequate mean pH_i should give the same intracellular P_{CO_2} as do cytosolic and mitochondrial pH.

$\Delta pH_{m-c} = 0.63$ → $C_c_{CO_2} = 5.46$, $C_m_{CO_2} = 21.53$ (mmol l^{-1})
$pH_{i,weak \ acid} = 7.35$ → $pH_c = 7.15$, $pH_m = 7.78$
  → $P_c_{CO_2} = 10.8$ mmHg $= P_m_{CO_2} = P_i_{CO_2}$

Conclusion: $pH_{i, \ weak \ acid}$ is required for $P_i_{CO_2}$ calculations.

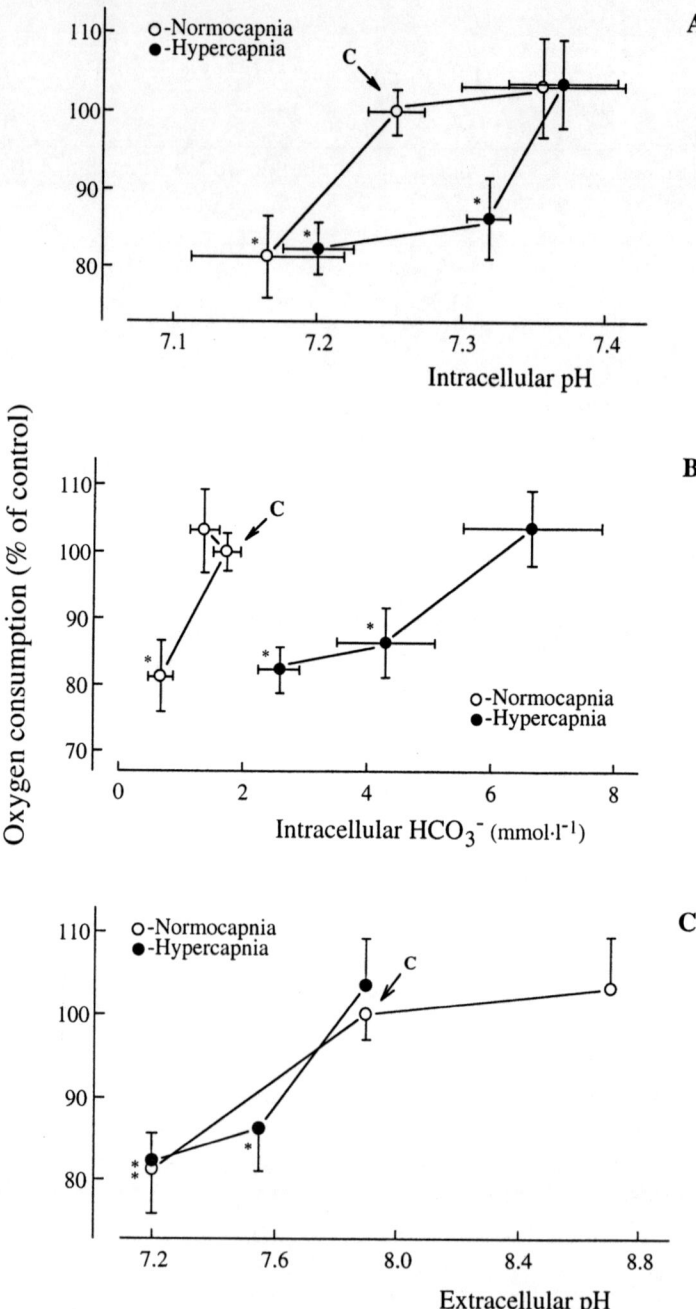

Fig. 2.

occurs from less ATP-efficient but more flexible acid–base transporters to more ATP-efficient exchange mechanisms (Reipschläger & Pörtner, 1996).

Cellular compartmentalisation

pH values in tissues poor in mitochondria, determined using homogenates, are in good agreement with the mean pH values obtained in DMO studies (cf. Pörtner *et al.*, 1990; Wood & Wang, this volume) and by ^{31}P-NMR (cf. Zange *et al.*, 1990). In cells containing large fractions of mitochondria, these values differ depending on the fraction of mitochondria and the pH gradient maintained between these organelles and the cytoplasm (the contribution of other organelles appears to be less relevant). The contribution of cellular compartments to determining the average pH_i (homogenate) follows their percentage contribution to cellular buffering and their relative volume (mixing two identical volumes of the same buffer values would yield the arithmetic mean of the two pH values). In mitochondria, pH is higher and, in consequence, total CO_2 levels are also higher than in the cytosol owing to pH-dependent distribution (Pörtner *et al.*, 1990; 1991a; 1991b; Tables 1 & 2). During the measurement procedure, all buffers (including total CO_2, which comprises the CO_2/bicarbonate buffers) are mixed in a closed system so that there is no exchange of gases, most importantly CO_2 between the homogenate and the air. This reduces the influence of the CO_2/bicarbonate buffer on homogenate pH. A much larger influence would result under open-system conditions when CO_2 leaves the mixture. (For an analysis of open system characteristics see Burton, 1973; Heisler, 1986a.) The pH dependence of the CO_2/bicarbonate buffer value and the fact that it is rather low at high mitochondrial pH also cause homogenate pH to be closer to cytosolic pH than expected when mean pH only depends upon the mixed volumes (see below).

Differences prevail between techniques for determining mean pH_i in

---

Fig. 2. O_2 consumption rates of isolated body wall musculature of *Sipunculus nudus* during normocapnia and during hypercapnia (1% CO_2) depicted as a function of pH_i values (**A**), intracellular HCO_3^- concentrations (**B**), and values of pH_e (**C**). Only plot **C** is consistent for both normocapnic and hypercapnic data and demonstrates that the O_2 consumption is significantly depressed (asterisks) below a threshold value of pH_e. Modified after Reipschläger & Pörtner (1996).

Table 2. *An estimate of mitochondrial (m) and cytosolic (c) pH, HCO_3^- levels and buffer values (β_{tot} = total, NB = non-bicarbonate) in the isolated perfused rat heart, based on the pH difference observed between weak acid-derived and homogenate (hom)-derived pH_i values*

Rat heart 37 °C

$pH_{i,hom}$ $= 6.90$ $C_iCO_2 = 11.76$ mmol l^{-1}, $P_iCO_2 \approx 40$ mmHg

$pH_{i,\ weak\ acid}$ $= 7.05$, $\Delta pH_{m-c} = 0.63$

($pK''' = 6.127$, $\alpha = 0.03134$ mmol/mmHg · l cell water)

Fm $= 0.19$, Fc $= 0.81$,

$\beta_{tot} = 44.7$ mmol/pH per l cell water (at $pH_{i,\ hom}$)

	[HCO_3^-]	C_{CO_2}	pH	mean pH	β_{tot}	β_{NB}
Cytosol	6.49	7.74	6.841		49.9	47.5
				6.961		
Mitochondria	27.71	28.96	7.471		22.0	19.2

See Pörtner *et al.*, 1990; for comparison and equations, see Table 1.

The accuracy of the calculation largely depends on a correct estimate of the mitochrondrial–cytosolic pH difference and the fraction of the mitochondrial matrix (Kauppinen *et al.*, 1980; see text and legend of Table 1; concentrations given in mmol/l cell or compartmental water). The buffer value of the rat heart was adopted from Hansen and Gesser (1980). The literature value was corrected for the accumulation of inorganic phosphate expected from the degradation of high-energy phosphates (mean pH: volume-weighted mean pH, which would result with identical β_{tot} values in cytosol and mitochondria, equation (e) in Table 1). Total closed system buffer value varies by a factor of 2.27 and non-bicarbonate buffer value by a factor of 2.5 lower in the mitochondria than in the cytosol. Similar conclusions arise from a comparison of homogenate and DMO pH_i values in the cardiac ventricle of a toad, *Bufo marinus* (Pörtner *et al.*, 1990).

their ability to weight all compartmental parameters such as volumes, buffer values and pH values (homogenate technique) or volumes and pH differences only (DMO). The different techniques must therefore lead to different values of pH_i when mitochondrial density is high. Moreover, weak acid distribution characteristics cause the DMO technique not only to ignore compartmental buffer values but also to deviate even further from the arithmetic mean according to volume fractions (see Table 1). The determination of pH_i by DMO distribution relies on a quantification of mean cellular DMO levels. DMO (DMO$^-$) accumulates at high pH in the mitochondria. A mitochondrial DMO level approximately ten times higher than in the cytosol is reached with a pH gradient of 1 ($[DMO^-]m/[DMO^-]_c = 10^{\Delta pH}$, see Table 1). Mixing equal volumes of cytosol and mitochondrial matrix assuming equal buffer values, the homogenate pH would be 0.5 pH units above cytosolic pH, whereas the DMO-derived pH would be as much as 0.74 pH units higher. Accordingly, DMO yields high values of mean pH_i in multicompartmental cells (rich in mitochondria; Pörtner *et al.*, 1990; Whiteley, Naylor & Taylor, 1995). However, the comparison of homogenate-derived and DMO-derived pH values yielded even larger differences than could be explained by the characteristics of the DMO technique and, thus, led to the conclusion that non-bicarbonate buffer values within mitochondria must be small (Pörtner *et al.*, 1990; 1991b; see below and Table 2).

In conclusion, homogenate analysis provides an estimate of mean pH_i emphasising cytosolic pH, whereas DMO distribution largely overestimates mean pH_i. When using ^{31}P-MRS, the localisation of pH values measured in such cells is not yet satisfactorily explained, but is also assumed to reflect cytosolic pH (Gadian *et al.*, 1982). Because the chemical shift of inorganic phosphate is a direct measure of pH (see Kinsey & Moerland, this volume), a mean value of pH should result which is close to the arithmetic mean of the chemical shift values for all NMR-visible phosphate, with the assumption that most mitochondrial phosphate is NMR invisible. In support of this hypothesis, pH values derived from inorganic phosphate and from the cytosolic marker phosphodeoxyglucose were found to be very close in radula protractor muscle of a marine whelk, with a mitochondrial content of about 15 per cent of the cell volume (Wiseman & Ellington, 1989; W.R. Ellington, personal communication). In squid mantle, a tissue with an average mitochondrial content of 14 per cent (see Pörtner *et al.*, 1991b), pH data obtained by homogenate and ^{31}P-NMR techniques were also found to be in close agreement (Pörtner *et al.*, 1996; H.O. Pörtner, D.M. Webber, P.G. Lee & M. Quast, unpublished), in accordance with the conclusion that both techniques emphasise cytosolic pH.

Some complication results from the fact that the various methods yield different mean values of pH_i. Which is the one to use for the calculation of mean intracellular P_{CO_2} (equation 3) in mitochondria-rich cells? Because C_{CO_2} distribution between cytosol and mitochondria (like DMO distribution) follows pH-dependent weak acid distribution characteristics, a mean cellular pH value determined by weak acid distribution analysis is required (see above; Table 1). Again, owing to the high inherent inaccuracy (see above), pH_i values determined in individual samples by the DMO technique are usually not suitable for this process such that, at present, model considerations of the general difference between homogenate and weak acid-derived mean pH_i values are required to obtain reasonable estimates of mean intracellular P_{CO_2} values in mitochondria-rich tissues (Pörtner et al., 1990; 1991a; 1991b; 1996).

Non-bicarbonate buffer values

As mentioned above, the response of cellular buffers in vivo to disturbances of the acid–base equilibrium comprises closed and open system characteristics. The latter is true for the CO_2/bicarbonate buffer, which reaches high buffer values in the open system (when P_{CO_2} can be adjusted), and is particularly relevant in animals with high internal P_{CO_2} levels such as air-breathing species. Accordingly, it appears appropriate to distinguish bicarbonate from non-bicarbonate (mostly protein and inorganic phosphate) buffers. Determinations of tissue non-bicarbonate buffer values by tonometry (Heisler, 1989a; Pörtner, 1990a) use CO_2 for the titration of buffers and thereby mimic respiratory changes in the acid–base status in vivo. Changes in bicarbonate levels with changes in pH reflect the action of non-bicarbonate buffers (Heisler & Piiper, 1971).

Regardless of whether buffer values are determined in intact tissue or homogenates, metabolic processes may respond to experimental changes in pH and may interfere with the measurements of buffer values in intact tissue and homogenates, leading to erroneous pH values during the titration procedure or the liberation of additional buffers (see Pörtner, 1989). This means that the measured buffer value does not actually correspond to any definite physiological state of the tissue, and definitely not that under control conditions. These effects are now being increasingly considered, and only recently were invasive (homogenate technique) and non-invasive approaches (by ^{31}P-NMR) for the determination of tissue buffer values brought together, yielding similar values in tissue preparations of low P_{CO_2} (Wiseman & Ellington, 1989; Pörtner, 1989; 1990a). The homogenate technique will yield non-bicarbonate buffer values under control conditions as required for quantitative treatments of the acid–base status, that is the determination of the proton quantities responsible for the observed changes in acid–base parameters through pH/bicarbonate analyses (see below; e.g. Pörtner, 1990a; Pörtner et al., 1991a; 1991b; Ferguson et al., 1993).

The concentrations and relative contributions of non-bicarbonate buffer substances are likely to be different in mitochondrial and cytosolic compartments. Cytosolic actomyosin comprises the major protein fraction in muscle tissue and was found in trout to be two to three times more important in cellular buffering than soluble protein (Abe *et al.*, 1985). Histidine-related compounds (carnosine, anserine, which are predominantly found in white muscle; Abe *et al.*, 1985) may also prevail in the cytosol. Inorganic phosphate levels are similar or lower in mitochondria than in the cytosol (Soboll & Bünger, 1981) but are thought to be bound to Ca^{2+} in the mitochondrial matrix and, therefore, may be inefficient in mitochondrial buffering. A low non-bicarbonate buffer value is expected, a conclusion supported by the observation that the mitochondrial influence on mean pH_i derived from homogenate analyses is small (see Table 2). During anaerobic exercise, cytosolic buffering is increased even further due to the release of inorganic phosphate from phosphagen and ATP.

As emphasised above, mitochondrial bicarbonate buffering may be substantial, even more so since bicarbonate buffering is considerably enhanced in an open system (Burton, 1973; Heisler, 1986a). Therefore, low mitochondrial non-bicarbonate buffer values in the living cell are compensated for to some extent by high bicarbonate levels. A low non-bicarbonate buffer value in mitochondria, however, facilitates the generation of a large pH gradient across the inner mitochondrial membrane because only a small amount of protons is then required for this purpose. The generation of the pH gradient plays an important role in mitochondrial ATP formation. These interrelationships have not been addressed before and call for further investigation of the quantitative contribution of mitochondria to cellular buffering.

pH/bicarbonate analyses

The pH/bicarbonate analysis (Fig. 3) follows these lines of thinking and quantifies respiratory (via changes in P_{CO_2}) and non-respiratory processes contributing to changes in the acid–base status. The pH/bicarbonate diagram illustrates the Henderson–Hasselbalch equation for the CO_2/bicarbonate system, that is the relationships between pH, bicarbonate levels and P_{CO_2}. Non-respiratory changes in a compartment include the net exchange of acid or base equivalents across epithelia or membranes and the influence of metabolic pathways. Also included in the non-respiratory processes are changes in the protonation of proteins which may be connected with the binding or release of ligands (e.g. Pörtner, 1990b). For this type of analysis, control values and associated changes in pH, bicarbonate concentration, P_{CO_2}, and non-bicarbonate buffer value β_{NB} must be known for the respective compartment. The respiratory component of the pH change is derived from the change in

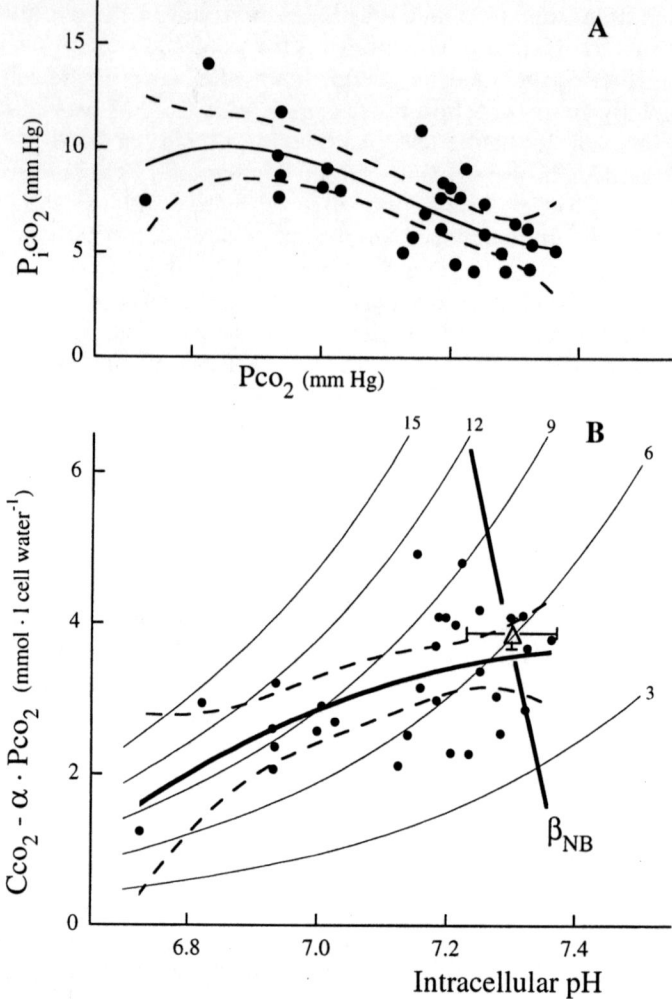

Fig. 3. **A.** Changes in intracellular P_{CO_2} (1 mmHg = 0.1333 kPa; 1 kPa = 7.502 mmHg) of squid (*L. brevis*) mantle musculature with pH_i during progressive fatigue from swimming at different velocities (r = 0.69, third order regression). The data originate from the same set of animals as in Figure 1. The depiction of individual data points for intracellular P_{CO_2} illustrates the variability introduced by the methodological procedure (see text). **B.** Changes in acid–base status of squid (*L. brevis*) mantle musculature during progressive fatigue displayed in a pH/HCO$_3^-$ diagram. The depiction illustrates the predominantly non-respiratory origin of the acidosis (see Fig. 1 and text)

P_{CO_2} along the buffer line, in accordance with the titration of non-bicarbonate buffers by CO_2. This process leads to changes in pH and in bicarbonate levels in opposite directions (pH falls, bicarbonate levels rise). The respiratory (resp.) proton quantities can be derived by pH/bicarbonate analysis as the bicarbonate increment along the buffer line β_{NB}, starting from a control data point specified by a combination of pH, bicarbonate and P_{CO_2} values (see Fig. 3).

$$\Delta pH_{resp.} \cdot -|\beta_{NB}| = \Delta[HCO_3^-]_{resp.} = -\Delta H^+_{resp.} \qquad (4)$$

In contrast, non-respiratory (non-resp.) changes in the acid–base status cause a unidirectional change in both pH and bicarbonate levels. Such a process would follow one of the P_{CO_2} isopleths in the diagram, again starting from a control point in the graph defined by pH, bicarbonate and P_{CO_2}. The respective proton quantities are calculated from the pH change and the change in bicarbonate concentration along the P_{CO_2} isopleth considering the non-bicarbonate buffer value (equation 4; tot = total).

$$\Delta H^+_{non-resp.} = (\Delta pH_{tot} \cdot -|\beta_{NB}|) - \Delta[HCO_3^-]_{tot} \qquad (5)$$

Equation 5 is still valid when the changes in pH and in bicarbonate are affected by respiratory processes, if a change in P_{CO_2} occurs. Because a respiratory change in pH is associated with a change in bicarbonate levels in the opposite direction (equation 4), respiratory changes always balance to 0 (equation 4 in the format of equation 6). Equation 6 is considered in equation 5 (ΔpH_{tot} includes $\Delta pH_{resp.}$ and $\Delta[HCO_3^-]_{tot}$ includes $\Delta[HCO_3^-]_{resp.}$). Any change in P_{CO_2} will influence ΔpH_{tot} and $\Delta[HCO_3^-]_{tot}$ (again, in opposite directions) but not the resulting $\Delta H^+_{non-resp.}$

$$(\Delta pH_{resp.} \cdot -|\beta_{NB}|) - \Delta[HCO_3^-]_{resp.} = 0 \qquad (6)$$

Prior to the advent of the homogenate technique, the only way to determine pH_i during muscle activity in the whole unrestrained animal was

---

Fig. 3 caption contd.

with a small respiratory contribution (r = 0.56, second order regression; Δ = control value, mean $\pm$ SD; β_{NB} = non-bicarbonate buffer line of muscle cell water). Solid lines were determined by regression analysis and delineate a significant relationship ($p<0.05$); dashed lines represent 95 per cent confidence intervals. Modified after Pörtner *et al.* (1996).

by calculation. The pH/bicarbonate analysis was used to calculate changes in pH_i from tissue metabolic changes, considering any exchange of acid–base equivalents between intracellular and extracellular space (Pörtner, 1987a; 1987b).

Interstitial pH

Most attention has focused on the properties of intracellular and extracellular bulk fluids when acid–base regulation of an animal is discussed. However, the interstitial fluid is the one in contact with the cell membrane and acts as a mediator for any extracellular signal transferred to the cell. In this context, it is important to note that a P_{CO_2} gradient may prevail between intracellular and extracellular fluids which is likely to cause a drop in interstitial pH below plasma pH, owing to minimal non-bicarbonate buffering in the interstitial fluid (Pörtner et al., 1991a; Pörtner, 1993). In support of this conclusion, studies in vertebrates have shown that a pH gradient may prevail from the cell surface to the venous blood or ambient medium (De Hemptinne & Huguenin, 1984). The presence of a P_{CO_2} gradient would explain why pH values are lower on the cell surface than in the surrounding medium or blood. The resulting pH gradient would be most evident with rapid CO_2 hydration, as expected from the action of extracellular, membrane-bound carbonic anhydrase (e.g. Henry, Wang & Wood, 1997).

The P_{CO_2} gradient and, thus, the difference between plasma and interstitial pH might be characteristic for the tissue in question. However, it was shown in an amphibian, *Bufo marinus*, that a P_{CO_2} gradient remains between the intracellular space and the plasma, which is approximately the same in gastrocnemius muscle and in the ventricle of the heart despite the differing metabolic rates. Moreover, intracellular P_{CO_2} reflects the drop in venous P_{CO_2} during long-term hyperventilation in hypoxia (Pörtner et al., 1991a). This observation complies with the fact that mitochondrial density and capillarisation of a tissue are interdependent (Weibel, 1984), explaining similar partial pressure gradients under steady state resting conditions in spite of differing rates of CO_2 production.

When taking the steady state P_{CO_2} gradient between intracellular and extracellular space under resting conditions into account, an interstitial pH can be determined that is between 0.1 and 0.15 pH units lower than that of the venous plasma. Such relationships have been interpreted to explain the deviation from a pH-dependent distribution pattern of lactate which becomes visible during situations of low metabolic rate, e.g. during long-term hypoxia (Pörtner et al., 1991a; Pörtner, 1993). In general, they are expected to influence the patterns of pH-dependent distribution between the intracellular and extracellular spaces not only of organic acids *via* pH-dependent carrier mechanisms like the lactate H^+ symporter (cf. Pörtner, 1993), but also of total CO_2 and ammonia, the volatile fractions of both of which are eliminated from the body following the laws of respiratory gas exchange (Heisler, 1995; Henry *et al.*, 1997; see also Wood & Wang, this volume).

Any deviation from steady state will influence such a pH-dependent distribution pattern, for example when net production or removal of the respective metabolite in metabolism occurs at a rate in the same order of magnitude as the transmembrane (facilitated) diffusion process. Some studies have revealed patterns of changes in intracellular and extracellular P_{CO_2} that differ especially during exercise in poorly perfused tissues like white muscle. Larger P_{CO_2} gradients between intracellular and extracellular space are likely to develop with any increase in metabolic rate such as during exercise (Pörtner *et al.*, 1991b; Boutilier *et al.*, 1993). In poorly perfused white muscle, this trend may be exacerbated by the glycolytic acidification and titration of bicarbonate stores, thus leading to even higher intracellular P_{CO_2} levels, steeper P_{CO_2} gradients and lower values of interstitial pH. These considerations emphasise the relevance of compiling P_{CO_2} estimates in various body compartments for a more complete picture, especially during non-steady state metabolic situations.

Temperature and pH$_i$

Reeves (1972) introduced the imidazole alphastat hypothesis, stating that pH regulation in poikilotherms maintains the degree of protonation (α) of imidazole groups in proteins despite changes in body temperature. A pH$_i$ change of around -0.018 pH units $/°C$ is expected to support the alphastat pattern and to ensure protein function at fluctuating temperature. Cameron (1989) refined this approach in proposing a 'Z-stat' model, emphasising that protein net charge Z is maintained rather than α in diverse histidine groups. This may be a more adequate treatment because $\Delta pK \, °C^{-1}$ depends upon local charge configurations in the environment of the imidazole group as well as on ionic strength and, therefore, ranges between -0.016 and $-0.024 \, °C^{-1}$ for histidine and free imidazole compounds and between -0.0010 and $-0.051 \, °C^{-1}$ for histidine residues in proteins (Heisler, 1986b).

In a comprehensive paper, Ultsch and Jackson (1996) recently reviewed the literature on the relationship between pH$_i$ and temperature in ectothermic vertebrates, concluding that the data in general support the concept of alphastat regulation, particularly within the normal temperature range of the species. With some exemptions, the alphastat pattern of pH_i regulation could also be confirmed for marine ectotherms (invertebrates and fish) exposed to various temperatures, both depending on the season and latitude (review by Pörtner *et al.*, 1998). However, alphastat regulation of pH is likely to be restricted to a temperature window between the critical temperature limits of a species (Sommer *et al.*, 1997).

The original hypothesis claimed that the observed changes in pH with temperature are only elicited by passive mechanisms, mostly proton binding or release from intracellular and extracellular buffers owing to the change in dissociation equilibria (pK values) of the buffer components. Later on, the alphastat hypothesis was extended, integrating an active component. In air breathers, active pH regulation is mainly due to the control of P_{CO_2} by means of ventilatory adjustments, whereas active ion exchange predominates in water breathers. The relative contributions of active and passive mechanisms to the adjustment of pH_i have been quantified by model calculations (Reeves & Malan, 1976; Heisler, 1984). However, this analysis is complicated by the variability of the $\Delta pK/\Delta T$ values (see above).

The homogenate technique offers the opportunity to distinguish active and passive elements in temperature-dependent pH regulation by direct analysis. For a quantification of passive mechanisms, tissues collected from animals exposed to control temperatures are analysed *in vitro* at different temperatures, whereas *in vivo* $\Delta pH/\Delta T$ values are determined in animals exposed to various temperatures for long enough time periods to allow active mechanisms to achieve and maintain new steady state values of pH_i.

In contrast to the original hypothesis, these authors found that the passive contribution to pH regulation was considerably below the alphastat value in some of the animals investigated (van Dijk *et al.*, 1997; Sartoris & Pörtner, 1997a; Pörtner *et al.*, 1998). The passive contribution accounted for only 35 per cent of the temperature-induced pH shift in white muscle of the eelpout *Zoarces viviparus*. In contrast, passive pH changes in an isolated mixture of intracellular proteins revealed a slope not significantly different from the *in vivo* slope. This discrepancy could not be ascribed to low molecular weight components with low $\Delta pK/\Delta T$ values like bicarbonate ($\Delta pK/\Delta T = -0.006$ °C^{-1}) or phosphate ($\Delta pK/\Delta T = -0.003$ °C^{-1}; van Dijk *et al.*, 1997). The nature of the buffers responsible for the passive contribution to alphastat and the reasons for the difference between the pH shifts of the protein mixture isolated from white muscle and of the passive components in the KF/NTA homogenate remain unexplained.

First results obtained in comparative studies suggest that the relative contribution of active and passive processes to the pH shift is largely influenced by the temperature regime of the animals' habitat. In all animal species investigated so far, the passive contribution to alphastat regulation was reduced and the active component was higher in eurythermal than in stenothermal species (Fig. 4). Lower passive pH shifts would, on the one hand, lead to more acidic pH values in the cold and leave a larger contribution to ion exchange mechanisms to accomplish alphastat pH regulation. On the other hand, a low passive slope allows flexible adjustments of pH according to metabolic requirements. Large passive slopes would require active pH regulation to compensate for their effect when more acidic pH values are to be maintained. In particular, animals exposed to large seasonal temperature variations exhibit low pH values at low temperatures in the winter (Thebault & Raffin, 1991; Spicer, Morritt & Taylor, 1994). The shrimp *Palaemon* tends to be inactive at temperatures below 10 °C, metabolic depression being reflected by a drop in pH_i below the alphastat pattern (Thebault & Raffin, 1991). Acidic pH_i values were also reported by Whiteley *et al.* (1995) for winter crayfish *Austrapotamobius pallipes*. Low passive slopes may support metabolic depression which should comprise the down-regulation of energy-consuming ion exchange mechanisms otherwise responsible for alphastat pH regulation. One might speculate that a capacity for metabolic depression in eurythermal animals is correlated with a reduced contribution of passive mechanisms to pH adjustment during temperature change.

As a corollary, these interspecies comparisons suggest that the temperature-dependent adjustment of pH_i mostly occurs by active mechanisms in eurythermal animals, whereas in stenothermal animals pH adjustment is mostly achieved by passive processes. A larger active than passive component of alphastat regulation may not only be a prerequisite to colonise shallow coastal waters, but may also allow for a variable adjustment of metabolic activity on a seasonal time scale.

Summary and conclusions

Invasive methodology can be used to study the relationships between acid–base and metabolic regulation in such a way that the role of acid–base parameters in modulating metabolic rate, the mode of catabolism and energetic parameters becomes evident. This includes an analysis of

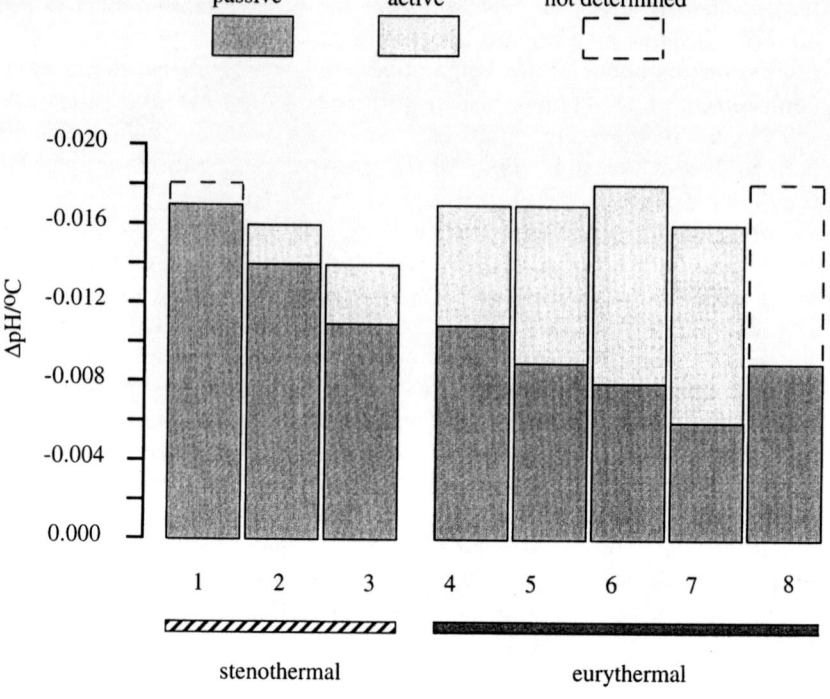

Fig. 4.

the full set of acid–base parameters such as pH_i and pH_e (as well as interstitial pH), bicarbonate and P_{CO_2} levels, considering the buffering characteristics of each compartment. The study of these relationships is especially interesting in marine invertebrates and lower vertebrates, many of which experience wide fluctuations in various environmental parameters such as temperature, CO_2 or O_2 levels. Specific patterns of acid–base regulation observed in different groups need to be considered in an attempt to evaluate unifying principles.

The specific features of the homogenate technique allow a precise evaluation of mean pH_i as required for a quantitative picture of the patterns of acid–base regulation in whole animals. It yields cellular P_{CO_2} and bicarbonate levels and quantifies non-bicarbonate buffers, which specifically respond to respiratory, but also to non-respiratory, acidification. The technique allows an estimate of extracellular/intracellular P_{CO_2} gradients and, thereby, interstitial pH. During temperature change, it is possible to distinguish between changes in pH_i elicited by physico-chemical buffering and those caused by a readjustment of the setpoints of proton equivalent ion exchange. The homogenate technique is a useful, easy to use and a readily available alternative to other methods of pH analysis, especially when the experimental design does not allow the use of online techniques. Since tissues can be stored away under

---

Fig. 4. Relative contributions of passive and active mechanisms to temperature-induced changes in pH_i in stenothermal (1–3) and eurythermal (4–8) marine ectotherms. *Tryphosella murrayi* is an Antarctic amphipod, *Pandalus borealis* is a deep water shrimp caught in Spitzbergen waters, and *Pachycara brachycephalum* is an Antarctic eelpout. These polar species were compared with more temperate zone species such as the lugworm *Arenicola marina*, the sand shrimp *Crangon crangon*, and the eelpout *Zoarces viviparus*. In the polar species, the temperature-induced passive pH change is close to the pH change observed *in vivo* (passive slope $= \Delta pH_i/\Delta T = -0.017$ in *Tryphosella murrayi*, -0.013 in *Pandalus borealis*, -0.011 in *Pachycara brachycephalum*). In species 1 and 8, only the passive contribution was determined; the dashed lines reflect expected *in vivo* values according to the alphastat pattern. Data originate from Sartoris and Pörtner (1997) for species 2, 5 and 6; from van Dijk *et al.* (1997) for species 7; data for species 1, 3, 4 and 8 are unpublished by F.J. Sartoris, A. Sommer, C. Tesch, I. Hardewig, P. van Dijk and H.O. Pörtner.

liquid nitrogen until analysis, the method is applicable to samples collected in the field. Samples can be collected from free-ranging, unrestrained and even exercising animals. The specific advantages of the homogenate technique are balanced to some extent by its limitations in that it does not allow changes to be recorded online and continuously in one preparation or animal and that it does not access compartmental pH in multicompartmental cells.

Acknowledgements

Supported by grants of the Deutsche Forschungsgemeinschaft to H.O.P. The authors appreciate the discussions with Chris Wood on the relevance of interstitial pH in the analysis of transmembrane distribution patterns.

References

Abe, H., Dobson, G.P., Hoeger, U. & Parkhouse, W.W. (1985). Role of histidine-related compounds to intracellular buffering in fish skeletal muscle. *American Journal of Physiology* **249**, R449–54.

Abele-Oeschger, D., Sartoris, F.J. & Pörtner, H.O. (1997). Hydrogen peroxide causes a decrease in aerobic metabolic rate and in intracellular pH in the shrimp *Crangon crangon*. *Comparative Biochemistry and Physiology* **117C**, 123–9.

Atkinson, D.E. & Bourke, E. (1995). pH homeostasis in terrestrial vertebrates; ammonium ion as a proton source. *Advances in Comparative and Environmental Physiology* **22**, 3–26.

Atkinson, D.E. & Camien, M.N. (1982). The role of urea synthesis in the removal of metabolic bicarbonate and the regulation of blood pH. *Current Topics in Cellular Regulation* **21**, 261–302.

Boutilier, R.G., Ferguson, R.A., Henry, R.P. & Tufts, B.L. (1993). Exhaustive exercise in the sea lamprey (*Petromyzon marinus*): Relationships between anaerobic metabolism and intracellular acid–base balance. *Journal of Experimental Biology* **178**, 71–88.

Branco, L.G.S., Pörtner, H.O. & Wood, S.C. (1993). Interaction between temperature and hypoxia in the alligator. *American Journal of Physiology* **265**, R1339–43.

Burton, R.F. (1973). The role of buffers in body fluids. Mathematical analysis. *Respiration Physiology* **18**, 34–42.

Cameron, J.N. (1989). Acid–base homeostasis: past and present perspectives. *Physiological Zoology* **62**, 845–65.

Chih, C.P. & Ellington, W.R. (1985). Metabolic correlates of intracellular pH change during rapid contractile activity in a molluscan muscle. *Journal of Experimental Zoology* **236**, 27–34.

Claiborne, J.B., Walton, J.S. & Compton-McCullough, D. (1994).

Acid–base regulation, branchial transfers and renal output in a marine teleost fish (the long-horned sculpin *Myoxocephalus octodecimspinosus*. *Journal of Experimental Biology* **193**, 79–95.

Day, N. & Butler, P.J. (1996). Environmental acidity and white muscle recruitment during swimming in the rainbow trout (*Salmo trutta*). *Journal of Experimental Biology* **199**, 1947–59.

De Hemptinne, A. & Huguenin, F. (1984). The influence of muscle respiration and glycolysis on surface and intracellular pH in fibres of the rat soleus. *Journal of Physiology* **347**, 581–92.

Ferguson, R.A., Kiefer, J.D. & Tufts, B.L. (1993). The effects of body size on the acid–base and metabolite status in the white muscle of rainbow trout before and after exhaustive exercise. *Journal of Experimental Biology* **180**, 195–207.

Gadian, D.G., Radda, G.K., Dawson, M.J. & Wilkie, D.R. (1982). pH$_i$ measurements of cardiac and skeletal muscle using ^{31}P-NMR. In *Intracellular pH: its measurement, regulation and utilization in cellular functions*, ed. R. Nuccitelli & D.W. Deamer, pp. 61–77. New York: Alan R. Liss.

Grieshaber, M.K., Hardewig, I., Kreutzer, U. & Pörtner, H.O. (1994). Physiological and metabolic responses to hypoxia in invertebrates. *Reviews of Physiology, Biochemistry and Pharmacology* **125**, 43–147.

Hansen, H.D. & Gesser, H. (1980). Relation between non-bicarbonate buffer value and tolerance to acidosis: a comparative study of myocardial tissue. *Journal of Experimental Biology* **84**, 161–7.

Hardewig, I., Addink, A.D.F., Grieshaber, M.K., Pörtner, H.O. & van den Thillart, G. (1991a). Metabolic rates at different oxygen levels determined by direct and indirect calorimetry in the oxyconformer *Sipunculus nudus*. *Journal of Experimental Biology* **157**, 143–60.

Hardewig, I., Kreutzer, U., Pörtner, H.O. & Grieshaber, M.K. (1991b). The role of phosphofructokinase in the glycolytic control in the facultative anaerobe *Sipunculus nudus*. *Journal of Comparative Physiology* **161B**, 581–9.

Hardewig, I., Pörtner, H.O. & Grieshaber, M.K. (1994). Interactions of anaerobic propionate formation and acid–base status in *Arenicola marina*: an analysis of propionyl-CoA-carboxylase. *Physiological Zoology* **67**, 892–909.

Häussinger, D., Meijer, A.J., Gerok, W. & Sies, H. (1988). Hepatic nitrogen metabolism and acid–base homeostasis. In *pH homeostasis: mechanisms and control*, ed. D. Häussinger, pp. 337–77. New York: Academic Press.

Heisler, N. (1984). Role of ion transfer processes in acid–base regulation with temperature changes in fish. *American Journal of Physiology* **246**, R441–51.

Heisler, N. (1986a). Buffering and transmembrane ion transport processes. In *Acid–base regulation in animals*, ed. N. Heisler, pp. 3–47. Amsterdam: Elsevier North Holland.

Heisler, N. (1986b). Comparative aspects of acid–base regulation. In *Acid–base regulation in animals*, ed. N. Heisler, pp. 397–450. Amsterdam: Elsevier North Holland.

Heisler, N. (1989a). Parameters and methods in acid–base physiology. In *Techniques in comparative respiratory physiology*, ed. C.R. Bridges & P.J. Butler, pp. 305–32. Cambridge: Cambridge University Press.

Heisler, N. (1989b). Interactions between gas exchange, metabolism, and ion transport in animals: an overview. *Canadian Journal of Zoology* **67**, 2923–35.

Heisler, N. (1995). Ammonia vs. ammonium: elimination pathways of nitrogenous wastes in ammoniotelic fishes. *Advances in Comparative and Environmental Physiology* **22**, 63–87.

Heisler, N. & Piiper, J. (1971). The buffer value of rat diaphragm muscle tissue determined by P_{CO_2} equilibration of homogenates. *Respiration Physiology* **12**, 169–78.

Henry, R.P., Wang, Y. & Wood, C.M. (1997). Carbonic anhydrase facilitates CO_2 and NH_3 transport across the sarcolemma of trout muscle. *American Journal of Physiology* **272**, R1754–61.

Hochachka, P.W. & Mommsen, T.P. (1983). Protons and anaerobiosis. *Science* **219**, 1391–8.

Hoppeler, H., Lindstedt, S.L., Claassen, H., Taylor, C.R., Mathieu, O. & Weibel, E.R. (1984). Scaling mitochondrial volume in heart to body mass. *Respiration Physiology* **55**, 131–7.

Kauppinen, R.A. (1983). Proton electrochemical potential of the inner mitochondrial membrane in isolated perfused rat hearts, as measured by exogenous probes. *Biochimica et Biophysica Acta* **725**, 131–7.

Kauppinen, R.A., Hiltunen, J.K. & Hassinen, I.E. (1980). Subcellular distribution of phosphagens in isolated perfused rat heart. *FEBS Letters* **112**, 273–6.

Kiefer, J.D., Currie S. & Tufts, B.L. (1994). Effects of environmental temperature on the metabolic and acid–base responses of rainbow trout to exhaustive exercise. *Journal of Experimental Biology* **194**, 299–317.

Krause, U. & Wegener, G. (1996). Exercise recovery in frog muscle: metabolism of PCr, adenine nucleotides, and related compounds. *American Journal of Physiology* **270**, R811–20.

Larsen, B.K., Pörtner, H.O. & Jensen, F.B. (1997). Extra- and intracellular acid–base balance and ionic regulation in cod (*Gadus morhua*) during combined and isolated exposures to hypercapnia and copper. *Marine Biology* **128**, 337–46.

Milligan, C.L. (1996). Metabolic recovery from exhaustive exercise in rainbow trout. *Comparative Biochemistry and Physiology* **113A**, 51–60.

Pörtner, H.O. (1987a). Contributions of anaerobic metabolism to pH regulation in animal tissues: theory. *Journal of Experimental Biology* **131**, 69–87.

Pörtner, H.O. (1987b). Anaerobic metabolism and changes in acid–base status: quantitative interrelationships and pH regulation in the marine worm *Sipunculus nudus*. *Journal of Experimental Biology* **131**, 89–105.

Pörtner, H.O. (1989). The importance of metabolism in acid–base regulation and acid–base methodology. *Canadian Journal of Zoology* **67**, 3005–17.

Pörtner, H.O. (1990a). Determination of intracellular buffer values after metabolic inhibition by fluoride and nitrilotriacetic acid. *Respiration Physiology* **81**, 275–88.

Pörtner, H.O. (1990b). An analysis of the effects of pH on oxygen binding by squid (*Illex illecebrosus, Loligo pealei*) haemocyanin. *Journal of Experimental Biology* **150**, 407–24.

Pörtner, H.O. (1993). Multicompartmental analyses of acid–base and metabolic homeostasis during anaerobiosis: invertebrate and lower vertebrate examples. In *Surviving hypoxia: mechanisms of control and adaptation*, ed. P.W. Hochachka, P.L. Lutz, T. Sick, M. Rosenthal & G. van den Thillart, pp. 139–56. Boca Raton, FL: CRC Press.

Pörtner, H.O. (1994). Coordination of metabolism, acid–base regulation and haemocyanin function in cephalopods. *Marine and Freshwater Behaviour and Physiology* **25**, 131–48.

Pörtner, H.O. (1995). pH homeostasis in terrestrial vertebrates: a comparison of traditional and new concepts. *Advances in Comparative and Environmental Physiology* **22**, 51–62.

Pörtner, H.O. (1997). Oxygen limitation of metabolism and performance in pelagic squid. In *The responses of marine organisms to their environments. Proceedings of the 30th European Marine Biology Symposium*, ed. L.E. Hawkins, S. Hutchinson with A.C. Jensen, M. Sheader & J.A. Williams, pp. 45–56. Chichester: RPM Reprographics.

Pörtner, H.O., Heisler, N. & Grieshaber, M.K. (1984). Anaerobiosis and acid–base status in marine invertebrates: a theoretical analysis of proton generation by anaerobic metabolism. *Journal of Comparative Physiology* **155B**, 1–12.

Pörtner, H.O., Boutilier, R.G., Tang, Y. & Toews, D.P. (1990). Determination of intracellular pH and P_{CO_2} after metabolic inhibition by fluoride and nitrilotriacetic acid. *Respiration Physiology* **81**, 255–74.

Pörtner, H.O., MacLatchy, L.M. & Toews, D.P. (1991a). Acid–base regulation in the toad *Bufo marinus* during environmental hypoxia. *Respiration Physiology* **85**, 217–30.

Pörtner, H.O., Webber, D.M., Boutilier, R.G. & O'Dor, R.K. (1991b). Acid–base regulation in exercising squid (*Illex illecebrosus, Loligo pealei*). *American Journal of Physiology* **261**, R239–46.

Pörtner, H.O., Webber, D.M., O'Dor, R.K. & Boutilier, R.G. (1993). Metabolism and energetics in squid (*Illex illecebrosus, Loligo pealei*) during muscular fatigue and recovery. *American Journal of Physiology* **265**, R157–65.

Pörtner, H.O., Finke, E. & Lee, P.G. (1996). Metabolic and energy correlates of intracellular pH in progressive fatigue of squid (*Lolliguncula brevis*) mantle muscle. *American Journal of Physiology* **271**, R1403–14.

Pörtner, H.O., Hardewig, I., Sartoris, F.J. & van Dijk, P. (1998). Energetic aspects of cold adaptation: critical temperatures in metabolic, ionic and acid–base regulation? In *Cold ocean physiology*, ed. H.O. Pörtner & R. Playle, pp. 88–120. Cambridge: Cambridge University Press.

Reeves, R.B. (1972). An imidazole alphastat hypothesis for vertebrate acid–base regulation: Tissue carbon dioxide content and body temperature in bullfrogs. *Respiration Physiology* **14**, 219–36.

Reeves, R.B. & Malan, A. (1976). Model studies of intracellular acid–base temperature responses in ectotherms. *Respiration Physiology* **28**, 49–63.

Reipschläger, A. & Pörtner, H.O. (1996). Metabolic depression during environmental stress: the role of extra- *versus* intracellular pH in *Sipunculus nudus*. *Journal of Experimental Biology* **199**, 1801–7.

Roos, A. & Boron, W.F. (1981). Intracellular pH. *Physiological Reviews* **61**, 296–434.

Sahlin, K., Harris, R.C. & Hultman, E. (1975). Creatine kinase equilibrium and lactate content compared with muscle pH in tissue samples obtained after isometric exercise. *Biochemical Journal* **152**, 173–80.

Sahlin, K., Harris, R.C., Nylind, B. & Hultman, E. (1976). Lactate content and pH in muscle samples obtained after dynamic exercise. *Pflügers Archiv* **367**, 143–9.

Sartoris, F.J. & Pörtner, H.O. (1997a). Temperature dependence of ionic and acid–base regulation in boreal and arctic *Crangon crangon* and *Pandalus borealis*. *Journal of Experimental Marine Biology and Ecology* **211**, 69–83.

Sartoris, F.J. & Pörtner, H.O. (1997b). Increased concentrations of haemolymph Mg^{2+} protect intracellular pH and ATP levels during temperature stress and anoxia in the common shrimp *Crangon crangon*. *Journal of Experimental Biology* **200**, 785–92.

Schmidt, H., Wichmann, A., Lamprecht, I. & Zerbst-Boroffka, I. (1996). Anaerobic metabolism in the leech (*Hirudo medicinalis* L.): direct and indirect calorimetry during severe hypoxia. *Journal of Comparative Physiology* **166**, 205–14.

Schulte, P.M., Moyes, C.D. & Hochachka, P.W. (1992). Integrating metabolic pathways in post-exercise recovery of white muscle. *Journal of Experimental Biology* **166**, 181–95.

Siggaard-Andersen, O. (1974). *The acid–base status of the blood.* Baltimore: William & Wilkins.

Smith, H.E. & Page, E. (1976). Morphometry of rat heart mitochondrial subcompartments and membranes: Application to myocardial cell atrophy after hypophysectomy. *Journal of Ultrastructural Research* **55**, 31–41.

Soboll, S. & Bünger, R. (1981). Compartmentation of adenine nucleotides in the isolated working guinea pig heart stimulated by noradrenaline. *Hoppe Seyler's Zeitschrift für Physiologische Chemie* **362**, 125–32.

Sommer, A., Klein, B. & Pörtner, H.O. (1997). Temperature induced anaerobiosis in two populations of the polychaete worm *Arenicola marina*. *Journal of Comparative Physiology* **167B**, 25–35.

Spicer, J.L., Morritt, D. & Taylor, A.C. (1994). Effect of low temperature on oxygen uptake and haemolymph ions in the sandhopper *Talitrus saltator* (Crustacea: Amphipoda). *Journal of the Marine Biological Association UK* **74**, 313–21.

Tang, Y. & Boutilier, R.G. (1991). White muscle intracellular acid–base and lactate status following exhaustive exercise: a comparison between freshwater- and seawater-adapted trout. *Journal of Experimental Biology* **156**, 153–71.

Thebault, M.T. & Raffin, J.P. (1991). Seasonal variations in *Palaemon serratus* abdominal muscle metabolism and performance during exercise, as studied by ^{31}P NMR. *Marine Ecology Progress Series* **74**, 175–83.

Ultsch, G.R. & Jackson, D.C. (1996). pH and temperature in ectothermic vertebrates. *Bulletin of the Alabama Museum of Natural History* **18**, 1–41.

van den Thillart, G. & van Waarde, A. (1996). Nuclear magnetic resonance spectroscopy of living systems: Applications in comparative physiology. *Physiological Reviews* **76**, 799–837.

van Dijk, P., Hardewig, I. & Pörtner, H.O. (1997). The adjustment of intracellular pH after temperature change in fish: relative contributions of passive and active processes. *American Journal of Physiology* **272**, R84–9.

Waddell, W.J. & Butler, T.C. (1959). Calculation of intracellular pH from the distribution of 5,5-dimethyl-2,4-oxazolidine-dione (DMO). Application to skeletal muscle of the dog. *Journal of Clinical Investigations* **38**, 720–9.

Walsh, P.J. & Milligan, C.L. (1989). Coordination of metabolism and intracellular acid–base status: ionic regulation and metabolic consequences. *Canadian Journal of Zoology* **67**, 2994–3004.

Wang, Y., Heigenhauser, G.J.F. & Wood, C.M. (1994). Integrated responses to exhaustive exercise and recovery in rainbow trout white muscle: Acid–base, phosphagen, carbohydrate, lipid, ammonia, fluid volume and electrolyte metabolism. *Journal of Experimental Biology* **195**, 227–58.

Wang, Y., Heigenhauser, G.J.F. & Wood, C.M. (1996a). Lactate and metabolic H^+ transport and distribution after exercise in rainbow trout white muscle. *American Journal of Physiology* **271**, R1239–50.

Wang, Y., Heigenhauser, G.J.F. & Wood, C.M. (1996b). Ammonia movement and distribution after exercise across white muscle cell membranes in rainbow trout. *American Journal of Physiology* **271**, R738–50.

Wasser, J.S., Lawler, R.G. & Jackson, D.C. (1996). Nuclear magnetic resonance spectroscopy and its applications in comparative physiology. *Physiological Zoology* **69**, 1–34.

Weibel, E.R. (1984). *The pathway for oxygen: structure and function of the mammalian respiratory system.* Cambridge, MA: Harvard University Press.

Whiteley, N.M., Naylor, J.K. & Taylor, E.W. (1995). Extracellular and intracellular acid–base status in the fresh water crayfish *Austropotamobius pallipes* between 1 and 12 °C. *Journal of Experimental Biology* **198**, 567–76.

Wiseman, R.W. & Ellington, W.R. (1989). Intracellular buffering capacity in molluscan muscle: superfused muscle vs. homogenates. *Physiological Zoology* **62**, 541–58.

Wollenberger, A., Ristau, D. & Schoffa, G. (1960). Eine einfache Technik der extrem schnellen Abkühlung größerer Gewebestücke. *Pflügers Archiv* **270**, 399–412.

Zange, J., Pörtner, H.O., Jans, A.W.H. & Grieshaber, M.K. (1990). The intracellular pH of a molluscan smooth muscle during a contraction–catch–relaxation cycle as estimated by the distribution of ^{14}C-DMO and by ^{31}P-NMR spectroscopy. *Journal of Experimental Biology* **150**, 81–93.

Zielinski, S. & Pörtner, H.O. (1996). Energy metabolism and ATP free-energy change of the intertidal worm *Sipunculus nudus* below a critical temperature. *Journal of Comparative Physiology* **166B**, 492–500.

CHRIS M. WOOD and YUXIANG WANG

Lactate, H⁺ and ammonia transport and distribution in rainbow trout white muscle after exhaustive exercise

Introduction

Since its introduction 40 years ago by E.C. Black and colleagues (Black, 1958; Black *et al.*, 1962), the exhaustively exercised rainbow trout has proven a popular model in fish exercise physiology. To date, there are well over 200 papers on this system, and progress has been reviewed at regular intervals (Driedzic & Hochachka, 1978; Wood & Perry, 1985; Wood, 1991; Moyes, Schulte & Hochachka, 1993; Moyes & West, 1995; Weber & Haman, 1996; Milligan, 1996). It is not the intention in this chapter to recap this extensive literature again, but rather to focus on several specific issues. Firstly, in keeping with the theme of this volume, methodology is assessed – specifically, how should blood and white muscle samples be taken, and how should they be processed and analysed so as to represent most faithfully the metabolic and acid–base status existing *in vivo*? Secondly, the major post-exercise responses in this model system are briefly described, emphasising those studies in which the best methodology has been employed. Thirdly, several theoretical issues or 'problems' of current understanding are described, one of which constitutes the focus of the remainder of this chapter – specifically, what mechanisms determine the distribution and movement of ammonia, lactate, and protons between the intracellular fluid of white muscle and blood plasma of rainbow trout at rest and after exhaustive exercise? Fourthly, a new *in vitro* preparation is described, the isolated–perfused tail-trunk, which we have developed to investigate these problems. Lastly, recent findings are summarized, based on both this preparation and *in vivo* studies, which cast new light on these topics. Unless otherwise noted, the information presented here is based on studies from our laboratory reported by Wang *et al.* (1994a; 1994b; 1996a; 1996b; 1997) and Henry, Wang and Wood (1997).

Methodology

For blood sampling in mechanistic studies, there is no alternative but to use chronic arterial catheterization (Soivio, Westman & Nyholm,

1972). Not only is disturbance minimized, but repetitive samples can be drawn from the same animal to track trends over time. A minimum 48-hour recovery period after cannulation and prior to the start of experiments is recommended. When blood and tissue sampling are combined, it is essential that blood samples be drawn prior to administration of the anaesthetic recommended below. Most blood parameters of importance (e.g. blood acid–base status, respiratory gases, catecholamines, electrolytes, ammonia) are drastically disturbed by anaesthesia and caudal or cardiac puncture sampling; a few (e.g. lactate, pyruvate) are reasonably resilient as long as struggling is avoided and the puncture sample is done rapidly, using a massive anaesthetic overdose rather than a cephalic blow. In general, artifacts caused by puncture sampling are greatest in resting fish, and less after exercise. In these authors' experience, the only measurement that is more reliable when taken by rapid caudal puncture is cortisol, which is routinely lower than in catheterized fish. Presumably, chronic low-level stress caused by the presence of the catheter perturbs the pituitary–interrenal axis.

This raises the question of whether it is better to take tissue (white muscle) samples from non-catheterized trout to avoid potential metabolic artifacts caused by cortisol. Wang *et al.* (1994b) examined this question and concluded that there were negligible differences in most muscle metabolites between cannulated and non-cannulated fish. It therefore seems preferable to take blood and tissue samples from the same fish (cannulated), both for internal consistency and to minimize animal usage.

Of much greater concern are the methods used to sample the white muscle, a matter first raised by Dobson and Hochachka (1987, for creatine phosphate) and Tang and Boutilier (1991, for lactate). For example, Dobson and Hochachka (1987) noted that only three to four 'tail flaps' prior to freeze-clamping will reduce white muscle creatine phosphate levels by 70 per cent. We therefore conducted a systematic comparison of methods, together with a critical survey of literature values for white muscle metabolites in resting trout. This clearly demonstrated the superiority of the sampling method pioneered by Tang and Boutilier (1991): rapid lethal overdose with neutralized MS-222 anaesthetic followed by immediate (10–15 seconds) freeze-clamping of the muscle filet in liquid N_2-cooled aluminum tongs. We favour 0.5 g/l MS-222 for greatest speed and to prevent any struggling. With this method, tissue levels of creatine phosphate and glycogen were higher, and levels of lactate and ammonia were lower than by various needle biopsy methods or by sampling methods that involved sacrifice by decapitation or cephalic blows (Wang *et al.*, 1994b).

An equally important issue is the subsequent processing of the sample. Routinely used methods are: (i) the traditional technique of glass homogenization of frozen tissue in ice-cold perchloric acid; (ii) mortar and pestle grinding of the frozen tissue to a fine powder *prior* to perchloric acid extraction; and (iii) freeze-drying of the latter powder from the mortar *prior* to perchloric acid extraction. For most metabolites, the three methods yielded similar values, but creatine phosphate levels were consistently higher and creatine levels lower by the freeze-drying method, reflecting better preservation of this highly labile substance. In addition to this benefit, freeze-drying is recommended because the lyophilized tissue is much more stable for long-term handling. A word of caution however: the negative pressure applied during freeze-drying causes evaporative loss of ammonia. The other methods are suitable for measuring ammonia, as long as critical care is taken that the tissue remains frozen up until the point of extraction. Prior thawing, even for a few seconds, will cause a large artifactual rise in tissue ammonia.

Based on these tests, we conclude that in most studies which did not use these recommended methods (i.e. most studies prior to this decade), resting white muscle lactate levels were too high, creatine phosphate and glycogen levels were too low, and ammonia levels were suspect. Curiously, ATP levels appear to be remarkably resistant to sampling or processing artifacts, an observation supported by the ^{31}P-nuclear magnetic resonance (^{31}P-NMR) study of van den Thillart *et al.* (1990) on goldfish. These workers reported that white muscle ATP levels were stable for 1 hour after excision, yet creatine phosphate levels declined by 36 per cent in only 6 seconds! Creatine phosphate preservation is particularly important for acid–base studies because its hydrolysis will cause a transient alkalinization (Hochachka & Mommsen, 1983; Spriet *et al.*, 1986). A useful check on creatine phosphate data is to measure the total creatine pool simultaneously. While the exact percentage phosphorylation in resting fish muscle is debatable (see Moyes & West, 1995), values less than 50 per cent of the total pool as creatine phosphate should be considered with suspicion.

Methods for intracellular acid–base measurement also deserve comment. Ultimately, the technique of choice may be ^{31}P-NMR (van den Thillart & van Waarde, 1996), but it cannot yet be applied to unrestrained fish, so most workers have used either [^{14}C]5,5- dimethyloxazolidine-2,4-dione ([^{14}C]DMO) (Waddell & Butler, 1959) or homogenate methods. The latter consist of freeze–thaw lysis followed by direct pH reading for red blood cells (Zeidler & Kim, 1977), and mortar and pestle grinding of freeze-clamped tissue, subsequent homogenization in a cocktail of metabolic inhibitors, and again final direct reading of pH for muscle (Pörtner *et al.*, 1990). In our hands, the [^{14}C]DMO and homogenate approaches yield identical values for intracellular pH (pH$_i$) in both red blood cells and white muscle, both at rest and after exercise (e.g. Milligan & Wood, 1985; 1986b *versus* Wang *et al.*, 1994a). Schulte, Moyes and Hochachka (1992) reached a similar conclusion. (Discrepancies are expected in other tissues, especially those that are rich in alkaline mitochondria or acidic lysosomes.) However, the homogenate techniques are recommended because they offer several advantages relative to the tedious [^{14}C]DMO method – directness, speed, simplicity, precision, and economy. Furthermore, for white muscle, the technique of Pörtner *et al.* (1990) facilitates the *direct* measurement of intracellular total CO_2, and therefore the calculation of P_{ICO_2} and [HCO_3^-]$_i$ for the same sample. Earlier studies that assumed equilibration of muscle P_{ICO_2} with blood P_{aCO_2} (so as to calculate tissue values) may have underestimated the post-exercise P_{ICO_2} and [HCO_3^-]$_i$ in white muscle after exercise (e.g. compare Milligan & Wood, 1986b, with Kieffer, Currie & Tufts, 1994), though this is not entirely certain (e.g. compare Milligan & Wood, 1986b, with Tang & Boutilier, 1991).

A final methodological concern is the necessity of measuring extracellular space accurately to allow calculation of intracellular fluid (ICF) space, and therefore expression of the concentration of intracellular ions and metabolites on the correct basis. Munger, Reid and Wood (1991) evaluated a number of different extracellular space labels and concluded that [^{3}H]polyethylene glycol-4000 (PEG-4000), allowed to equilibrate for about 12 hours prior to measurement, was the marker of choice for fish tissues. Low-molecular-weight markers such as mannitol should be avoided because they tend to penetrate into the ICF. In situations in which it is not practical to use [^{3}H]PEG-4000 (e.g. the perfused tail-trunk preparation), calculation of the Cl^-–K^+ space as an extracellular fluid (ECF) estimate (see Conway, 1957, for formula) appears to be a reasonable alternative in trout white muscle. This requires measurement of electrolyte levels in plasma (or perfusate) and whole muscle, and assumes a Donnan distribution of K^+ and Cl^- between the ECF and ICF. An added benefit is that with the additional measurement of extracellular and tissue Na^+, the white muscle membrane potential (E_m) can be calculated using the Goldman–Hodgkin–Katz equation (Hodgkin & Horowicz, 1959), as explained by Wang *et al.* (1996a).

Post-exercise responses of the rainbow trout

Exhaustive exercise is achieved most easily by vigorously chasing the fish in a circular tank for a 6-minute period. (Electric shocks used in early studies are unnecessary with rainbow trout.) Swimming the fish at progressively higher speeds in a swim tunnel up to the maximum sustainable velocity (U_{Crit}), followed by repeated burst swimming at higher velocity to ensure exhaustion, yields similar results. The endpoint is an inability to perform further burst exercise, though slow swimming by aerobic red muscle usually persists. The procedure is undoubtedly stressful, resulting in a rapid catecholamine and slower cortisol mobilization, and is probably better termed 'exhaustive exercise stress' (Wood, 1991). With a focus on white muscle physiology, key features of the response are summarized below. Keeping in mind the methodological issues outlined above, this description relies on data judiciously selected from Turner and Wood (1983), Turner, Wood and Clark (1983), Neumann, Holeton and Heisler (1983), Primmett *et al.* (1986), Milligan and Wood (1986a; 1986b), Dobson and Hochachka (1987), Dobson, Parkhouse and Hochachka (1987), Milligan and McDonald (1988), Mommsen and Hochachka (1988), Wood (1988), Wood *et al.* (1990), Tang and Boutilier (1991), Pagnotta and Milligan (1991), Scarabello, Heigenhauser and Wood (1991), Schulte *et al.* (1992), Moyes, Schulte and Hochachka (1992), Milligan and Girard (1993), Pagnotta, Brooks and Milligan (1994), Kieffer *et al.* (1994), Wang *et al.* (1994a), West, Schulte and Hochachka (1994), Kieffer and Tufts (1996), and Eros and Milligan (1996).

Exhaustive exercise causes large declines in white muscle levels of the three 'anaerobic' fuels – ATP, creatine phosphate, and glycogen – and accumulation of creatine, inorganic phosphate, lactate and glycolytic intermediates in the intracellular compartment (Fig. 1). Lactate accumulation, which may exceed 40 mmol/l ICF, is often almost equimolar to the depletion of glycogen (in glucosyl equivalents; Fig. 1A). Similarly, due to the AMP deaminase reaction, accumulation of inosine monophosphate (IMP) and ammonia is often equimolar to the depletion of ATP (Fig. 1B). Muscle pH_i drops markedly (Table 1), due to the production of H+ ('metabolic acid' = H+m) associated with lactate (and pyruvate) production and ATP hydrolysis, with a small contribution from H+ due to P_{CO_2} build-up ('respiratory acid'). These changes are at their greatest either immediately or within half an hour after the end of exercise. With the exception of creatine phosphate, creatine (Fig. 1C) and 'respiratory acidosis', which are restored/corrected by 0.5–2 hours, recovery of other parameters occurs slowly, with completion taking 4–24 hours. Ammonia and IMP clearance occur at similar rates, and generally mirror ATP restoration (Fig. 1B), presumably reflecting the reaminating arm of the purine nucleotide cycle. Lactate clearance is coincident with, but slightly greater than, glycogen replenishment (Fig. 1A). Later in recovery (post 4 hours), lactate disappearance corresponds more closely to glycogen restoration.

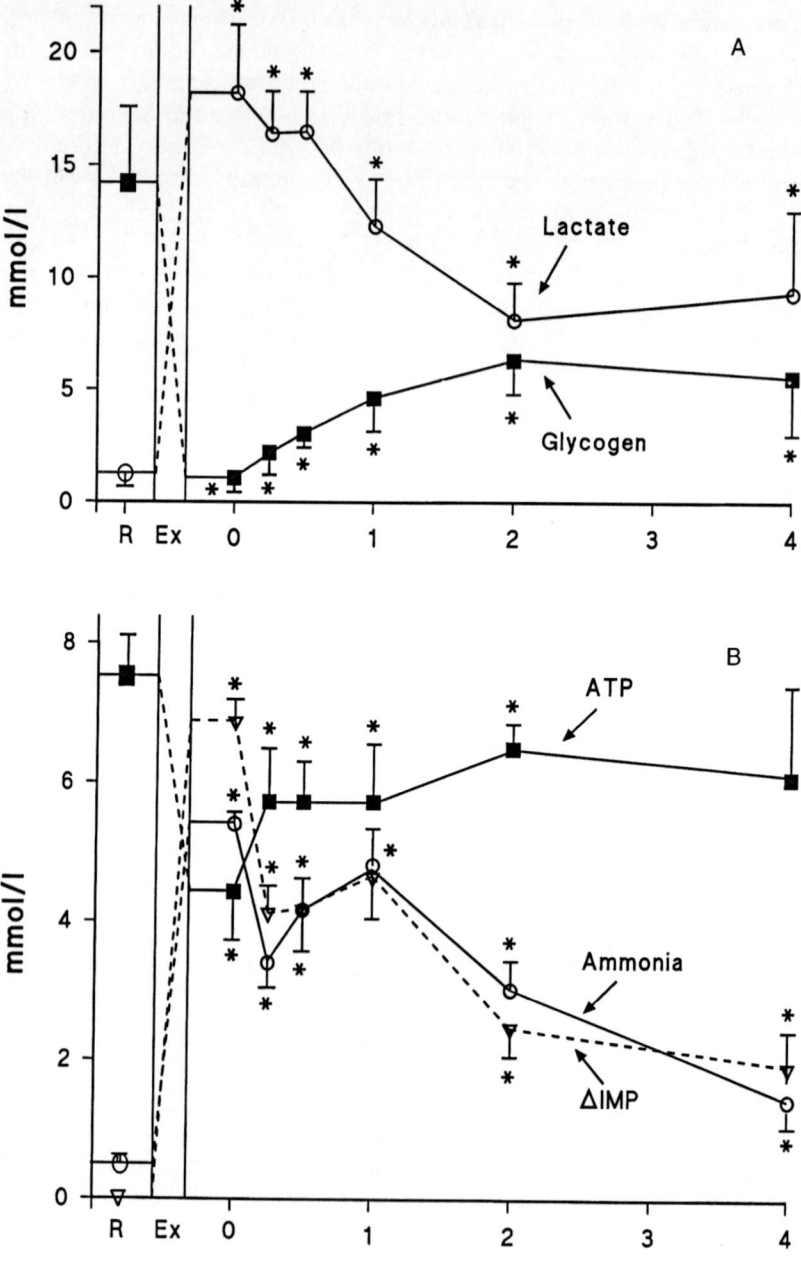

Fig. 1.

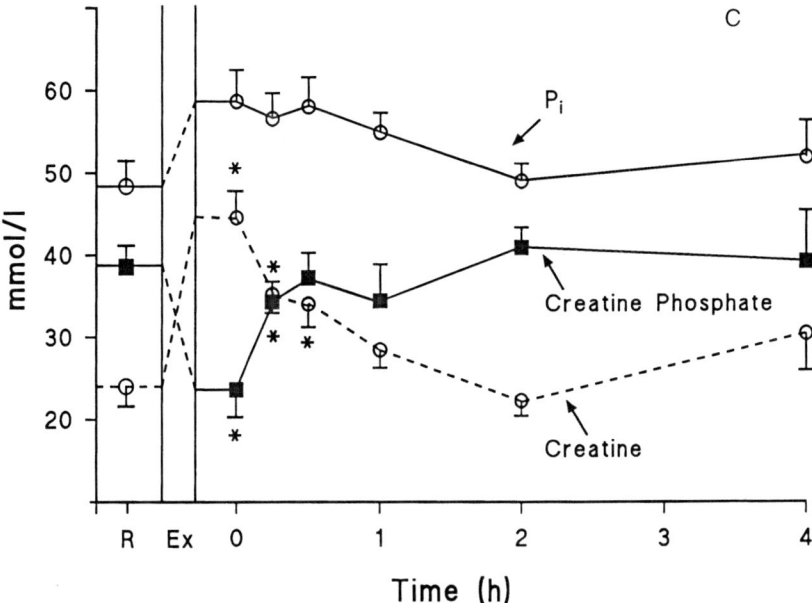

Fig. 1. Changes in metabolite concentrations in the intracellular fluid of white muscle after exhaustive exercise in rainbow trout. **A.** Glycogen and lactate, both expressed in glucosyl equivalents for ease of comparison. **B.** ATP, total ammonia, and ΔIMP (resting levels subtracted for ease of comparison on the same scale). **C.** Creatine, creatine phosphate, and inorganic phosphate (Pi). Within each panel, note the general equivalence of the changes, suggesting that the white muscle functions as a closed unit during recovery from exhaustive exercise. Means ± 1 SEM (n = 8–13). Asterisks indicate significant differences ($p < 0.05$) from resting values. Data from Wang *et al.* (1994a).

Most of these changes are reflected in the blood, with elevations in plasma P_{CO_2}, lactate, pyruvate, H^+m, inorganic phosphate and ammonia, and depressions in plasma pH (pHa = pH$_e$; Table 1). With the exception of P_{CO_2} ('respiratory acid'), which is again cleared in 1–2 hours, these changes lag behind those in muscle, reaching their peak at 1–3 hours post-exercise and declining thereafter. The pHa and H^+m response in the plasma is confounded by the β-adrenergic extrusion of H^+ from erythrocytes triggered by the mobilization of catecholamines. As a result, red cell pH$_i$ undergoes little change and blood O_2 transport is protected at a time of extracellular acidosis. An initial export of H^+

Table 1. *Membrane potential (E_m) and pH gradients between extracellular (pH_e = arterial pH_a) and intracellular (pH_i) compartments in white muscle of rainbow trout in vivo after exhaustive exercise. Means ± 1 SEM (n = 8–13)*

	E_m (mV)	pH_e	pH_i
Rest	-95 ± 7	7.96 ± 0.02	7.22 ± 0.03
Post-exercise			
0 h	-102 ± 6	$7.50 \pm 0.03^*$	$6.83 \pm 0.11^*$
0.25 h	-98 ± 6	$7.56 \pm 0.02^*$	$6.72 \pm 0.05^*$
0.50 h	-95 ± 6	$7.61 \pm 0.04^*$	$6.75 \pm 0.05^*$
1.0 h	-93 ± 4	$7.65 \pm 0.04^*$	$6.84 \pm 0.08^*$
2.0 h	-90 ± 6	$7.67 \pm 0.03^*$	7.04 ± 0.06
4.0 h	-85 ± 4	7.94 ± 0.06	7.05 ± 0.12

$^*p < 0.05$ relative to rest value.
Data from Wang *et al.* (1994a).

to the external environment and a later re-uptake, both mediated by manipulation of ionic exchange processes on the gills, help to reduce changes in plasma pHa and H⁺m levels in the blood.

A point of key importance is that not only do extracellular changes in metabolite concentrations lag behind intracellular changes, they are *much* smaller (only 5–20 per cent) in absolute magnitude. This difference is accentuated when the smaller size (approximately 40 per cent) of the extracellular compartment relative to the white muscle intracellular compartment is taken into account. The simplest interpretation is that the white muscle functions largely as a closed compartment during the post-exercise period, a conclusion which is now supported by a large weight of evidence from both salmonids and other teleosts; this information has been summarized by Wood (1991) and Milligan (1996). This takes place despite the fact that blood flow to white muscle appears to increase substantially at this time (Wardle, 1978; Neumann *et al.*, 1983). A word of caution, however, is that to date there are no white muscle blood flow measurements in trout subjected to chasing-type exhaustive exercise; the data of Neumann *et al.* (1983) were collected from fish exercised by electrical stimulation.

In particular, the evidence indicates that the Cori cycle – lactate release to liver, hepatic conversion to glucose, glucose transport to muscle – plays a negligible role in glycogen resynthesis, and that direct gluconeogenesis from lactate occurs in white muscle. The great bulk of the lactate and H⁺m loads is found in the white muscle, not the extracellular fluid, at all times after exercise. The measured capacity of the liver to perform gluconeogenesis from lactate is low, as are the measured lactate and H^+m release rates, glucose uptake rates, and rates of glucose incorporation into glycogen of white muscle. Labelled lactate itself, however, is taken up by muscle and is incorporated into glycogen at a much higher rate. Functional removal of the liver actually accelerates lactate clearance and glycogen resynthesis in white muscle, suggesting that the small amount of lactate 'leakage' and subsequent hepatic processing which normally occurs are actually a drain on the muscle recovery process! In a similar vein, blockade of cortisol mobilization accelerates muscle recovery and lessens lactate and metabolic acid appearance in the blood. This observation suggests that the white muscle system is designed to retain as much lactate and H^+m in the muscle as possible, and that the normal post-exercise cortisol surge acts slightly to lessen this retention, thereby ensuring that small amounts are released as fuel for other tissues early in the recovery period.

Problems in current understanding

Several theoretical problems must be overcome before this theory of lactate and metabolic acid retention in white muscle for glyconeogenesis *in situ* can be fully accepted. Firstly, what biochemical pathway is employed – specifically, how is pyruvate converted back to phosphoenolpyruvate since, based on *in vitro* tests, pyruvate kinase is normally thought to be an irreversible enzyme? Pyruvate carboxylase and phosphenolpyruvate carboxykinase (PEPCK), which do this job in liver, have not been detected in fish muscle; this would also seem to rule out another possibility, the combination of malic enzyme with PEPCK. However, one possibility is reversal of pyruvate kinase; several papers have suggested that this may be possible under *in vivo* conditions, but to date, proof is lacking (Schulte *et al.*, 1992; Moyes *et al.*, 1992; Moyes & West, 1995).

Secondly, if lactate and H^+m stay in the white muscle, why are they not burned in aerobic respiration to meet the considerable demands of excess post-exercise oxygen consumption (EPOC; Scarabello *et al.*, 1991)? Certainly, white muscle is capable of lactate oxidation (Bilinski & Jonas, 1972), but other tissues (e.g. heart, red muscle, liver, red blood cells) have much higher capacities and a significant portion of the lactate and metabolic acid which does leak out of muscle (perhaps 30–40 per cent; Milligan & Girard, 1993) is consumed by oxidation. These factors undoubtedly explain the small stoichiometric discrepancy between lactate clearance and glycogen reappearance in white muscle early in recovery (Fig. 1A). However, the major portion of muscle lactate appears to be spared from an oxidative fate so that it can be used as the substrate for *in situ* gluconeogenesis. Recent circumstantial data suggest that an activation of lipid uptake and oxidation in white muscle occurs at this time (Dobson & Hochachka, 1987; Moyes *et al.*, 1992; Milligan & Girard, 1993; Wang *et al.*, 1994a), thereby providing the ATP needed to power both glycogen resynthesis and the other major components of EPOC. The control point may be at the pyruvate dehydrogenase complex (PDH), which funnels C units from lactate into the Krebs cycle, because Moyes *et al.* (1992) reported that increased availability of free fatty acids allosterically inhibited PDH activity in isolated trout white muscle mitochondria. Clearly, more work is needed to confirm this theory.

The third theoretical challenge forms the focus of the remainder of this chapter. If white muscle really does function as a largely closed system during post-exercise recovery, despite greatly increased blood flow at this time (Wardle, 1978; Neumann *et al.*, 1983), what are the mechanisms which prevent key metabolites from leaking out to the blood plasma? While this question can clearly be directed at lactate and H^+m, it is also relevant to ammonia, a molecule which plays an important role in intracellular acid–base regulation and which is usually considered to be highly diffusible. Figure 2 illustrates the issue. If ammonia and lactate obey the classic behaviour of weak acids and bases (Jacobs & Stewart, 1936), then only the uncharged forms (NH_3, undissociated lactic acid = HLac respectively) should be permeable across cell membranes and, at equilibrium, the transmembrane distribution ratio should be dictated by the transmembrane pH gradient (i.e. pH_e–pH_i). Diffusive movement will be driven by concentration gradients of NH_3 (actually P_{NH_3} because NH_3 is a gas) and HLac. If, however, the charged forms (NH_4^+, Lac^- respectively) are also permeable, then the situation becomes more complex. The distribution ratio may now be influenced by the membrane potential (E_m) as well, and diffusive movements may be influenced by the net electrochemical driving force (i.e. E_m – Nernst potential) on the charged form. Depending on the relative permeability of the cell membrane to charged and uncharged forms, the actual distribution may lie anywhere between bounds determined by the pH gradient and the E_m. If the distribution lies outside these bounds, then it cannot be explained by solely passive factors. The theory for this situation has been derived by Boron and Roos (1976) and explained by Wright, Randall and Wood (1988), to which the reader is referred.

Figure 2 presents the measured distribution ratios ([ICF]/[ECF]) for lactate and ammonia in trout white muscle at rest and after exhaustive

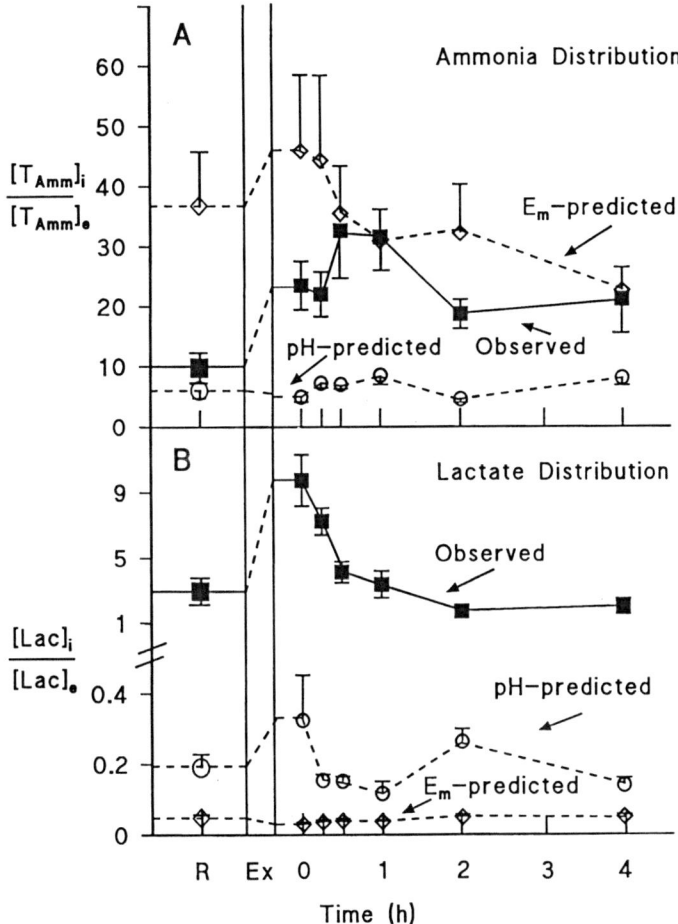

Fig. 2. Comparisons of the observed [ICF]/[ECF] distribution ratios of (**A**) ammonia ($[T_{Amm}]i/[T_{Amm}]e$) and (**B**) lactate ($[Lac]i/[Lac]e$) with theoretical ratios in rainbow trout at rest (R) and after exhaustive exercise (Ex). The observed ratios (solid squares) are compared with ratios calculated on the basis of passive distribution according to the measured transmembrane pH gradient (open circles) and according to the measured membrane potential (E_m, open diamonds). In (**A**), note that the observed $[T_{Amm}]i/[T_{Amm}]e$ moves from a value close to the pH-predicted ratio at rest to values close to the E_m-predicted ratios after exercise, but remains within the theoretical bounds. In (**B**), note that the observed $[Lac]i/[Lac]e$ remains greatly above both theoretical ratios throughout the experiment. Means ± 1 SEM ($n = 8$–13). Data from Wang *et al.* (1994a).

exercise, and compares them with the bounds predicted by theory from measured pH gradients (0.6–0.8 pH units) and E_m (−85 to −102 mV) as tabulated in Table 1. E_m was calculated from measured ionic distributions *via* the Goldman–Hodgkin–Katz equation (Wang *et al.*, 1994a). From this it is clear that ammonia distribution lies between the two bounds, but changes markedly from a situation closer to a pH-dictated distribution at rest to a situation closer to an E_m-dictated distribution during post-exercise recovery (Fig. 2A). Therefore, there is no need to invoke active transport mechanisms, but this result suggests a change in relative membrane permeability to NH_3 *versus* NH_4^+ after exercise. In contrast, lactate distribution is held well out of passive equilibrium with either the pH gradient or the E_m (Fig. 2B). This is true even at rest, and the discrepancy increases after exercise, suggesting that active processes are in some way involved in maintaining intracellular [lactate] at much higher concentrations than explicable from passive forces. Based on both pH and E_m, intracellular lactate concentration should be well below the extracellular concentration! Finally, it is not possible to distinguish between H^+ of 'metabolic' *versus* 'respiratory' origin in this sort of analysis. Nevertheless, if H^+ were distributed passively, then H^+ ions distribution would follow the same 'E_m-predicted line' as for ammonia in Fig. 2A; of course it does not, but rather follows the 'pH-predicted' line, illustrating the well-known fact (Roos & Boron, 1981) that H^+ are held well out of electrochemical equilibrium by active extrusion mechanisms in most vertebrate cells.

Clearly, to understand the situation better, it would be useful to manipulate the pH and electrical gradients across white muscle cell membranes, and to apply various pharmacological inhibitors of suspected carrier-mediated processes for lactate, ammonia, and H^+ transport. It would also be useful to measure the actual fluxes of these substances, rather than just their distribution. Such studies are difficult or impossible to do *in vivo*, so there is an obvious need for new *in vitro* approaches.

The isolated-perfused trout tail-trunk preparation

With these goals in mind, we have recently developed this new *in vitro* preparation, which is described in greater detail and shown diagrammatically in Fig. 1 of Wang *et al.* (1996a). Large trout (0.6–1.0 kg) are sacrificed either at rest or immediately after exhaustive exercise using an overdose of neutralized MS-222 (as above), and transected at the level of the anus. Section at this level yields a short tail-trunk of about 80–130 g, with discrete arterial inflow (caudal artery) and outflow (caudal vein) vessels. These are immediately cannulated, perfusion is started at a rate of 2 ml/100 g tail weight/min, and the preparation is submerged in a saline bath at 15.0 ± 0.5 °C. White muscle makes up about 90 per cent of the total soft tissue volume, the remainder being mainly discrete bands of red muscle along the lateral line; there is no contaminating kidney tissue. The basic perfusate consists of Cortland salmonid saline (Wolf, 1963) supplemented with 50 i.u./ml heparin and 3% bovine serum albumin (important in providing oncotic pressure) and gassed with 0.25% CO_2/balance O_2 (P_{CO_2} ~ 2 torr, P_{O_2} > 500 torr), which yields a control pH of approximately 7.9, HCO_3^- of approximately 7 mmol/l. Perfusate pH can be easily adjusted by manipulating the $NaHCO_3$ concentration (nominally 2 and 18 mmol/l for pH 7.4 and 8.4), and partial depolarization of E_m (from about −90 mV to −60 mV) is achieved by raising the KCl concentration from 3 to 15 mmol/l; for both manipulations, the concentration of NaCl is reciprocally adjusted. In practice, most experiments last 60 minutes, with the first 30 minutes used to ensure stabilization and complete washout of erythrocytes, and the second 30 min used for experimental measurements. Flux rates of substances (respiratory gases, metabolites, electrolytes) into and out of the muscle are calculated *via* the Fick principle based on exact gravimetric measurements of the perfusion flow rate and differences in measured concentration between arterial inflow and venous outflow samples. At the end of the experiment, muscle samples are taken by freeze-clamping immediately after the last perfusate samples have been drawn. Tissue composition at this time may be compared with pre-perfusion muscle samples taken at the time of sacrifice from the region immediately anterior to the point of transection. Extracellular fluid volume is estimated by the Cl^-–K^+ space method and E_m by the Goldman–Hodgkin–Katz equation (as above).

The preparation exhibits a stable M_{O_2} of approximately 800 μmol/kg tissue per h, equivalent to about 25 per cent of the whole animal M_{O_2} measured *in vivo* in trout of this size, which seems reasonable given the low mitochondrial density of white muscle. Ammonia efflux (M_{Amm}) is approximately 50 μmol/kg tissue/h (Fig. 3A), again equivalent to about 25 per cent of the whole animal value; while net lactate efflux is about 300 μmol/kg tissue/h (Fig. 3B), comparable to resting *in vivo* rates estimated from radiolabelled turnover (Milligan & McDonald, 1988; Weber, 1991). Control experiments with resting preparations demonstrated no change in key intracellular metabolites over 60 minutes of perfusion (Table 2), apart from an increase in ATP levels. Note in particular the preservation of high stable creatine phosphate concentrations. There is some gain in water content, but this remains within the normal

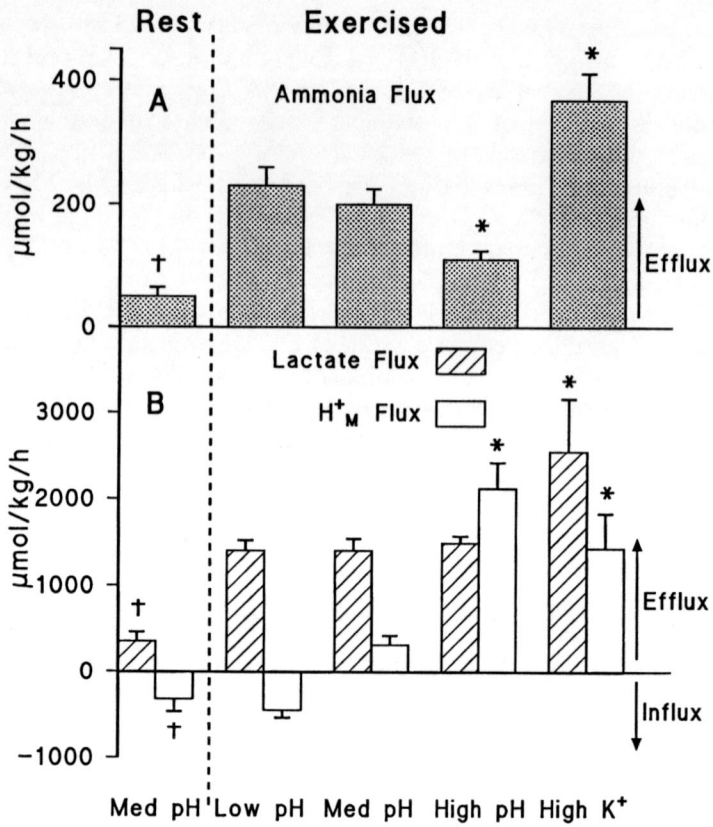

Fig. 3. **A**. Fluxes of ammonia and **B**. lactate and metabolic acid (H^+m) across the perfused tail-trunk preparation of rainbow trout at rest or 60 minutes after exhaustive exercise. Low pH (7.4), medium pH (7.9), and high pH (8.4) perfusates were employed, as well as a medium pH perfusate with high [K^+] to elicit partial depolarization. Note that ammonia and H^+m fluxes respond to changes in pH and electrical gradients in the predicted fashion, while lactate fluxes do not. Daggers indicate significant differences ($p < 0.05$) between resting values and post-exercise values at medium pH; asterisks indicate significant differences ($p < 0.05$) relative to the medium pH value after exercise. Means ± 1 SEM ($n = 6$–11). Data from Wang *et al.* (1996a; 1996b) and Henry *et al.* (1997).

Table 2. *Metabolite levels and other parameters in white muscle of the resting isolated–perfused tail-trunk preparation after 60 minutes of perfusion under control conditions, compared with values measured immediately prior to the start of perfusion. Means ± 1 SEM (n = 5–8)*

	Pre-perfusion	60-minute perfusion
pH_i	7.25±0.02	7.22±0.03
$[CO_2]_i$ (mmol/kg)	4.69±0.75	6.21±0.61
Lactate (mmol/kg)	3.97±0.86	3.04±0.59
ATP (mmol/kg)	7.51±0.56	10.08*±0.10
Creatine phosphate (mmol/kg)	38.57±3.44	39.30±1.80
Ammonia (µmol/kg)	323.4±36.8	258.8±57.0
Water (%)	74.5±0.8	77.1*±0.5

Values expressed per litre of intracellular water.
*$p \leq 0.05$ relative to respective pre-perfusion value.
Data from Wang *et al.* (1996b) and Henry *et al.* (1997).

range and no obvious oedema occurs. The E_m stays at approximately –90 mV.

When the preparation is made from a trout which has been exhaustively exercised, E_m remains largely unchanged, as *in vivo*. M_{O_2} also remains largely unchanged, whereas M_{Amm} increases to about 200 µmol/ kg tissue/h (Fig. 3A) and net lactate efflux to about 1500 µmol/kg tissue/h (Fig. 3B), the latter two comparable to changes seen *in vivo* (Wood, 1988; Milligan & McDonald, 1988). Intracellular lactate (approximately 50–75 mmol/l ICF) and ammonia (approximately 9 mmol/l ICF) are greatly elevated and pH_i greatly depressed (to about 6.6) in these large trout relative to resting levels (see Table 2) and remain stable or increase slightly over the 60 minutes of perfusion. *In vivo*, small decreases would be expected over this period (see Fig. 1). Overall, the preparation provides a useful simplifying model for studying the mechanisms of metabolite and acid–base transfer across the white muscle of rainbow trout.

Ammonia, H⁺m, and lactate distribution and transport in white muscle

Changes in perfusate pH_a from the control level of 7.9 to either 7.4 or 8.4 had no effect on pH_i in the tail-trunk after exhaustive exercise; pH_i remained stable at about 6.6, whereas partial depolarization with high

K^+ medium caused a significant decrease to about 6.4. Total intracellular ammonia levels were also similar and stable in all groups, at about 9 mmol/l, providing an informative set of conditions for analysis of ammonia transport. Increases in pH_e at constant pH_i thereby produced decreases in the P_{NH_3} gradient between ICF and ECF. If NH_3 permeability across white muscle cell membranes is significant, one would predict an accompanying fall in ammonia efflux, which was exactly the response observed (Fig. 3A). Conversely, if NH_4^+ permeability is significant, one would predict that partial depolarization of E_m from -90 mV to -60 mV would increase ammonia efflux. Indeed, ammonia efflux almost doubled (Fig. 3A), despite a 65 per cent drop in the P_{NH_3} gradient, which suggests that NH_4^+ permeability is also substantial, and perhaps even more important than NH_3 permeability in post-exercise muscle. Notably, amiloride (10^{-4} mol), a potent cation exchange blocker (Benos, 1982), had no effect on ammonia efflux, thereby eliminating any direct or indirect role of Na^+/H^+ or Na^+/NH_4^+ exchangers in ammonia transport in this preparation.

Comparison of the measured distribution ratios ([ICF]/[ECF]) for ammonia with those predicted from distribution according only to pH gradients (i.e. NH_3 permeability dominant) or according only to E_m (i.e. NH_4^+ permeability dominant) proved particularly informative (Fig. 4A). $[T_{Amm}]i/[T_{Amm}]e$ changed from a resting level only slightly above that dictated by the pH gradient to a much greater post-exercise level similar to that dictated by E_m and greatly above that dictated by the pH gradient. Partial depolarization decreased $[T_{Amm}]i/[T_{Amm}]e$ in accord with the predicted effect of the electrical gradient on NH_4^+ distribution. Nevertheless, increases in pH_e (which thereby increased the pH gradient and decreased the P_{NH_3} gradient) elevated $[T_{Amm}]i/[T_{Amm}]e$ in parallel to but above the predictions of a pH-dictated distribution, showing that significant NH_3 permeability persists.

The clear conclusion from these perfusion studies is that both NH_3 and NH_4^+ permeability are important, with a shift towards a greater role of the latter after exercise. This is in accord with our in vivo studies (see Fig. 1A), and with earlier in vivo experiments by Tang, Lin and Randall (1992) in which apparent post-exercise 'washout' of ammonia in trout was influenced by pH_e manipulations, but the $[T_{Amm}]i/[T_{Amm}]e$ distribution ratio in muscle remained close to that predicted from E_m. The in vivo data of both Wright and Wood (1988) and Tang et al. (1992) suggested that E_m determined $[T_{Amm}]i/[T_{Amm}]e$ at rest as well as after exercise, whereas the present data suggest a shift away from pH control towards E_m control between rest and exercise. The energetic advantages of such a shift may be considerable. Earlier, Heisler (1990)

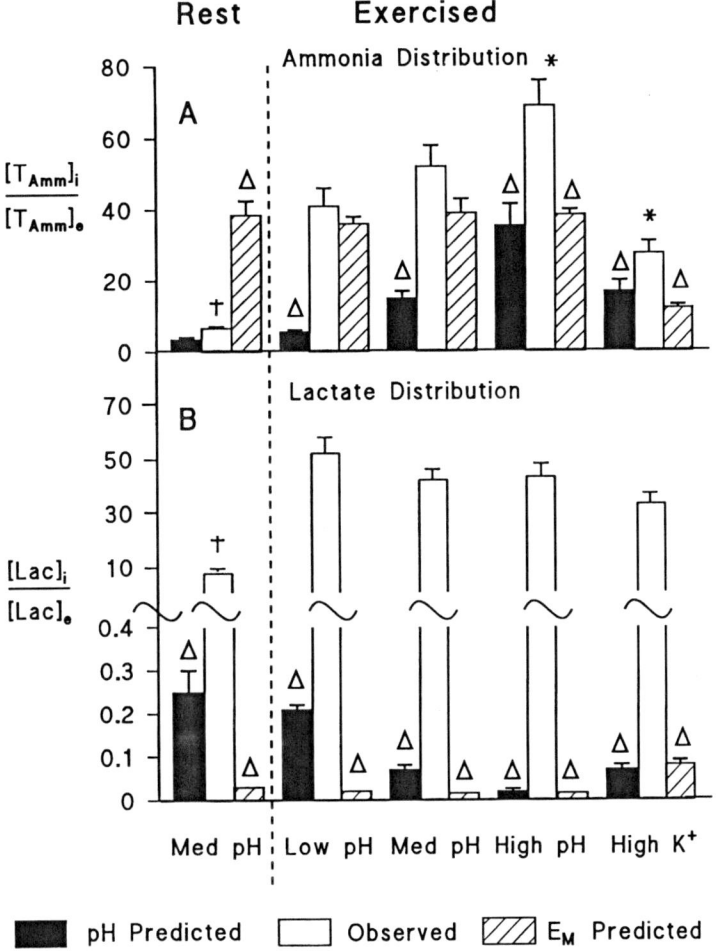

Fig. 4. Comparisons of the observed [ICF]/[ECF] distribution ratios of (A) ammonia ([T_{Amm}]i/[T_{Amm}]e) and (B) lactate ([Lac]i/[Lac]e) with theoretical ratios in the perfused tail-trunk preparation of the rainbow trout at rest and after exhaustive exercise. The observed ratios (open bars) are compared with ratios calculated on the basis of passive distribution according to the measured transmembrane pH gradient (black bars) and according to the measured membrane potential (E_m, cross-hatched bars). Note the general accord with the *in vivo* data of Figure 2. (See the legend of Figure 3 for details of perfusate composition.) Triangles indicate significant differences ($p < 0.05$) between theoretical and observed values; daggers indicate significant differences ($p < 0.05$) between resting values and post-exercise values at medium pH; asterisks indicate significant differences ($p < 0.05$) relative to the medium pH value after exercise. Means ± 1 SEM ($n = 6$–11). Data from Wang *et al.* (1996a; 1996b) and Henry *et al.* (1997).

had pointed out that an E_m-dictated distribution of ammonia at rest would be costly because it would create an inward H^+ shuttle, taxing the active H^+ extrusion mechanisms of the cell. A pH-dictated distribution at rest would avoid this shuttle, whereas an E_m-dictated distribution after exercise would help retain much higher levels of ammonia in the muscle for ATP resynthesis by the purine nucleotide cycle (Mommsen & Hochachka, 1988) and for buffering intracellular acidosis (Dobson & Hochachka, 1987). No active transport mechanism for retention is necessary, and the muscle functions as a more or less closed system. In this regard, note that the measured elevation in ammonia efflux to about 200 µmol/kg tissue/h during the first hour after exhaustive exercise in the perfused tail-trunk (Fig. 3A), while comparable to the elevation *in vivo* (Wood, 1988), will have negligible impact on the load of total ammonia built up in the white muscle mass (5–9 mmol/l ICF).

The mechanism behind this shift would obviously involve a change in the relative permeability of the white muscle sarcolemma to NH_4^+ *versus* NH_3 (see Boron & Roos, 1976, Wright *et al.*, 1988, for theory), though the proximate stimulus is unknown; perhaps acidosis itself plays a role. It is noteworthy that in mammalian physiology, the traditional view that ammonia is always distributed as a weak base according to the pH gradient (Mutch & Bannister, 1983) has been challenged by recent exercise data (Graham *et al.*, 1990; 1995).

This same perfusion approach also illuminates the mechanisms of metabolic acid (H^+m) and lactate transport in trout white muscle. Clearly, the net movements of H^+m and lactate after exercise are dissociated in magnitude. Net lactate efflux, while elevated four- to five-fold relative to resting levels, was completely unaltered by large changes in the pH_e–pH_i gradient (Fig. 3B), whereas net H^+m flux changed from actual influx at low pH_e to an efflux greater in magnitude than that of lactate at high pH_e! Thus, net H^+m movement was clearly responsive to the pH gradient. H^+m movement also appeared responsive to the electrochemical gradient in the expected manner, because partial depolarization (high $[K^+]$) accelerated H^+m efflux. In these arguments, it must be appreciated that H^+m does not represent free protons but rather is a composite acid–base measure. Base efflux and acid uptake are indistinguishable from one another in the H^+m measurement, a fact which has no significance in terms of net acid–base balance, but clearly may in terms of mechanism. In this regard, it is noteworthy that at low pH_e (the physiologically 'realistic condition' after exhaustive exercise), where H^+m influx into the muscle occurred (Fig. 3B), this phenomenon was significantly depressed by amiloride (10^{-4} M) and accelerated by

4-acetomido-4'-isothiocyanatostilbene-2,2'-disulphonic acid (SITS, 5×10^{-4} M). These data suggest the involvement of classic cation (e.g. Na^+/H^+) and anion (Cl^-/HCO_3^-) exchange processes ('antiporters'). Overall, the data are consistent with the 'equilibrium limitation' model first proposed by Holeton and Heisler (1983), such that H^+m efflux falls as pH_e declines, eventually reaching a point of reversal. By this mechanism, extracellular acidosis will automatically limit the amount of H^+m exported from the muscle after exercise, thereby promoting its retention for glyconeogenesis.

In addition to being unresponsive to the pH gradient, net lactate efflux from the perfused trunk increased rather than decreased in the presence of partial depolarization (Fig. 3B), the opposite to the response expected from the change in electrochemical gradient. The distribution ratio ($[Lac]_i/[Lac]_e$) remained at a high level (about 40), regardless of pH or E_m manipulation, far out of equilibrium with either pH or E_m (Fig. 4B). As *in vivo* (Fig. 2B), this discrepancy increased markedly after exercise. Overall, the best correlation for net lactate efflux was with the simple concentration gradient of total lactate from ICF to ECF. The post-exercise efflux rate in the perfused tail-trunk was very comparable to the low rate seen *in vivo*, which can account for no more than 10 per cent of the observed post-exercise lactate clearance from white muscle (Milligan & McDonald, 1988). These data reinforce the conclusion that active processes are in some way involved in maintaining intracellular high lactate concentration against all prevailing gradients, as first suggested by Turner and Wood (1983) based on much less rigorous evidence.

We have used pharmacological tools (Poole & Halestrap, 1993; Gladden, 1996) and kinetic analysis to examine possible carrier mechanisms for net lactate transport across the white muscle sarcolemma in the perfused tail-trunk. CIN (5×10^{-3} M α-cyano-4-hydroxycinnamic acid) is considered a specific competitive blocker of the Lac^--H^+ symporter and a non-competitive inhibitor of the Lac^-/HCO_3^-, Cl^- antiporter ('band 3'); SITS (5×10^{-4} mol) is a competitive blocker of the anion antiporter only; and amiloride (10^{-4} M) is a competitive blocker of Na^+/H^+ antiporter. D(+)-lactate, the non-native stereoisomer of regular L(+)-lactate, was also employed to separate the stereospecific symport pathway (Lac^--H^+ co-transport) from the non-stereospecific antiport (Lac^-/HCO_3^-, Cl^-) and free diffusion pathways. Uptake experiments were performed in resting preparations and efflux experiments in post-exercise preparations, i.e. in the direction of the simple concentration gradients in each case.

L(+)-lactate uptake kinetics (rates *versus* concentration) were curvilinear, indicative of carrier mediation, while D(+)-lactate uptake kinetics were strictly linear. As summarized in Figure 5, net lactate uptake rates in resting preparations were almost double net efflux rates in post-exercise preparations, despite a much more favourable electrochemical driving force in the latter. This in itself suggests a system that is poised to take up rather than lose lactate. The various pharmacological tests indicated that Lac$^-$–H$^+$ symport and the Lac$^-$/HCO$_3^-$, Cl$^-$ antiport each contributed about 35–40 per cent to the uptake process, with free diffusion accounting for the balance. In contrast, for the smaller net efflux, only the symport (approximately 40 per cent) and diffusion

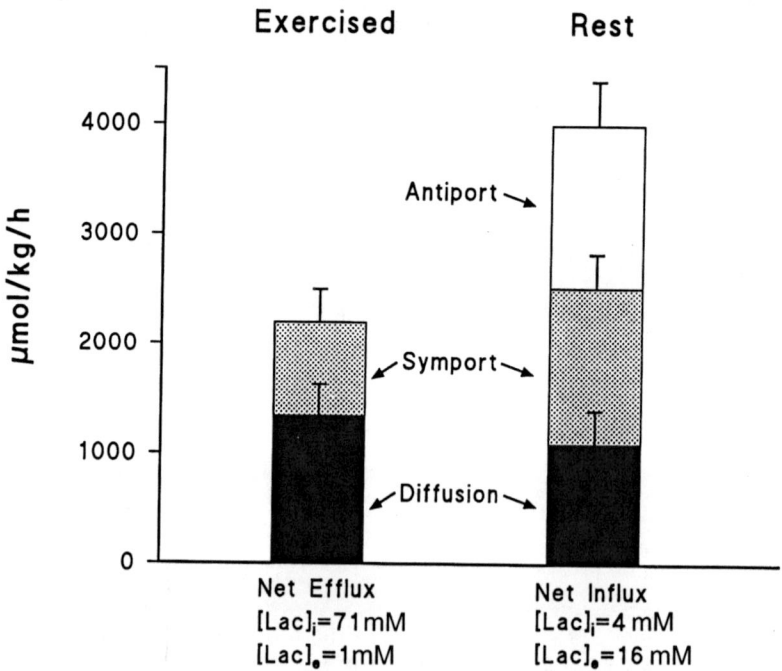

Fig. 5. Net efflux rates of lactate after exhaustive exercise and net influx rates of lactate at rest in the perfused tail-trunk of rainbow trout. The results of a pharmacological analysis of the various component mechanisms are shown; see text for details. The measured intracellular and extracellular concentrations of lactate are given. Note the twofold higher influx rate compared to efflux rate, despite the much greater driving gradients in the case of the latter, and the presence of three components to influx *versus* only two for efflux. Means ± 1 SEM ($n = 7$–11). Data from Wang *et al.* (1997).

(approximately 60 per cent) appeared to be involved. These results, together with *in vivo* indications of vigorous lactate uptake into white muscle (Batty & Wardle, 1979; Milligan & Girard, 1993), are in accord with a model in which active lactate uptake mechanisms largely counteract lactate efflux (Turner & Wood, 1983), thereby retaining lactate in the white muscle cells for glyconeogenesis *in situ*. The challenge now is to understand the energetic linkage of this active retention mechanism, and the mechanism(s) of its regulation.

Acknowledgements

The authors' original research, reported in this chapter, was supported by an NSERC Canada research grant to CMW.

References

Batty, R.S. & Wardle, C.S. (1979). Restoration of glycogen from lactic acid in the anaerobic swimming muscle of plaice, *Platichthys stellatus* L. *Journal of Fish Biology* **15**, 509–19.

Benos, D.J. (1982). Amiloride: a molecular probe of Na⁺ transport in tissue and cells. *American Journal of Physiology* **242**, C131–45.

Bilinski, E. & Jonas, R.E.E. (1972). Oxidation of lactate to carbon dioxide by rainbow trout (*Salmo gairdneri*) tissues. *Journal of the Fisheries Research Board of Canada* **29**, 1467–71.

Black, E.C. (1958). Hyperactivity as a lethal factor in fish. *Journal of the Fisheries Research Board of Canada* **15**, 245–51.

Black, E.C., Robertson, A.C., Lam, K.-C. & Chiu, W.-G. (1962). Changes in glycogen, pyruvate and lactate in rainbow trout (*Salmo gairdneri*) during and following muscular activity. *Journal of the Fisheries Research Board of Canada* **19**, 409–36.

Boron, W.F. & Roos, A. (1976). Comparison of microelectrode, DMO, and methylamine methods for measuring intracellular pH. *American Journal of Physiology* **231**, 799–801.

Conway, E.J. (1957). Nature and significance of concentration relations of potassium and sodium ions in skeletal muscle. *Physiological Reviews* **37**, 84–132.

Dobson, G.P. & Hochachka, P.W. (1987). Role of glycolysis in adenylate depletion and repletion during work and recovery in teleost white muscle. *Journal of Experimental Biology* **29**, 125–40.

Dobson, G.P., Parkhouse, W.S. & Hochachka, P.W. (1987). Regulation of anaerobic ATP-generating pathways in trout fast-twitch skeletal muscle. *American Journal of Physiology* **253**, R186–94.

Driedzic, W.R. & Hochachka, P.W. (1978). Metabolism in fish during exercise. In *Fish physiology*, Vol. VII, ed. W.S. Hoar & D.J. Randall, pp. 503–43. New York: Academic Press.

Eros, S.K. & Milligan, C.L. (1996). The effect of cortisol on recovery from exhaustive exercise in rainbow trout (*Oncorhynchus mykiss*): potential mechanisms of action. *Physiological Zoology* **69**, 1196–214.

Gladden, L.B. (1996). Lactate transport and exchange during exercise. In *Exercise: regulation and integration of multiple systems, Handbook of Physiology, Sec. 12.*, ed. L.B. Rowell & J. T. Shepherd, pp. 614–48. New York: Oxford University Press.

Graham, T.E., Bangsbo, J., Gollnick, P.D., Juel, C. & Saltin, B. (1990). Ammonia metabolism during intense dynamic exercise and recovery in humans. *American Journal of Physiology* **259**, E170–6.

Graham, T.E., Rush, J. W. E. & MacLean, D.A. (1995). Skeletal muscle amino acid metabolism and ammonia production during exercise. In *Exercise metabolism*, ed. M. Hargreaves, pp. 131–75. Champaign, IL: Human Kinetics.

Heisler, N. (1990). Mechanisms of ammonia elimination in fishes. In *Animal nutrition and transport processes 2. Transport, respiration, and excretion: comparative and environmental aspects*, ed. J-P. Truchot & B. Lahlou, pp. 137–51. Basel: Karger.

Henry, R.P., Wang, Y. & Wood, C.M. (1997). Carbonic anhydrase facilitates CO_2 and NH_3 transport across the sarcolemma of trout white muscle. *American Journal of Physiology* **272**, R1754–61.

Hochachka, P.W. & Mommsen, T.P. (1983). Protons and anaerobiosis. *Science* **219**, 1391–7.

Hodgkin, A.L. & Horowicz, P. (1959). The influence of potassium and chloride ions on the membrane potential of single muscle fibres. *Journal of Physiology* **148**, 127–60.

Holeton, G.F. & Heisler, N. (1983). Contribution of net ion transfer mechanisms to acid–base regulation after exhausting activity in the larger spotted dogfish (*Scyliorhinus stellaris*). *Journal of Experimental Biology* **103**, 31–46.

Jacobs, M.H. & Stewart, D.R. (1936). The distribution of penetrating ammonium salts between cells and their surroundings. *Journal of Cellular Comparative Physiology* **7**, 351–65.

Kieffer, J.D., Currie, S. & Tufts, B.L. (1994). Effects of environmental temperature on the metabolic and acid–base responses of rainbow trout to exhaustive exercise. *Journal of Experimental Biology* **194**, 299–317.

Kieffer, J.D. & Tufts, B.L. (1996). The influence of environmental temperature on the role of the rainbow trout gill in correcting the acid–base disturbance following exhaustive exercise. *Physiological Zoology* **69**, 1301–23.

Milligan, C.L. (1996). Metabolic recovery from exhaustive exercise in rainbow trout. *Comparative Biochemistry and Physiology* **113A**, 51–60.

Milligan, C.L. & Girard, S.S. (1993). Lactate metabolism in rainbow trout. *Journal of Experimental Biology* **180**, 175–93.

Milligan, C.L. & McDonald, D.G. (1988). *In vivo* lactate kinetics at rest and during recovery from exhaustive exercise in coho salmon (*Oncorhynchus kisutch*) and starry flounder (*Platichthys stellatus*). *Journal of Experimental Biology* **135**, 119–31.

Milligan, C.L. & Wood, C.M. (1985). Intracellular pH transients in rainbow trout tissues measured by dimethadione distribution. *American Journal of Physiology* **248**, R668–73.

Milligan, C.L. & Wood, C.M. (1986a). Intracellular and extracellular acid–base status and H⁺ exchange with the environment after exhaustive exercise in the rainbow trout. *Journal of Experimental Biology* **123**, 93–121.

Milligan, C.L. & Wood, C.M. (1986b). Tissue intracellular acid–base status and the fate of lactate after exhaustive exercise in the rainbow trout. *Journal of Experimental Biology* **123**, 123–44.

Mommsen, P. & Hochachka, P.W. (1988). The purine nucleotide cycle as two temporarily separated metabolics units: a study on trout muscle. *Metabolism* **37**, 552–6.

Moyes, C.D., Schulte, P.M. & Hochachka, P.W. (1992). Recovery metabolism of trout white muscle: role of mitochondria. *American Journal of Physiology* **262**, R295–304.

Moyes, C.D., Schulte, P.M. & Hochachka, P.W. (1993). Burst exercise recovery metabolism in fish white muscle. In *Surviving hypoxia: mechanisms of control and adaptation*, ed. P.W. Hochachka, P.L. Lutz, T. Sick, M. Rosenthal & G. van den Thillart, pp. 527–39. Boca Raton, FL: CRC.

Moyes, C.D. & West, T.G. (1995). Exercise metabolism of fish. In *Biochemistry and molecular biology of fishes*, Vol. 4, ed. P.W. Hochachka & T.P. Mommsen, pp. 367–92. Amsterdam: Elsevier.

Munger, R.S., Reid, S.D. & Wood, C.M. (1991). Extracellular fluid volume measurements in tissues of the rainbow trout (*Oncorhynchus mykiss*) *in vivo* and their effects on intracellular pH and ion calculations. *Fish Physiology and Biochemistry* **9**, 313–23.

Mutch, J.C. & Bannister, E.W. (1983). Ammonia metabolism in exercise and fatigue: a review. *Medical Science and Sports Exercise* **15**, 41–50.

Neumann, P., Holeton, G.F. & Heisler, N. (1983). Cardiac output and regional blood flow in gills and muscles after exhaustive exercise in rainbow trout (*Salmo gairdneri*). *Journal of Experimental Biology* **105**, 1–14.

Pagnotta, A., Brooks, L. & Milligan, C.L. (1994). The potential regulatory roles of cortisol in recovery from exhaustive exercise in rainbow trout. *Canadian Journal of Zoology* **72**, 2136–46.

Pagnotta, A. & Milligan, C.L. (1991). The role of blood glucose in the restoration of muscle glycogen during recovery from exhaustive

exercise in rainbow trout (*Oncorhynchus mykiss*) and winter flounder (*Pseudopleuronectes americanus*). *Journal of Experimental
Biology* **161**, 489–508.

Poole, R.C. & Halestrap, A.P. (1993). Transport of lactate and other
monocarboxylates across mammalian plasma membranes. *American Journal of Physiology* **264**, C761–82.

Pörtner, H.O., Boutilier, R.G., Tang, Y. & Toews, D.P. (1990). Determination of intracellular pH and PCO_2 after metabolic inhibition by
fluoride and nitrilotriacetic acid. *Respiration Physiology* **81**, 255–74.

Primmett, D.R.N., Randall, D.J., Mazeaud, M. & Boutilier, R.G.
(1986). The role of catecholamines in erythrocyte pH regulation
and oxygen transport in rainbow trout (*Salmo gairdneri*) during
exercise. *Journal of Experimental Biology* **122**, 139–48.

Roos, A. & Boron, W.F. (1981). Intracellular pH. *Physiological
Reviews* **61**, 296–434.

Scarabello, M., Heigenhauser, G.H.J. & Wood, C.M. (1991). The
oxygen debt hypothesis in juvenile rainbow trout after exhaustive
exercise. *Respiration Physiology* **84**, 245–59.

Schulte, P.M., Moyes, C.D. & Hochachka, P.W. (1992). Integrating
metabolic pathways in post-exercise recovery of white muscle.
Journal of Experimental Biology **166**, 181–95.

Soivio, A., Westman, K. & Nyholm, K. (1972). Improved method of
dorsal aorta catheterization: haematological effects followed for
three weeks in rainbow trout (*Salmo gairdneri*). *Finnish Fisheries
Research* **1**, 11–21.

Spriet, L.L., Soderlund, K., Thomson, J.A. & Hultman, E. (1986). pH
measurement in human skeletal muscle samples: effect of phosphagen hydrolysis. *Journal of Applied Physiology* **61**, 1949–54.

Tang, Y. & Boutilier, R.G. (1991). White muscle intracellular acid–
base and lactate status following exhaustive exercise: a comparison
between freshwater- and seawater-adapted rainbow trout. *Journal
of Experimental Biology* **156**, 153–71.

Tang, Y., Lin. H. & Randall, D.J. (1992). Compartmental distributions
of carbon dioxide and ammonia in rainbow trout at rest and following exercise, and the effect of bicarbonate infusion. *Journal of
Experimental Biology* **169**, 235–49.

Turner, J.D. & Wood, C.M. (1983). Factors affecting lactate and
proton efflux from pre-exercised, isolated-perfused rainbow trout
trunks. *Journal of Experimental Biology* **105**, 395–401.

Turner, J.D., Wood, C.M. & Clark, D. (1983). Lactate and proton
dynamics in the rainbow trout (*Salmo gairdneri*). *Journal of Experimental Biology* **104**, 247–68.

van den Thillart, G., van Waarde, A., Muller, H.J., Erkelens, C. &
Lugtenburg, J. (1990). Determination of high energy phosphate
compounds in fish muscle: ^{31}P NMR spectroscopy and enzymatic
methods. *Comparative Biochemistry and Physiology* **95B**, 789–95.

van den Thillart, G. & van Waarde, A. (1996). Nuclear magnetic resonance spectroscopy of living systems: applications in comparative physiology. *Physiological Reviews* **76**, 799–837.

Waddell, W.J. & Butler, T.C. (1959). Calculation of intracellular pH from the distribution of 5,5-dimethyl-2,4-oxazolidine dione (DMO): application to skeletal muscle of the dog. *Journal of Clinical Investigation* **38**, 720–9.

Wang, Y., Heigenhauser, G.J.F. & Wood, C.M. (1994a). Integrated responses to exhaustive exercise and recovery in rainbow trout white muscle: acid–base phosphagen, carbohydrate, lipid, ammonia, fluid volume, and electrolyte metabolism. *Journal of Experimental Biology* **195**, 227–58.

Wang, Y., Heigenhauser, G.J.F. & Wood, C.M. (1996a). Ammonia movement and distribution after exercise across white muscle cell membranes in rainbow trout. *American Journal of Physiology* **271**, R738–50.

Wang, Y., Heigenhauser, G.J.F. & Wood, C.M. (1996b). Lactate and metabolic H⁺ transport and distribution after exercise in rainbow trout white muscle. *American Journal of Physiology* **271**, R1239–50.

Wang, Y., Wilkie, M.P., Heigenhauser, G.J.F. & Wood, C.M. (1994b). The analysis of metabolites in rainbow trout white muscle: a comparison of different sampling and processing methods. *Journal of Fish Biology* **45**, 855–73.

Wang, Y., Wright, P.M., Heigenhauser, G.J.F. & Wood, C.M. (1997). Lactate transport by rainbow trout white muscle: kinetic characteristics and sensitivity to inhibitors. *American Journal of Physiology* **272**, R1577–87.

Wardle, C.S. (1978). Non-release of lactic acid from anaerobic swimming muscle of plaice, *Pleuronectes platessa* L.: a stress reaction. *Journal of Experimental Biology* **77**, 141–55.

Weber, J-M. (1991). Effect of endurance swimming on the lactate kinetics of rainbow trout. *Journal of Experimental Biology* **158**, 463–76.

Weber, J-M. & Haman, F. (1996). Pathways for metabolic fuels and oxygen in high performance fish. *Comparative Biochemistry and Physiology* **113A**, 33–8.

West, T.G., Schulte, P.M. & Hochachka, P.W. (1994). Implications of hyperglycemia for post-exercise resynthesis of glycogen in trout skeletal muscle. *Journal of Experimental Biology* **189**, 69–84.

Wolf, K. (1963). Physiological salines for freshwater teleosts. *The Progressive Fish-Culturist* **25**, 135–40.

Wood, C.M. (1988). Acid–base and ionic exchanges at gills and kidney after exhaustive exercise in the rainbow trout. *Journal of Experimental Biology* **136**, 461–81.

Wood, C.M. (1991). Acid–base and ion balance, metabolism, and their interactions, after exhaustive exercise in fish. *Journal of Experimental Biology* **160**, 285–308.

Wood, C.M. & Perry, S.F. (1985). Respiratory, circulatory, and metabolic adjustments to exercise in fish. In *Circulation, respiration, and metabolism*, ed. R. Gilles, pp. 2–22. Berlin: Springer-Verlag.

Wood, C.M., Walsh, P.J., Thomas, S. & Perry, S.F. (1990). Control of red blood cell metabolism in rainbow trout after exhaustive exercise. *Journal of Experimental Biology* **154**, 491–507.

Wright, P.A., Randall, D.J. & Wood, C.M. (1988). The distribution of ammonia and H^+ between tissue compartments in lemon sole (*Parophrys vetulus*) at rest, during hypercapnia, and following exercise. *Journal of Experimental Biology* **136**, 149–75.

Wright, P.A. & Wood, C.M. (1988). Muscle ammonia stores are not determined by pH gradients. *Fish Physiology and Biochemistry* **5**, 159–62.

Zeidler, R. & Kim, D.H. (1977). Preferential hemolysis of postnatal calf red cells induced by internal alkalinization. *Journal of General Physiology* **70**, 385–401.

NORBERT HEISLER

Limiting factors for acid–base regulation in fish: branchial transfer capacity versus diffusive loss of acid–base relevant ions

Introduction

Metabolic energy production, indispensable for the maintenance of life, relies essentially on the regulation of pH as a central parameter for the involved processes. Deviation by more than a few tenths of a pH unit may considerably reduce metabolic flux or even completely inhibit further energy production on the basis of pronounced pH activity optima of certain key enzymes (such as phosphofructokinase) (see Heisler, 1990b). Tight regulation of pH is accordingly indispensable for conservation of homeostasis.

In fish, pH regulation is extensively challenged by changes of the environment: utilization of water as the gas exchange medium frequently subjects fish to large and rapid changes in water O_2 and CO_2 concentrations, in temperature and in environmental electrolytes, and to various other natural or human-induced stress factors that air breathers never encounter to any comparable extent. This chapter focuses on the description of characteristics, sites and mechanisms of fish integumentary acid–base regulation, with particular emphasis on the limitations of active and passive components contributing to the overall regulatory pattern. Due to limitations of space, the reader is referred to review articles wherever appropriate.

Regulatory mechanisms

Mechanisms for acid–base regulation are common to all living organisms and are essentially the same for fish as for terrestrial animals. The extent to which individual mechanisms are utilized, however, is quite different. The situation for fish is characterized by the intimate contact with an aqueous environment, generally including continuous immersion and utilization of water as the gas exchange medium. While buffering has little importance in fish, due to the low buffer values (see below), and water immersion provides severe restrictions for the regulation of body fluid P_{CO_2}, the intimate contact with the aqueous environment strongly supports ion transfer mechanisms affecting acid–base regulation.

Buffering

Buffering is a valuable mechanism for transient acid–base regulation in fish, but its role is rather limited. While terrestrial animals generally normalize pH only with removal of the original stress factor (for example with return to environmental normocapnia or aerobic metabolic processing of lactic acid), acid–base regulation in fish is characterized by early recovery of pH before the original stress factor is removed. The amount of buffered H^+ is a direct function of the shift in pH and the process is accordingly incapable of restoring the original pH. Thus, with exceptions to the rule (Heisler, 1982a; Heisler et al., 1982), the role of buffering is limited to early periods of acid–base disturbances (for details, see Heisler, 1986a).

Non-bicarbonate buffering (mainly at protein residues such as histidine, cysteine and terminal NH_2-groups with pK' values close to the physiological pH) takes place in a system closed for acid and base forms of the buffer (for definitions see Brønsted, 1923; Heisler, 1986a), at constant total buffer concentration. The buffer values (β; Van Slyke, 1922) in the intracellular compartments of fish are generally lower than in terrestrial vertebrates (by factors of 1.5–4; Heisler, 1986a), but higher than in blood and extracellular space. The largest proportion of the buffer capacity ($\kappa = \beta \cdot V$; V = volume; Heisler, 1986a) is accordingly located in the intracellular body compartments. The efficiency of the bicarbonate buffer system (consisting of CO_2, H_2CO_3, HCO_3^- and CO_3^-) is based on *in vivo* adjustment of the volatile acid-anhydride (CO_2) by respiratory gas exchange (for details see Woodbury, 1965; Heisler, 1986a, 1989, 1990a). Due to the limited scope of P_{CO_2} adjustment, this is hardly possible in fish (see below; Heisler, 1986a, 1989, 1990a). Also, the concentration of the buffer is smaller than in air breathers as a consequence of the low P_{CO_2} in fish (Heisler, 1984; Heisler, 1986b, 1986c).

Adjustment of P_{CO_2}

The low O_2 capacitance of water as compared to air (approximately 0.03, depending on temperature) is unfavourable for fish homeostatic control. Metabolic demand can be satisfied only with a large specific ventilation (volume per unit O_2 consumption), resulting in much lower

P_{CO_2} values than in terrestrial animals. This is related to the much larger capacitance of pure water for CO_2 than for O_2 (30–40 times), with even higher capacitance ratios at very low water P_{CO_2} (and correspondingly high pH) as a result of carbonate and other non-bicarbonate buffering of CO_2. Accordingly, the high relative rate of gill ventilation (about 5 ml/μmol O_2 consumption *versus* 0.6 ml pulmonary ventilation/μmol O_2 in humans) leads to much lower P_{CO_2} differences between arterial blood and respiratory medium in fish (ΔP_{CO_2} 1–4 mmHg) than in terrestrial vertebrates (typically 30–45 mmHg) (Heisler, 1989).

Such small environmental–arterial P_{CO_2} differences hardly leave any scope for respiratory modulation of P_{CO_2}, a mechanism frequently exploited in mammals (Woodbury, 1965), particularly with part of ΔP_{CO_2} being independent of gill ventilation. Factors such as blood or water shunting past the gas exchange surface (Heisler, 1989), or incomplete CO_2 hydration and disequilibrium as a result of insufficient carbonic anhydrase activity on the mucosal side of the gill epithelium (e.g. Henry, Smatresk & Cameron, 1988; Heisler, 1990a), contribute to ΔP_{CO_2}. A significant reduction of arterial P_{CO_2} can accordingly be achieved only in fish with relatively low ventilation and large ΔP_{CO_2}, such as carp (3–4 mmHg; Claiborne & Heisler, 1984, 1986).

Any regulatory effect of hyperventilation is small in terms of the environmental hypercapnia fish may encounter. Seawater P_{CO_2} usually reflects equilibrium with air (0.15–0.3 mmHg), but may rise by 5–10 mmHg due to anaerobic metabolism of micro-organisms or related to mixing of water currents of different [HCO_3^-] at depths of 200–500 m (Harvey, 1974). Natural freshwater P_{CO_2} ranges from extremely low, photosynthesis-related values (less than atmospheric, approximately 0.26 mmHg) to as high as 60 mmHg (Heisler *et al.*, 1982) due to the inhibition of surface gas exchange combined with CO_2 release from bicarbonate by anaerobic microbial metabolism and mixing with more acidic water. Therefore, changes in ΔP_{CO_2} by adjustment of gill ventilation cannot effectively ameliorate the impact of hypercapnia. In fish, an increase of 10 mmHg in inspired P_{CO_2} will cause arterial P_{CO_2} to rise by a factor of three to six, and arterial pH to fall by about 0.4–0.5 pH units (Heisler, 1986a). In contrast, mammals will hardly be affected after reducing the inspired/alveolar P_{CO_2} difference from 40 mmHg to 30 mmHg and readjusting arterial P_{CO_2} by a mere 30 per cent of hyperventilation. Hence, readjustment of ventilation is barely exploited in fish during environmental hypercapnia: the response is generally small and transient, and not related to the actual time course of pH normalization (e.g. Randall, Heisler & Drees, 1976; Dejours, 1981).

Bicarbonate adjustment by ionic transfer

pH in a fluid compartment can be affected by changes in the two variables P_{CO_2} and $[HCO_3^-]$ of the Henderson–Hasselbalch equation for the CO_2/HCO_3^- buffer system (Heisler, 1986c, 1989, 1990a). Because changes in P_{CO_2} are too small for any significant contribution with respect to acid–base regulation (owing to the limitations provided by water as the gas exchange medium, see above), fish have to rely much more than terrestrial animals on mechanisms affecting the $[HCO_3^-]$ in the body fluids. Transfer of H^+, OH^- and HCO_3^- directly affect $[HCO_3^-]$ in the body fluids (H^+ and OH^- *via* relevant buffer equilibria; Heisler, 1986c, 1989, 1990a). Acid–base relevant effects may also result from acid or base forms of a buffer system (Brønsted, 1923; Heisler, 1986a) changing ionization during transfer between fluid compartments of different pH.

Transmembrane and transepithelial ion transfer processes are the only mechanisms capable of permanent elimination of acid–base relevant ions produced as non-volatile metabolic end-products. The capacities of transmembrane mechanisms are larger than those at the epithelial interface by a factor of more than ten compared on a body-weight basis (Heisler, 1986b, 1992). The mechanisms for transepithelial transfer of acid–base relevant ions are much more efficient in fish than in terrestrial animals and compensate for the poor efficiency of respiratory adjustment. The involved processes generally include transfer of co-ions and counter-ions in order to maintain electroneutrality (for details, see below).

Characteristics of acid–base regulation in fish

In fish, acid–base homeostasis is continuously challenged by various endogenous and exogenous factors. While large amounts of CO_2 are exchanged with the environment by non-ionic diffusion, non-volatile acid–base relevant ions generated as end-products of metabolism are continuously eliminated by acid–base relevant ion transfer. This relatively small endogenous load may be tremendously enhanced during times of extensive muscular activity and extreme hypoxia, when lactic acid is produced by anaerobic glycolysis. Changes in environmental temperature, O_2 and CO_2 concentrations, as well as low water pH and associated toxic effects of substances such as aluminium on epithelial surfaces, represent exogenous disturbances to be dealt with by epithelial acid–base regulation. This chapter focuses on the typical pattern of acid–base regulation during hypercapnia as a basis for the discussion of branchial acid–base regulation. For the effects of other challenges of acid–base regulation, such as changes in temperature, lactacidosis, low water pH and associated toxicity the reader is referred to excellent and extensive literature reviews (e.g. Fromm, 1980; Spry, Wood & Hodson, 1981; Wood & McDonald, 1982, 1987; Heisler, 1984, 1986b, 1989, 1900a, 1990b; Wood & Soivio, 1991).

Environmental hypercapnia

Fish, in particular freshwater species, are frequently exposed to elevated environmental levels of CO_2 (see above), which are readily transmitted to the arterial blood by the large gill surface area. The equivalent rise in plasma P_{CO_2} results in a fall in plasma pH and pH_i values (Fig. 1). With some notable exceptions (see below), pH recovers towards control values soon after initiation of hypercapnia by elevation of the $[HCO_3^-]$, partially or completely compensating for the rise in P_{CO_2}. During the initial phase of hypercapnia with large deflections of pH, extracellular HCO_3^- is elevated on the basis of non-bicarbonate buffering of CO_2, particularly in well-buffered intracellular muscle compartments (Fig. 2). With rising $[HCO_3^-]$ and increasing degree of pH recovery towards control values, the fraction of HCO_3^- originating from buffering falls, approaching zero when pH compensation is complete (Fig. 2). Mobilization of carbonate from the osteal structures does not contribute significantly during short-term hypercapnia in fish (directly measured: Cameron, 1985; estimated from changes in plasma $[Ca^{2+}]$ and $[Mg^{2+}]$: N. Heisler & N.A. Andersen, unpublished data).

After an initial phase, the accumulation of HCO_3^- in the body fluids mainly originates from acid–base relevant transepithelial ion transfer mechanisms. The amount of HCO_3^- equivalents gained from the environment during hypercapnia (1% CO_2) is in the range of 3–6 mmol/ kg body weight) (e.g. Heisler, Weitz & Weitz, 1976; Claiborne & Heisler, 1984, 1986; Heisler 1986b), an amount exceeding the total HCO_3^- pool at control conditions. Activation of transepithelial transfer upon environmental hypercapnia is often delayed by a short time (approximately 15–30 minutes; Figs. 1 & 2; e.g. *Scyliorhinus stellaris*: Heisler *et al.*, 1976; *Conger conger*: Toews, Holeton & Heisler, 1983). During this period, transmembrane transfer of acid–base relevant ions from the well-buffered intracellular fluid compartments contributes to an early elevation of plasma concentration. At the same time, HCO_3^- is often lost to the environment, a maladaptive response indicating still incomplete adjustment of the ion-retaining structures to elevated plasma HCO_3^- levels, analogous to the initial release of HCO_3^- equivalents from intracellular to extracellular space. Bicarbonate net loss from intra-

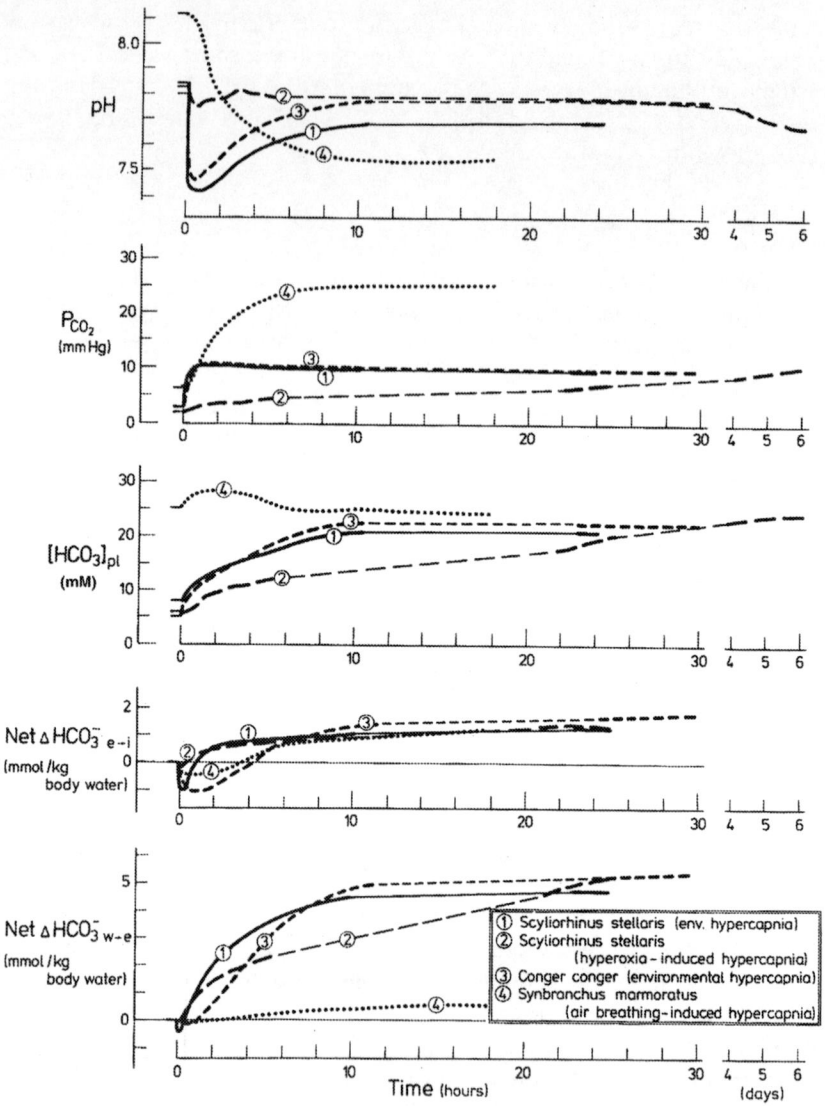

Fig. 1. Extracellular acid–base parameters (plasma pH, P_{CO_2} and [HCO_3^-]) in response to hypercapnia (environmental, hyperoxia-induced, or induced by transition from water to air breathing) as well as net HCO_3^- equivalent ion transfer ($\Delta HCO_3^-{}_{e \to i}$: extracellular to intracellular; $\Delta HCO_3^-{}_{w \to e}$: environmental water to extracellular space). Note differences in the time course of P_{CO_2} elevation and in the amount of HCO_3^- gained from the environment. Based on data from Heisler *et al.* (1976, 1981, 1988; *Scyliorhinus stellaris*), Heisler (1982a; *Synbranchus marmoratus*) and Toews *et al.* (1983; *Conger conger*); also see text.

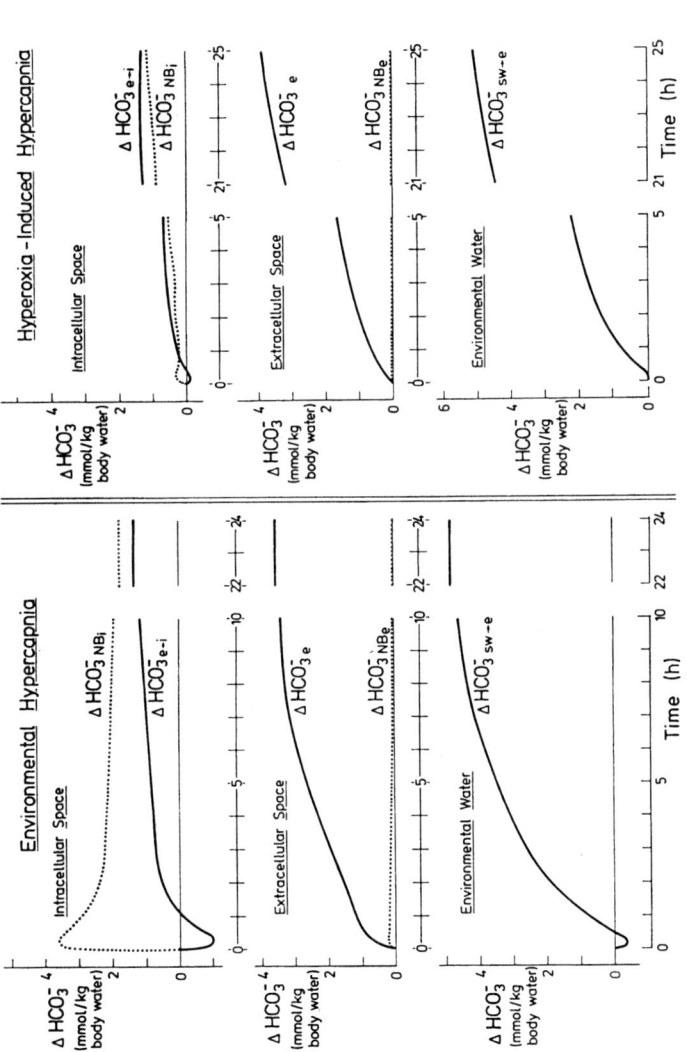

Fig. 2. Mechanisms contributing to the elevation of [HCO₃⁻] for the compensation of hypercapnia (left: environmentally induced; right: hyperoxia-induced) in *Scyliorhinus stellaris*. Plasma HCO_3^- (index 'e'), HCO_3^- produced by extracellular (NB$_e$) and intracellular non-bicarbonate buffering (NB$_i$), and net transfer of HCO_3^- equivalents from extracellular to intracellular fluid compartments (e→i) or seawater to extracellular space (sw→e). Note that although the shift in pH is very small, a sizeable amount of HCO_3^- is produced by steep intracellular non-bicarbonate buffer curves. Based on data from Heisler et al. (1976, 1981, 1988); also see text.

cellular and extracellular fluid compartments during this time period is later rectified when the transepithelial transfer is fully activated (Fig. 2). Then, the gain of HCO_3^- equivalents from the environment generally results in almost complete pH compensation in both the extracellular and the intracellular compartments.

The largest fraction of accumulated HCO_3^- is received by the extracellular space in spite of its small relative volume (20–25 per cent of the body water: e.g. Heisler, 1978, 1982a; Cameron, 1980), whereas only a relatively small amount of HCO_3^- (approximately one-third; see Fig. 2) is transferred to the large intracellular space (80–75 per cent of body water). Nevertheless, this amount is sufficient to effect almost complete pH compensation and generally much tighter restoration of pH to normocapnic controls than in the extracellular fluid (Heisler, 1984, 1986a, 1988a). A certain amount of HCO_3^- transferred to the intracellular space is more effective than it is in the extracellular compartments, because of the larger relative change in HCO_3^- with respect to intracellular control HCO_3^- levels, which are smaller by factors between four and eight. Intracellular compensation is clearly also supported by the high capacity of intracellular non-bicarbonate buffers (Heisler, 1984, 1986a, 1986b).

Hyperoxia-induced and air-breathing-induced hypercapnia

Two other sources of hypercapnia – hyperoxia and air breathing – are originated endogenously on the basis of the primarily O_2-oriented regulation of ventilation in fish (Dejours, 1981). This type of regulation is characterized by adjustment in order to provide sufficient amounts of O_2 for metabolism (in contrast, terrestrial animals regulate ventilation aiming for constant arterial P_{CO_2}). In view of the sparse availability of O_2 in the water, this is certainly advantageous in terms of the supply of O_2, but carries the inherent disadvantage of endogenous respiratory acid–base disturbances. Accordingly, elevation of the O_2 content of the gas exchange medium results in substantially reduced ventilation with simultaneously increased ΔP_{O_2} and ΔP_{CO_2} (e.g. Dejours, 1973; Heisler, Holeton & Toews, 1981; Heisler, Toews & Holeton, 1988; Wilkes *et al.*, 1981; Heisler, 1982a; Höbe, Wood & Wheatly, 1984; for review, see Heisler, 1986b).

Upon exposure to environmental hyperoxia, which is mainly found in natural environments during periods of photosynthesis, particularly in tropical freshwater habitats (see Heisler, 1984, 1986b), arterial P_{CO_2} rises, being compensated for by equivalent changes in $[HCO_3^-]$. In contrast to environmental hypercapnia, in which CO_2 is taken up rapidly from the ambient water, hyperoxia-induced hypercapnia develops slowly by accumulation of endogenously produced CO_2 (Heisler *et al.*, 1988), more slowly, in fact, than expected on the basis of endogenous CO_2 production and the large amount of O_2 available in the environment.

In the dogfish *Scyliorhinus stellaris*, at least two factors contribute to this phenomenon. With rising P_{CO_2}, large amounts of endogenously produced CO_2 are converted to HCO_3^- by non-bicarbonate buffering, slowly filling the tissue CO_2 stores. A more important factor is that the reduction of ventilation is less than equivalent to the elevated O_2 availability. This is evidenced by arterial P_{O_2} values exceeding the levels for complete blood saturation for a number of days after the onset of hyperoxia. Steady state values for arterial P_{CO_2} (11 mmHg at environmental P_{O_2} of about 600 mmHg) and P_{O_2} are only attained after five days of hyperoxia, and the rise in P_{CO_2} is well matched to the accumulation of plasma HCO_3^-. Arterial pH is hardly affected during adaptation to the new environment in spite of rising P_{CO_2} ($< - 0.08$ units; Heisler *et al.*, 1988). Indicating a switch from primarily O_2-oriented towards pH-stat regulation during hyperoxia (Heisler *et al.*, 1988), the adjustment of gill ventilation tracks (with some regulatory oscillations) the rate of HCO_3^- accumulation by ion transfer from the environment. The resulting disparity between O_2 demand and ventilation is obvious from excessively high arterial P_{O_2} values (initially more than 250 mmHg), returning to normoxic controls below 100 mmHg only after five days of hyperoxia. During this time, plasma $[HCO_3^-]$ rises steadily to values of almost 25 mM, concurrent with, and almost completely compensating, the elevation of P_{CO_2} (see Fig. 1; Heisler *et al.*, 1988). This pattern is related to elimination of the primary stimulus for ventilatory drive (low availability of O_2) such that control of ventilation closely follows plasma pH. Only when the reduction in ventilation reapproaches levels just sufficient to satisfy the metabolic O_2 demand is ventilatory control returned to the O_2-related drive, indicated by deviations of pH from the controls becoming larger than during the initial phases of hyperoxia-induced hypercapnia (Heisler *et al.*, 1988).

According to the lack of appreciable pH deviations, there is little production of HCO_3^- from non-bicarbonate buffering (see Fig. 2). Also, the rate of HCO_3^- equivalent gain is much lower compared to environmental hypercapnia. A rate of HCO_3^- gain close to the maximal epithelial transfer capacity of 15 μmol/min per kg of this species (Heisler *et al.*, 1976; Heisler & Neumann, 1977; Holeton & Heisler, 1983; Heisler,

1988b) occurs during the first hours, but the transfer rate is reduced to one-third, with plasma pH recovering to less than −0.03 after three to five hours, and falling to 2 µmol/min per kg after 21–25 hours of hyperoxia. Accumulation of the same quantity of HCO_3^- takes accordingly much longer during hyperoxia-induced than during environmental hypercapnia. This reduction of transfer rates and the partial switch to adjustment of pH by ventilatory means may be related to the lack of appreciable pH deviations, but may also be energetically less costly because the usual duration of hyperoxia is too short for complete pH compensation on the basis of ionic transfer anyway. The observed pattern saves the animal the largest fraction of HCO_3^- accumulation by ionic transfer as well as the reverse process after the end of maximally 12 hours of hyperoxia.

Air-breathing-induced hypercapnia

The mechanism of the rise in arterial P_{CO_2} in facultative air-breathing fish switching from water breathing to air breathing is essentially the same as in water hyperoxia (see above): air contains more O_2 (20.9 Vol %) than water equilibrated with pure O_2 (about 3 Vol % at 20 °C), or saturated with air (about 0.6 Vol %). The ventilatory drive of the animals is readjusted according to the enhanced availability of O_2, to quite a considerable extent in the tropical teleost fish *Synbranchus marmoratus*. Upon switching from water to air breathing, P_{CO_2} is elevated from almost 6 mmHg to 25 mmHg within 10 to 72 hours, causing plasma pH to drop from 8.15 to 7.5 (see Fig. 1; Heisler, 1982a). In contrast to the typical response of water-breathing fish to hypercapnia, however, pH in this species remains deflected, with no sign of compensation. Bicarbonate is adjusted, after some initial elevations due to blood non-bicarbonate buffering, at essentially the same level as during water breathing. *Synbranchus* does not gain significant amounts of HCO_3^- from the environment during air-breathing-induced hypercapnia (see Fig. 1; Heisler, 1982a). A similar lack of extracellular compensation is also observed in other air breathers such as *Channa argus* (Ishimatsu & Itazawa, 1983) and *Amia calva* (Daxboeck, Barnard & Randall, 1981). Dehydration may lead to a slight rise in HCO_3^- in *Protopterus*, as indicated by overproportionate concentration increases of other ions (Delaney, Lahiri & Fishman, 1974; DeLaney *et al.*, 1977).

In contrast to the complete lack of extracellular compensation, the pH_i of white and heart muscle of *Synbranchus* is almost completely restored (Heisler, 1982a). The required (small) elevation in intracellular HCO_3^- is primarily caused by transfer of HCO_3^-, originally produced by blood non-bicarbonate buffering. This type of preferential intracellular regulation is very efficient: the small amount of HCO_3^- produced by extracellular buffering would not effect any significant extracellular compensation in view of very high control levels of HCO_3^- (24 mM), but serves for almost complete pH_i compensation on the basis of much lower intracellular HCO_3^- levels (Heisler, 1982a).

Conditions for homeostasis are relatively adverse in the extracellular space of *Synbranchus*. Complete compensation of this compartment would require a fivefold rise in $[HCO_3^-]$ in order to offset the rise in P_{CO_2} completely (Heisler, 1986c). The high basal level of $[HCO_3^-]$, however, would require an equimolar reduction in extracellular $[Cl^-]$ in order to maintain electroneutrality; even a fraction of this could barely be tolerated (Heisler, 1982a), even if such $[HCO_3^-]$ could be maintained by the ion-retaining structures of the fish. Furthermore, sufficient amounts of HCO_3^- could not be gained from the environmental water. Filling the buccal cavity with air for respiratory gas exchange prevents gill water irrigation and reduces the contact time of branchial ion-transporting epithelia with water to short flush periods every 5 to 35 minutes (Heisler, 1982a). Also, the environmental conditions for these fish are extremely unfavourable with respect to ion exchange with the environment (Heisler, 1982a; see also below).

Mechanisms and sites of transfer

The preceding section indicates that fish rely heavily on ion transfer processes for pH regulation. Transfer of acid–base relevant substances takes places not only between intracellular and extracellular body compartments, but also to a great extent between fish and environmental water. A number of different epithelial sites may be involved in this process: the kidneys, the urinary bladder, the skin and, in elasmobranchs, rectal glands and abdominal pores, in addition to the gill epithelium with its large interface area between plasma and water. The few studies on the relative role of extrabranchial sites for acid–base regulation in fish indicate that the branchial epithelium is mainly responsible (Heisler, 1984, 1986b, 1988b).

Several branchial ion-transfer mechanisms have been depicted, which are generally considered to be best described as active electroneutral ion exchange processes of H^+ or NH_4^+ against Na^+, and of HCO_3^- against Cl^-, although this may not characterize the details of the processes (Fig. 3; Maetz, 1974; Evans, 1984, 1986; Heisler, 1986b, 1988a). Also, an electrogenic H^+ pump in the apical membrane of branchial chloride cells has recently been postulated, being linked only indirectly

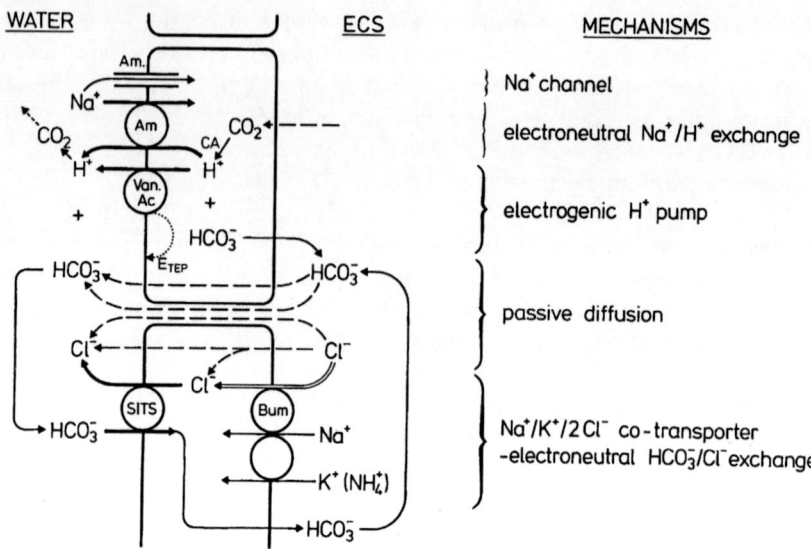

WATER   ECS   MECHANISMS

Fig. 3. Model of transepithelial ion transfer. H^+ may be exchanged against Na^+ by an amiloride-sensitive (Am) electroneutral ion exchange, or extruded by an electrogenic vanadate-sensitive (Van) and acetozolamide-sensitive (Ac) H^+ pump. For maintenance of electroneutrality, Na^+ moves through amiloride-sensitive (Am) Na^+ channels, or Cl^- through either paracellular pathways or the cytoplasm. HCO_3^- may be exchanged 1:1 against Cl^- at the apical membrane (SITS-sensitive). $K^+(NH_4^+)$, Na^+ and $2Cl^-$ may be translocated by a bumetanide-sensitive (Bum) electroneutral co-transporter. CA = carbonic anhydrase; dotted lines indicate the effect of the electrogenic transfer of H^+ on the transepithelial potential (E_{TEP}). See text.

to movement of Na^+ or other cations (Avella & Bornancin, 1989; Lin & Randall, 1991). Evidence has been provided for all of these mechanisms operating under certain conditions (Evans, 1984, 1986; Lin & Randall, 1991). However, not all of them seem to be involved under physiological conditions. Recent evidence from unidirectional Na^+ and Cl^- tracer ion fluxes in *Cyprinus carpio* indicates that Na^+-related mechanisms play no role in the transfer of acid–base relevant ions for compensation of hypercapnic acidoses (Heisler, 1990a, 1990b). Under these conditions, however, unidirectional Cl^- flux was directly related. This strategy may be based on the fact that the HCO_3^-/Cl^- ion exchange is osmotically neutral, whereas all other mechanisms (e.g. H^+/Na^+) lead to osmotic stress by transfer of osmotically active molecules to or from body fluids of the animal (Heisler, 1989). This factor may also prohibit utilization of co-transfer processes. The often observed lack of relationship between the transfer of acid–base relevant ions and Na^+ movements is in line with an electrogenic proton pump mechanism (Lin & Randall, 1991).

The role of the secondary circulatory system

Although the role of the gills as the main site of acid–base regulation in fish is by now well established, the precise location of ionic transfer within the branchial epithelium is still a matter of discussion. Mitochondria-rich cells (chloride cells) are generally claimed to be responsible on the basis of circumstantial evidence. With some variability, chloride cells are mainly located at the base of the secondary lamellae on the primary filament, whereas the apical fractions of the secondary lamellae are mainly covered by so-called respiratory cells (Laurent, 1984).

These areas are drained by two different circulatory systems. The respiratory cells on the secondary lamellae are in close connection with the bloodstream of the primary circulation through the branchial gas exchange area. The chloride cells are mainly located in close proximity to the central venous sinus in the primary filament of the fish gills, which drains the basal portion of the secondary lamellae as part of the secondary circulatory system in fishes (Laurent, 1984). Differences between the composition of fluids from the two circulatory circuits in the gills, together with the actual flow rates, are accordingly expected to reflect the fractional contribution of epithelia in juxtaposition to perfused areas.

The perfusion rate in the primary gill circulation (the respiratory, apical parts of the secondary lamellae) is close to cardiac output (Ishimatsu, Iwama & Heisler, 1988, 1995), owing to the need for a high rate of respiratory gas exchange. Venous–arterial (ventral to dorsal aortic) differences in ionic composition will accordingly be small and difficult to detect. Furthermore, uncertainties related to the correction for blood non-bicarbonate buffering, and to changes in HCO_3^- by oxygenation of haemoglobin (Haldane effect; up to 1.5 mmol/l) render quantification of ion transfer processes by the Fick principle (ionic transfer rate = ventral aortic − dorsal aortic concentration difference × flow rate) an inadequate approach in the range of the primary gill circulation.

Flow in the secondary circulation of the gills is much lower (Ishimatsu *et al.*, 1988), providing an ideal background on which to determine the contribution to ionic transfer of this circulatory system.

Measurement of flow rate of the central venous sinus and access to its effluent fluid, however, are rather difficult. The central venous sinus circuit is fed from the efferent filamental and branchial arteries and drains into the branchial vein (Laurent, 1984), with little contamination by blood from gill nutritional vessels (Ishimatsu et al., 1988). The central venous sinus flow rate was recently estimated in rainbow trout to be less than 7 per cent of cardiac output, using a special microcannulation technique of the branchial vein and an endogenous indicator distribution method employing the haemoglobin concentrations in central venous sinus fluid and dorsal aortic blood (Ishimatsu et al., 1988). A subsequent study focused on quantitative assessment of acid–base relevant transfer processes in branchial epithelia localized in juxtaposition to the secondary circulation, utilizing the central venous sinus flow rate and the dorsal aortic–branchial vein concentration differences for an application of the Fick principle (Iwama, Ishimatsu & Heisler, 1993).

Central venous sinus and dorsal aortic $[HCO_3^-]$ were the same during normocapnia, although pH was slightly lower and P_{CO_2} slightly higher in central venous sinus fluid, owing to the venous nature of the central venous sinus system. Upon environmental hypercapnia, $[HCO_3^-]$ rose significantly higher in the central venous sinus fluid than in dorsal aortic plasma, as a result of acid–base relevant transfer from the environmental water. Because the central venous sinus circuit is fed in rainbow trout from the efferent filamental and branchial arteries (Laurent, 1984), differences in $[HCO_3^-]$ between dorsal aorta and branchial vein (central venous sinus) are the result of ionic transfer across the central venous sinus-related epithelium, and, to a minor extent, to non-bicarbonate buffering (Fig. 4). Calculated on the basis of the Fick principle from the concentration differences at two or eight hours of hypercapnia, respectively, and the relative central venous sinus flow rate, about 3.1 or 2.3 mmol/kg body weight of HCO_3^- would have been accumulated during the entire time of exposure (Iwama et al., 1993). Because the transepithelial transfer rate during hypercapnia in fish is always highest at the largest pH deflections during the first hour of exposure (see Figs. 1 & 2 for data and references), the actual amount transferred during eight hours of exposure will be close to the two hour-based estimate. The estimated amount of transferred HCO_3^- compares favourably with that accumulated in the extracellular space, 2.4 mmol/kg body weight $(\Delta[HCO_3^-]_{pl} \times$ extracellular volume $= 12$ mmol/l $\times 0.2$ l/kg), taking into account that typically about one-third of HCO_3^- gained by ionic transfer from the environment is further transferred to the intracellular space (see Figs. 1 & 2; Heisler, 1984, 1986b). Accordingly, the largest fraction of additional HCO_3^- was evidently gained by action of epithelia

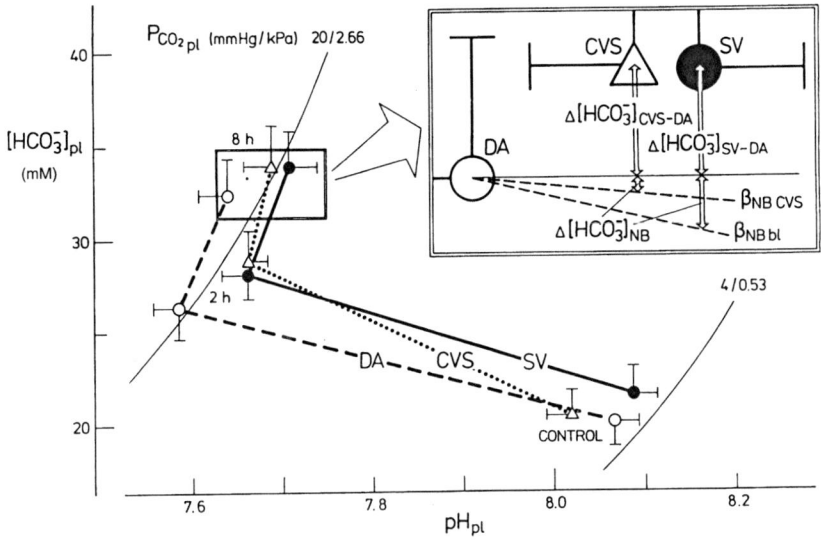

Fig. 4. Acid–base parameters of fluid from the central venous sinus (CVS) and blood of the dorsal (DA) and ventral aorta (VA) of rainbow trout during normocapnia (control) and exposed to environmental hypercapnia (2 and 8 h). Transfer from the environment ($\Delta HCO_3^-{}_{CVS-DA}$) and to a small extent also non-bicarbonate buffering ($\Delta HCO_3^-{}_{NB}$) are responsible for the difference in [HCO_3^-] between CVS and DA during hypercapnia. Differences between sinus venosus (SV) and dorsal aorta ($\Delta HCO_3^-{}_{SV-DA}$) are related to oxygenation of haemoglobin (Haldane effect). Based on data from Iwama *et al.* (1993); also see text.

related to the central venous sinus circuit (Iwama *et al.*, 1993). These data support the notion that mitochondria-rich cells at the base of the secondary lamellae are mainly responsible for ion transfer relevant for acid–base regulation in fish.

Ion transfer mechanisms of the secondary circulatory system

In response to the rise in dorsal aortic [HCO_3^-], the [Cl^-] of the extracellular fluid compartment was reduced equivalently, following the requirements of electroneutrality. Without any indication of an osmotically active transfer mechanism, these data suggest at first glance the operation of a 1:1 electroneutral ion exchange of HCO_3^-/Cl^-. However, no difference in [Cl^-] occurred during passage of the secondary circulation between dorsal aorta and central venous sinus (branchial vein), as would be expected on the basis of a HCO_3^-/Cl^- ion exchange mechanism in this system, and the observed net difference in $[HCO_3^-]$ between dorsal aorta and central venous sinus (Iwama et al., 1993). Although the expected concentration difference of about 2 mmol/l can well be distinguished on the background of plasma $[Cl^-]$, the data did not indicate any sign of a trend, even with paired analysis. Also, $[Na^+]$ was not enhanced during passage of the secondary circulatory system, as expected for Na^+-related H^+ extrusion, but rather was reduced (Iwama et al., 1993; Ishimatsu et al., 1995).

Measurement of unidirectional efflux from the secondary circulatory system using radioisotopes is impossible, but radioisotope influx into the central venous sinus did not indicate any significant counter-ionic or co-ionic movement (A. Ishimatsu, G. Iwama & N. Heisler, unpublished data). Together with the lack of net fluxes for Cl^- or Na^+, these data exclude the involvement of any Na^+-related or Cl^--related ion-transfer mechanisms in the secondary circulatory system (Iwama et al., 1993). On the basis of a large influx of acid–base relevant ions in the range of the central venous sinus circuit, the data indicate that correlated co-transfer or counter-transfer of ions required for electroneutrality takes place at other epithelial sites. According to present knowledge, this combination is compatible only with operation of an electrogenic H^+ pump, as proposed by Avella and Bornancin (1989) and Lin and Randall (1991). Readjustment of the transepithelial potential may accordingly drive passive ion movement, re-establishing electroneutrality. Because this is not linked to any specific location, as for a 1:1 carrier-mediated exchange process, the diffusive translocation will occur at the site of least diffusive resistance (Fig. 5), which is presented by the much larger surface area of the respiratory part of the secondary lamellae in the primary circulatory system (Iwama et al., 1993).

Limitations of transfer

Evidently, branchial acid–base relevant ion transfer is the most important mechanism for fish acid–base regulation (see above). The involved processes operate at a rate much higher than that of comparable mechanisms in terrestrial animals, with typical transfer rates per unit standard metabolic rate in the range of 0.13–0.47 (μmol ion transfer per μmol O_2 consumption) for fish as compared to 0.02–0.03 for typical mammals such as dog and man (Heisler, 1988b). An important factor in attaining this efficiency is the direct contact of the large surface area of the main transfer site, the gill epithelium, with a large volume of environmental

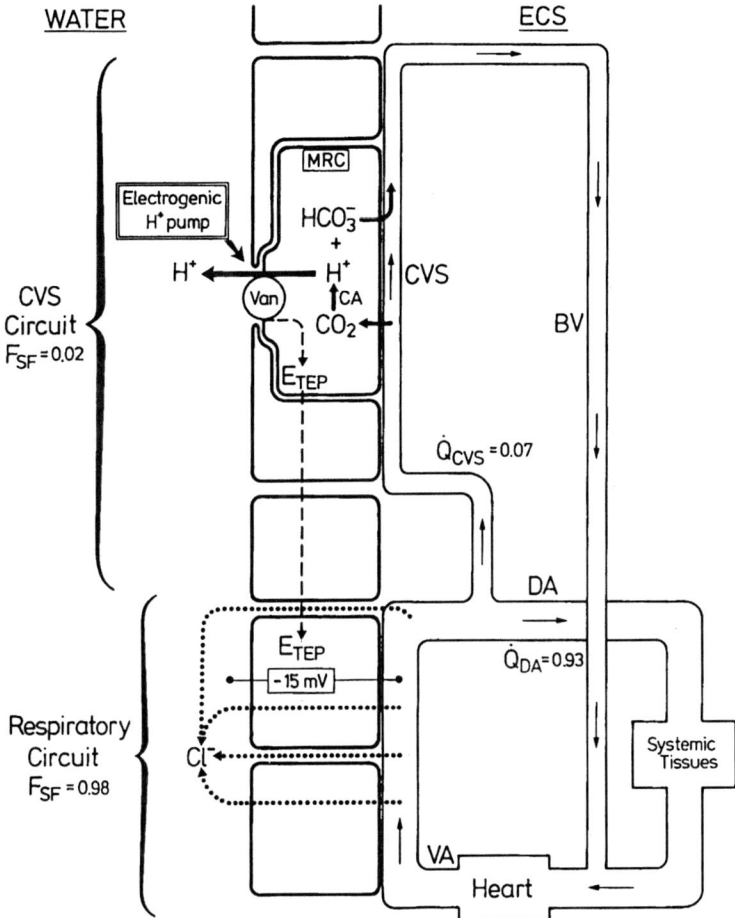

Fig. 5. Model of ionic transfer in trout during hypercapnia. Changes in $[HCO_3^-]$ are effected by an electrogenic H^+ pump in the epithelium of the primary filament at the base of the secondary lamellae (convective transport by central venous sinus (CVS) of the secondary branchial circulation, into the branchial vein (BV) at a flow rate of 0.07 of cardiac output). Note transfer of H^+ is bound to occur at the site of mitochondria-rich cells, whereas diffusive transfer of Cl^- following the changes in transepithelial potential will take place distributed across the epithelial surface at the site of least resistance, mainly the respiratory epithelium. E_{TEP} = transepithelial potential; F_{SF} = relative surface area of the two perfusion circuits. See text for details and references.

water (high specific ventilation, see above). Resting water flow past the gill epithelium in fish is in the range of 5000–20 000 ml/kg per hour) (e.g. Randall *et al.*, 1976) and larger than the typical urine flow rate in fish (1–10 ml/kg per hour; Hunn, 1982) and mammals (e.g. man, 1.1 ml/kg per hour) by factors of 10^3 to 10^4. The buffer capacity (i.e. the product of buffer value and volume) of urine will always be smaller by several orders of magnitude and negligible as compared to the buffer capacity of the water (at least about $2.3 \times [HCO_3^-] \times V$) flowing past the gill epithelium, even with extremely high urine titratable acidity and ammonia concentrations. Avoiding the necessity to establish large ionic gradients, such as in the mammalian renal tubular system, the high water flow facilitates ion-transfer processes to a much greater extent, provided the ionic composition of the environmental water is favourable.

Water ionic composition

Branchial ion-transfer mechanisms are directly affected by changes in ionic composition of the environment. This became evident from different time courses of compensation during hypercapnia (Heisler, 1982b). Trout in dilute water ($[Na^+] < 0.1$ mM; $[HCO_3^-] \sim 0.050$ mM) require much longer (72 hours) to arrive at the same level of pH compensation (Janssen & Randall, 1975) than trout in water of higher ionic concentration ($[Na^+]$ 0.5 mM; $[HCO_3^-]$ 3 mM: 24 hours, Eddy *et al.*, 1977) and *Conger* in seawater ($[Na^+]$ 480 mM; $[HCO_3^-]$ 2.5 mM: 10 hours; Toews *et al.*, 1983). Also, trout at different environmental salinities, as well as carp at different environmental $[HCO_3^-]$, dealt with compensation of environmental hypercapnia more rapidly at higher salinities (Iwama & Heisler, 1991) or $[HCO_3^-]$ (N.A. Andersen & N. Heisler, unpublished data), respectively. Genetically identical trout populations achieved much larger branchial HCO_3^- equivalent flux rates in seawater than in freshwater (Tang & Boutilier, 1988; Tang, McDonald & Boutilier, 1989).

Evaluation of the transfer conditions for acid–base relevant ions during hypercapnic compensation revealed a complicated pattern of ion transfer rates. In *Scyliorhinus stellaris*, net resorption of HCO_3^- from the environment was constant in a range of plasma (pl)/seawater (sw) concentration ratios ($[HCO_3^-]_{pl}/[HCO_3^-]_{sw}$) of 0.3 to 4, but was reduced in an apparently linear fashion at higher $[HCO_3^-]_{pl}/[HCO_3^-]_{sw}$ ratios, attaining zero at $[HCO_3^-]_{pl}/[HCO_3^-]_{sw} = 13$. At even higher ratios, HCO_3^- was lost to the water. On the basis of all other water ions remaining essentially constant, these data suggest water $[HCO_3^-]$ as a limiting factor for acid–base relevant ion transfer in this species (Heisler & Neumann, 1977; Heisler, 1988b). A similar pattern to that of *Scyliorhinus stellaris* is presented by preliminary data recently obtained for carp (see Fig. 7; N. Heisler, R. Neumann & N.A. Andersen, unpublished data).

Plasma bicarbonate threshold

In spite of large differences in time course, the extent of compensation achieved during hypercapnia is independent of the environmental ionic composition. This is illustrated by the regulatory pattern observed during hyperoxia-induced hypercapnia in *Scyliorhinus stellaris*. After some initial regulatory oscillations, pH_e is kept compensated to within 0.03 pH units below the normoxic controls, until plasma $[HCO_3^-]$ exceeds 20 mM. With still rising P_{CO_2} values, $[HCO_3^-]$ is not elevated equivalently, such that compensation becomes increasingly incomplete. Even after six days of hyperoxia, plasma $[HCO_3^-]$ does not rise above 25 mM, but pH deviates much more from the controls than during the initial phases of the experiment (see Fig. 1; Heisler *et al.*, 1988). Similar maximal $[HCO_3^-]$ values are also attained during environmental hypercapnia (Heisler *et al.*, 1976).

Literature data on the compensation of hypercapnia indicate that quite a number of aquatic lower vertebrates never exceed $[HCO_3^-]$ of 23–33 mM (Heisler, 1988b). With such a limit or threshold for plasma $[HCO_3^-]$, the degree of compensation that can be achieved must generally be lower in animals with high control $[HCO_3^-]$, due to the inability to elevate HCO_3^- by the same factor as P_{CO_2} (Heisler, 1986c). Due to the high arterial P_{CO_2} of about 4 mmHg, *Cyprinus carpio* maintains a comparatively high control $[HCO_3^-]$ of about 13 mM – much higher than other water breathers (trout = approximately 4 mM: Holeton, Neumann & Heisler, 1983; *Conger* = approximately 5 mM: Toews *et al.*, 1983; *Scyliorhinus* = approximately 7 mM: Heisler *et al.*, 1976; Heisler, 1988). In fact, carp did not raise plasma $[HCO_3^-]$ above 25 mM during exposure to hypercapnia of 1% or 5% CO_2, in spite of incomplete pH compensation (80 or 45 per cent), even after exposure for up to three weeks (Claiborne & Heisler, 1984, 1986). Also, the lack of any compensation in the tropical air breather *Synbranchus* (see above; Heisler, 1982a) may at least be partially related to the fact that the control $[HCO_3^-]$ in this species is as high as 24 mM, which may represent, or be close to, a limiting $[HCO_3^-]$ threshold. In this species, however, alternative factors such as the limited exposure time of the ion-transporting epithelia to water during air breathing, or adversely low environmental ion concentrations, may also be involved (Heisler, 1982a). Interestingly, the limit does not seem to be directly related to problems in gaining HCO_3^- from the water, but may rather represent the inability to maintain high plasma $[HCO_3^-]$. In at least two species of aquatic animals, *Cyprinus carpio* and the urodele amphibian *Siren lacertina*, HCO_3^- infused into the bloodstream during environmental hypercapnia is not retained, but is quantitatively released to the environment (Claiborne & Heisler, 1986; Heisler *et al.*, 1982). These data, in concert with the bulk of literature reports, support the notion of an epithelial threshold for retention and net resorption of HCO_3^- equivalents (Fig. 6). However, a few experiments in three fish species have indicated that much higher $[HCO_3^-]$ can actually be attained during compensation of hypercapnia (Fig. 6; Børjeson, 1976, 1977; Jensen & Weber, 1982; Cameron and Iwama, 1987; Dimberg, 1988).

Environmental Ca^{2+}

As evident from HCO_3^- equivalent resorption as a function of water $[HCO_3^-]$ in *Scyliorhinus* and *Cyprinus carpio*, water $[HCO_3^-]$ is an important factor in this matter (see above). Preliminary data indicate that also in carp HCO_3^- equivalent resorption is enhanced with rising water $[HCO_3^-]$ (Fig. 6; N.A. Andersen & N. Heisler, unpublished), but even at 50 mmol, carp fail to exhibit any degree of compensation comparable to the tremendous compensatory capacity of Uppsala trout (Børjeson, 1976, 1977; Dimberg, 1988), Odense tench (Jensen & Weber, 1982) and Port Aransas channel catfish (Cameron & Iwama, 1987), specimens acquiring up to 50 mmol plasma $[HCO_3^-]$. Long-term acclimation may seem to be an obvious factor, particularly for the Uppsala trout, which were adapted to elevated levels of environmental P_{CO_2} for several months before the experiments – much longer than the three weeks of adaptation during which carp did not attain levels of plasma $[HCO_3^-]$ higher than 26 mM (Claiborne & Heisler, 1986). In catfish, however, plasma $[HCO_3^-]$s of 50 mM were attained after only five days of hypercapnia (Cameron & Iwama, 1987).

A factor more likely to contribute is the environmental $[Ca^{2+}]$. All experiments in which high plasma $[HCO_3^-]$ levels were accomplished have been conducted at environmental $[Ca^{2+}]$ close to or higher than 2 mM, combined with relatively high levels of HCO_3^- in the range of 3–4 mM. The extraordinary capacity to compensate hypercapnic acidoses may accordingly be related to a combination of the supportive effect of high environmental $[HCO_3^-]$ with the sealing effect of Ca^{2+} on paracellular pathways, and the associated general reduction of unidirectional

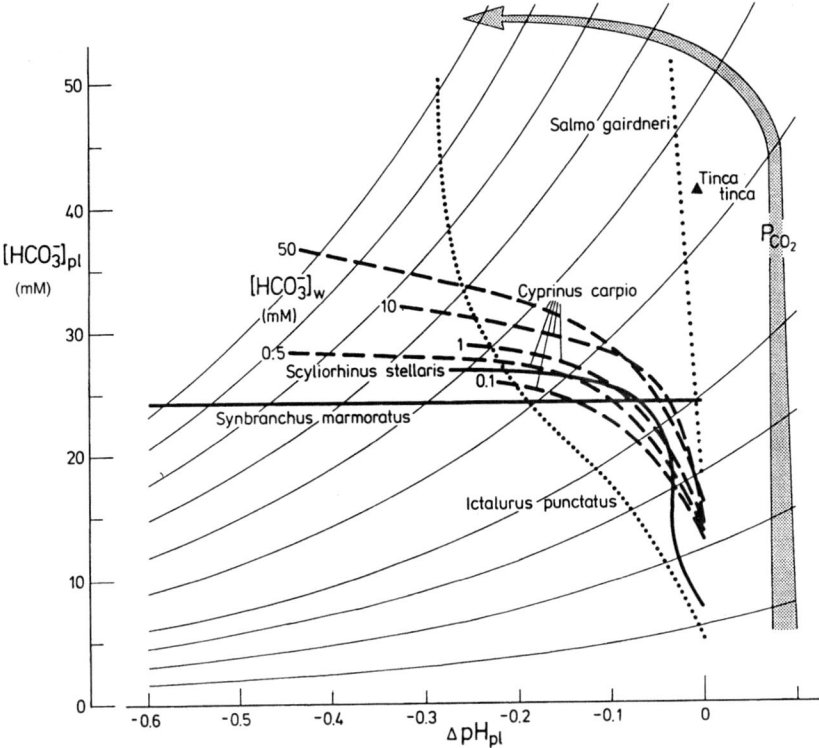

Fig. 6. Steady state plasma [HCO₃⁻] during hypercapnia and plasma pH as a function of water P_{CO_2} and [HCO₃⁻]. In most studied fish species, pH compensation is reduced (indicated by no further rise in [HCO₃⁻] and larger shifts in pH with rising P_{CO_2}) after arriving at certain plasma HCO₃⁻ levels (e.g. *Scyliorhinus stellaris, Cyprinus carpio* and *Synbranchus marmoratus*). Some experiments indicate an improved capacity for pH compensation and HCO₃⁻ accumulation (*Salmo gairdneri, Ictalurus punctatus* and *Tinca tinca* – see text). P_{CO_2} isobars are relative (with increasing values indicated by shaded arrow), due to the standardization of individual steady-state pH values. For references and details, see text.

flux rates (including a reduction in the rate of transepithelial HCO₃⁻ leakage). Even a direct effect of Ca²⁺ on the ion-transporting structures cannot be excluded. Preliminary data from closer examination of the interrelationships between net HCO₃⁻ resorption rate and environmental [Ca²⁺] in *Cyprinus carpio* indicate that Ca²⁺ is indeed a significant factor in this matter (Fig. 7). The shifts of the zero intercept of the inter-relationship between net HCO_3^- gain and the plasma/water $[HCO_3^-]$ ratio with rising $[Ca^{2+}]$ are suitable to explain a large proportion of the elevation in $[HCO_3^-]$ threshold observed in different experiments (Fig. 7).

Changes in transfer capacity *versus* diffusive loss

Two fundamentally different processes are probably counteracting each other: the active net uptake of HCO_3^- equivalents and the passive diffusive loss of accumulated HCO_3^- from the extracellular space (see Fig. 3). The change in $[Ca^{2+}]$ may have direct effects on turnover rate and/ or molecular capacity of an active carrier-mediated exchange process, thus stimulating the net uptake of HCO_3^-, and/or may diminish the leakage rate of HCO_3^- by changing the conductance of cellular and paracellular water passages. Little is known about modulation of such mechanisms, but on the basis of present knowledge, changes in both parameters are capable of explaining the observed phenomena.

Isolated direct determination of either the active uptake or the passive

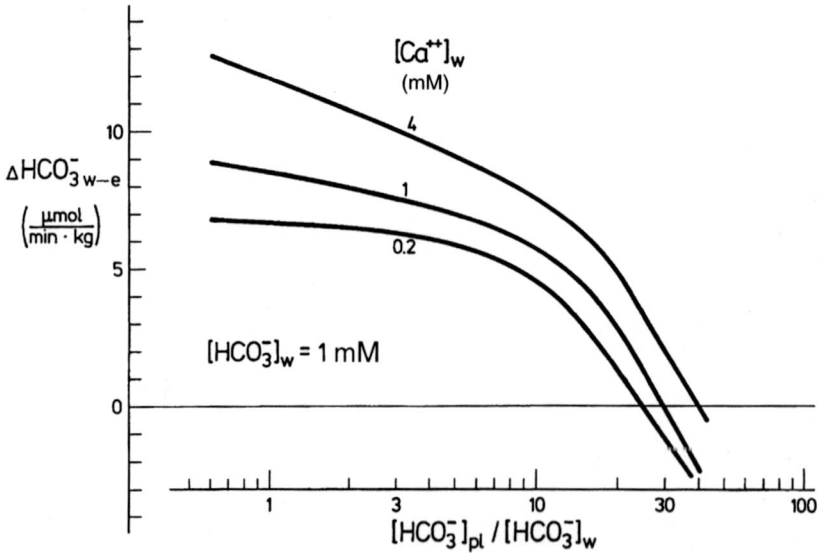

Fig. 7. Transepithelial HCO_3^- equivalent transfer rate in *Cyprinus carpio* as a function of the ratio between plasma and environmental bicarbonate concentration ($[HCO_3^-]_{pl}/[HCO_3^-]_w$) at three different environmental $[Ca^{2+}]$. Based on preliminary data of N. Heisler, R. Neumann and N. A. Andersen; also see text.

loss of acid–base relevant ions is technically impossible to date. The situation is complicated by the fact that not only would diffusive loss occur of HCO_3^- (untraceable because of conversions in its buffer equilibria), but also other acid–base relevant ions might be involved as well. The nature of such ions, their apparent pK' values and the exact interface conditions with respect to pH, ionic strength, ligating ions and diffusional pathways of secondary products originating from changes in ionization of acid–base relevant ions on the way out or into the animal, are unknown. Accordingly, the problem can only be attacked using a simplified and descriptive first approach.

Changes of diffusive acid–base relevant conductance of the integument (mainly the branchial epithelium) will affect the conductance of Cl^- as the main anion in the extracellular space of fish as well. Although molecular size and mobility of Cl^- through paracellular and cytoplasmic diffusional pathways (mainly in the gill area) may not ideally reflect the movement of HCO_3^- and other acid–base relevant ions, changes will still provide a relatively close first estimate. Determination of unidirectional efflux of tracer $^{36}Cl^-$ in specimens of *Cyprinus carpio* indicated that the Cl^- diffusional mobility was actually a function of the environmental $[Ca^{2+}]$. In addition to Ca^{2+}, the efflux of Cl^- was also affected by plasma $[HCO_3^-]$ (Fig. 8).

Taking the results of this study as indicative of the changes in passive

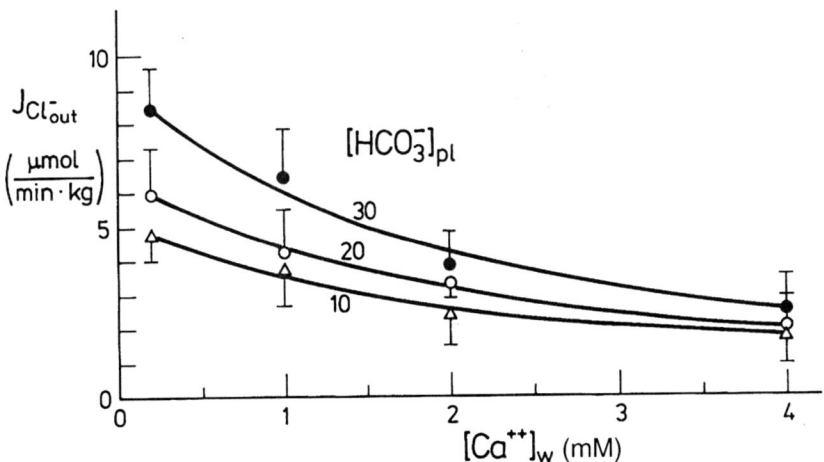

Fig. 8. Unidirectional efflux of Cl^- from *Cyprinus carpio* as a function of environmental $[Ca^{2+}]$ at three levels of plasma $[HCO_3^-]$. Based on preliminary data of R. Neumann, C. Schorer and N. Heisler; also see text.

permeability of acid–base relevant ions, a large fraction of the change in the plasma [HCO$_3^-$] threshold during hypercapnia can be explained on the basis of changes in diffusional conductance by high environmental [Ca^{2+}]. The active HCO$_3^-$ gain (resulting from the difference between apparent net flux and the apparent passive conductance obtained on the basis of the data of Fig. 8) is similar at the zero intercept of net HCO$_3^-$ flux for different environmental [Ca^{2+}] (Fig. 9). This suggests that in this range, critical for the absolute extent of pH compensation, the active component of net acid–base relevant transfer is less affected by environmental [Ca^{2+}] than by passive diffusional loss of acid–base relevant ions. However, differences in the shape of the relationship between net HCO$_3^-$ influx and the plasma/water HCO$_3^-$ ratio (Figs. 8 & 9) in the range of insignificant diffusional loss indicate that the active component of the process is also affected to some extent. Many questions are raised

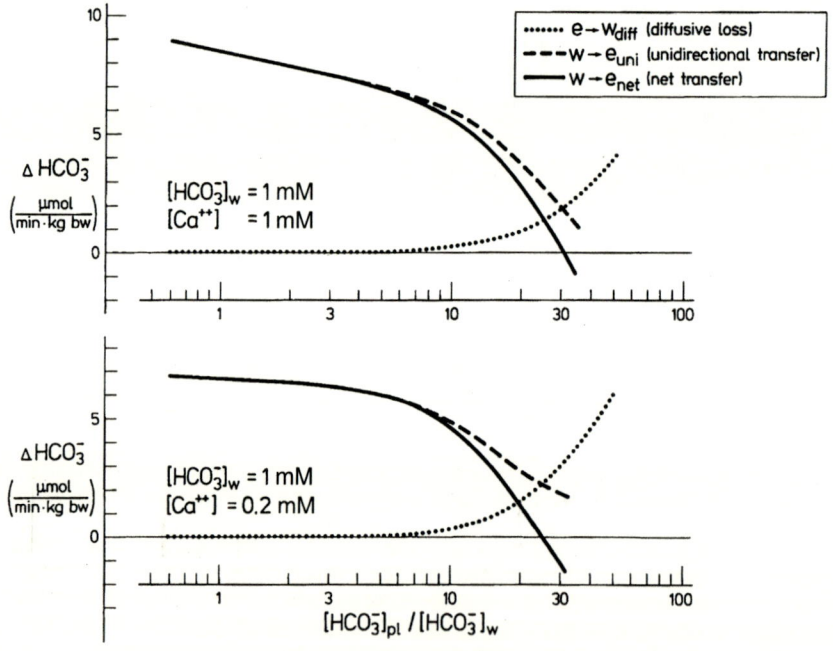

Fig. 9. Model of the contribution of active transfer of HCO$_3^-$ equivalents and passive diffusional loss of acid–base relevant ions (HCO$_3^-$) to the observed net rate of HCO$_3^-$ gain from the environment during ambient hypercapnia. Note the intersect of the net transfer and thus the maximal plasma [HCO$_3^-$] is enhanced by about 8 mM at an elevated [Ca^{2+}]$_w$ of 1 mM. Based on data in Figures 7 and 8; also see text.

by the complexity of this intriguing system and will be the subjects of future studies.

Conclusion

In water-breathing fish, respiratory regulation is impaired by the limitations of water as the gas exchange medium and the resulting low levels of P_{CO_2}. Lacking the possibility for respiratory compensation, fish rely mainly on regulation of the acid–base status by epithelial ion transfer, operating at a much larger capacity than in air-breathing animals. The high capacity to net gain $[HCO_3^-]$ from the environment generally facilitates normalization of the acid–base status before the original stress factor is removed, with preferential regulation of the intracellular body compartments. In the only species studied for this aspect, epithelial acid–base relevant transfer is mainly performed by the action of chloride cells located in juxtaposition to the secondary circulatory system of the central venous gill sinus. Acid–base relevant ion transfer mechanisms of fish are limited by low environmental ion concentrations, in particular of $[HCO_3^-]$ and $[Ca^{2+}]$, affecting integumentary ion conductance and, to a lesser extent, possibly also modulating active transfer mechanisms.

References

Avella, M. & Bornancin, M. (1989). A new analysis of ammonia and sodium transport through the gills of the freshwater rainbow trout (*Salmo gairdneri*). *Journal of Experimental Biology* **142**, 155–75.

Børjeson, H. (1976). Some effects of high carbon dioxide tension on juvenile salmon (*Salmo salar* L.). *Acta Universitatis Upsaliensis* **383**, 1–35.

Børjeson, H. (1977). Effects of hypercapnia on the buffer capacity and haematological values in *Salmo salar* (L.). *Journal of Fish Biology* **11**, 133–42.

Brønsted, J.N. (1923). Einige Bemerkungen über den Begriff der Säuren und Basen. *Recueil des Travaux Chimique des Pays-Bas* **42**, 718–28.

Cameron, J.N. (1980). Body fluid pools, kidney function, and acid–base regulation in the freshwater catfish *Ictalurus punctatus*. *Journal of Experimental Biology* **86**, 171–85.

Cameron, J.N. (1985). The bone compartment in a teleost fish, *Ictalurus punctatus*: size, composition and acid–base response to hypercapnia. *Journal of Experimental Biology* **117**, 307–18.

Cameron, J.N. & Iwama, G.K. (1987). Compensation of progressive hypercapnia in channel catfish and blue crabs. *Journal of Experimental Biology* **133**, 183–97.

Claiborne, J.B. & Heisler, N. (1984). Acid–base regulation in the carp (*Cyprinus carpio*) during and after exposure to environmental hypercapnia. *Journal of Experimental Biology* **108**, 25–43.

Claiborne, J.B. & Heisler, N. (1986). Acid–base regulation and ion transfers in the carp (*Cyprinus carpio*): pH compensation during graded long- and short-term environmental hypercapnia and the effect of bicarbonate infusion. *Journal of Experimental Biology* **126**, 41–61.

Daxboeck, C., Barnard, D.K. & Randall, D.J. (1981). Functional morphology of the gills of the bowfin, *Amia calva* L. with special reference to their significance during air exposure. *Respiratory Physiology* **43**, 349–64.

Dejours, P. (1973). Problems of control of breathing in fishes. In *Comparative physiology*, eds. L. Bolis, K. Schmidt-Nielsen & S.H.P. Maddrell, pp. 117–33. Amsterdam: North Holland.

Dejours, P. (1981). *Principles of comparative respiratory physiology*, 2nd edition. Amsterdam: North Holland.

DeLaney, R.G., Lahiri, S. & Fishman, A.P. (1974). Aestivation of the African lungfish *Protopterus aethiopicus*: Cardiovascular and respiratory functions. *Journal of Experimental Biology* **61**, 111–28.

DeLaney, R.G., Lahiri, S., Hamilton, R. & Fishman, A.P. (1977) Acid–base balance and plasma composition in the aestivating lungfish (*Protopterus*). *American Journal of Physiology* **232**, R10–17.

Dimberg, K. (1988). High blood CO_2 levels in rainbow trout exposed to hypercapnia in bicarbonate-rich hard fresh water – a methodological verification. *Journal of Experimental Biology* **134**, 463–6.

Eddy, F.B., Lomholt, J.P., Weber, R.E. & Johansen, K. (1977). Blood respiratory properties of rainbow trout (*Salmo gairdneri*) kept in water of high CO_2 tension. *Journal of Experimental Biology* **67**, 37–47.

Evans, D.H. (1984). The role of gill permeability and transport mechanisms in euryhalinity. In *Fish physiology*, Vol. XB, ed. W.S. Hoar & D.J. Randall, pp. 315–401. Orlando: Academic Press.

Evans, D.H. (1986). The role of branchial and dermal epithelia in acid–base regulation in aquatic animals. In *Acid–base regulation in animals*, ed. N. Heisler, pp. 139–72. Amsterdam: Elsevier Science Publishers.

Fromm, P.O. (1980). A review of some physiological and toxicological responses of freshwater fish to acid stress. *Environmental Biology Fish.* **5**, 79–93.

Harvey, H.W. (1974). *The chemistry and fertility of sea waters.* London, New York: Cambridge University Press.

Heisler, N. (1978). Bicarbonate exchange between body compartments after changes of temperature in the larger spotted dogfish (*Scyliorhinus stellaris*). *Respiration Physiology* **33**, 145–60.

Heisler, N. (1982a). Intracellular and extracellular acid–base regulation in the tropical freshwater teleost fish *Synbranchus marmoratus* in response to the transition from water breathing to air breathing. *Journal of Experimental Biology* **99**, 9–28.

Heisler, N. (1982b). Transepithelial ion transfer processes as mechanisms for fish acid–base regulation in hypercapnia and lactacidosis. *Canadian Journal of Zoology* **60**, 1108–22.

Heisler, N. (1984). Acid–base regulation in fishes. In *Fish physiology*, Vol. XA, ed. W.S. Hoar & D.J. Randall, pp. 315–401. New York, London: Academic Press.

Heisler, N. (1986a). Buffering and transmembrane ion transfer processes. In *Acid–base regulation in animals*, ed. N. Heisler, pp. 3–47. Amsterdam: Elsevier Science Publishers.

Heisler, N. (1986b). Acid–base regulation in fishes. In *Acid–base regulation in animals*, ed. N. Heisler, pp. 309–56. Amsterdam: Elsevier Science Publishers.

Heisler, N. (1986c). Comparative aspects of acid–base regulation. In *Acid–base regulation in animals*, ed. N. Heisler, pp. 397–450. Amsterdam: Elsevier Science Publishers.

Heisler, N. (1988a). Acid–base regulation in elasmobranch fishes. In *Physiology of elasmobranch fishes*, ed. T.J. Shuttleworth, pp. 215–52. Heidelberg: Springer.

Heisler, N. (1988b). Ion transfer processes as mechanisms for acid–base regulation. In *Lung biology in health and disease – comparative pulmonary physiology: current concepts*, ed. S.C. Wood, pp. 539–83. New York: Marcel Dekker.

Heisler, N. (1989). Acid–base regulation in fishes I. Mechanisms. In *Acid toxicity and aquatic animals*, ed. R. Morris, E.W. Taylor, D.J.A. Brown & J.A. Brown, pp. 85–97. Society of Experimental Biology Seminar Series. Cambridge: Cambridge University Press.

Heisler, N. (1990a). Acid–base regulation: interrelationships between gaseous and ionic exchange. In *Vertebrate gas exchange from environment to cell*, ed. R.G. Boutilier, pp. 211–51. Series: Advances in Environmental and Comparative Physiology Series, Vol. 6. Heidelberg, Berlin, New York: Springer.

Heisler, N. (1990b). Interaction between gas exchange, metabolism and ion transport in animals: an overview. *Canadian Journal of Zoology* **67**, 2923–35.

Heisler, N. (1992). Mode of transmembrane and transepithelial ammonia transfer. In *The vertebrate gas transport cascade: adaptations to environment and mode of life*, ed. J.E. Bicudo, pp. 1–11. Orlando: Academic Press.

Heisler, N., Forcht, G., Ultsch, G.R. & Anderson, J.F. (1982). Acid–base regulation in response to environmental hypercapnia in two aquatic salamanders, *Siren lacertina* and *Amphiuma means*. *Respiration Physiology* **49**, 141–58.

Heisler, N., Holeton, G.F. & Toews, D.P. (1981). Regulation of gill ventilation and acid–base status in hyperoxia-induced hypercapnia in the larger spotted dogfish (*Scyliorhinus stellaris*). *Physiologist* **24**, 58.

Heisler, N. & Neumann, P. (1977). Influence of sea water pH upon bicarbonate uptake induced by hypercapnia in an elasmobranch (*Scyliorhinus stellaris*). *Pflügers Archiv* **368**, Suppl., R19.

Heisler, N., Toews, D.P. & Holeton, G.F. (1988). Regulation of ventilation and acid–base status in the elasmobranch *Scyliorhinus stellaris* during hyperoxia-induced hypercapnia. *Respiration Physiology* **71**, 227–46.

Heisler, N., Weitz, H. & Weitz, A.M. (1976). Hypercapnia and resultant bicarbonate transfer processes in an elasmobranch fish. *Bulletin Européen de Physiopathologie Respiratoire* **12**, 77–85.

Henry, R.P., Smatresk, N.J. & Cameron, J.N. (1988). The distribution of branchial carbonic anhydrase and the effects of gill and erythrocyte carbonic anhydrase inhibition in the channel catfish, *Ictalurus punctatus*. *Journal of Experimental Biology* **134**, 201–18.

Höbe, H., Wood, C.M. & Wheatly, M.G. (1984). The mechanisms of acid–base and ionoregulation in the freshwater rainbow trout during environmental hyperoxia and subsequent normoxia. I. Extra- and intracellular acid–base status. *Respiration Physiology* **55**, 139–54.

Holeton, G.F. & Heisler, N. (1983). Contribution of net ion transfer mechanisms to the acid–base regulation after exhausting activity in the larger spotted dogfish (*Scyliorhinus stellaris*). *Journal of Experimental Biology* **103**, 31–46.

Holeton, G.F., Neumann, P. & Heisler, N. (1983). Branchial ion exchange and acid–base regulation after strenuous exercise in rainbow trout (*Salmo gairdneri*). *Respiratory Physiology* **51**, 303–18.

Hunn, J.B. (1982). Urine flow rate in freshwater salmonids: A review. *Progress in Fish-Culture* **44**, 119–25.

Ishimatsu, A. & Itazawa, Y. (1983). Blood oxygen levels and acid–base status following air exposure in an air-breathing fish, *Channa argus*: The role of air ventilation. *Comparative Biochemistry Physiology A* **74A**, 787–93.

Ishimatsu, A., Iwama, G.K. & Heisler, N. (1988). *In vivo* analysis of partitioning of cardiac output between systemic and central venous sinus circuits in rainbow trout: a new approach using chronic cannulation of the branchial vein. *Journal of Experimental Biology* **137**, 75–88.

Ishimatsu, A., Iwama, G.K. & Heisler, N. (1995). Physiological roles of the secondary circulatory system in fish. In *Mechanisms of systemic regulation in lower vertebrates*, Vol. 21: *Respiration and circulation*, pp. 215–36. Advances in Environmental and Comparative Physiology. Heidelberg: Springer.

Iwama, G.K. & Heisler, N. (1991). Effect of environmental water salinity on the acid–base regulation during environmental hypercapnia in the rainbow trout (*Salmo gairdneri*). *Journal of Experimental Biology* **158**, 1–18.

Iwama, G.K., Ishimatsu, A. & Heisler, N. (1993). Site of acid–base relevant ion transfer in the gills of rainbow trout (*Oncorhynchus mykiss*) exposed to environmental hypercapnia. *Fish Physiology & Biochemistry* **12**, 269–80.

Janssen, R.G. & Randall, D.J. (1975). The effect of changes in pH and P_{CO_2} in blood and water on breathing in rainbow trout, *Salmo gairdneri*. *Respiration Physiology* **25**, 235–45.

Jensen, F.B. & Weber, R.E. (1982). Respiratory properties of tench blood and hemoglobin. Adaptation to hypoxic–hypercapnic water. *Molecular Physiology* **2**, 235–50.

Laurent, P. (1984). Gill internal morphology. In *Fish physiology*, Vol. XA, ed. W.S. Hoar & D.J. Randall, pp. 73–183. New York: Academic Press.

Lin, H. & D.J. Randall (1991). Evidence for the presence of an electrogenic proton pump on the trout gill epithelium. *Journal of Experimental Biology* **161**, 119–34.

Maetz, J. (1974). Adaptation to hyper-osmotic environments. *Biochemical and Biophysical Perspectives in Marine Biology* **1**, 91–149.

Randall, D.J., Heisler, N. & Drees, F. (1976). Ventilatory response to hypercapnia in the larger spotted dogfish *Scyliorhinus stellaris*. *American Journal of Physiology* **230**, 590–4.

Spry, D.J., Wood, C.M. & Hodson, P.V. (1981) The effects of environmental acid on freshwater fish with particular reference to the softwater lakes in Ontario and the modifying effects of heavy metals. A literature review. *Canadian Technical Report of Fisheries and Aquatic Sciences* **999**.

Tang, Y. & Boutilier, R.G. (1988). Clearance of lactate and protons following acute lactacidosis: a comparison between seawater- and freshwater-adapted rainbow trout (*Salmo gairdneri*). *Journal of Experimental Biology* **48**, 41–4.

Tang, Y., McDonald, D.G. & Boutilier, R.G. (1989). Acid–base regulation following exhaustive exercise: A comparison between freshwater- and seawater-adapted rainbow trout (*Salmo gairdneri*). *Journal of Experimental Biology* **141**, 407–18.

Toews, D.P., Holeton, G.F. & Heisler, N. (1983). Regulation of the acid–base status during environmental hypercapnia in the marine teleost fish *Conger conger*. *Journal of Experimental Biology* **107**, 9–20.

Van Slyke, D.D. (1922). On the measurement of buffer values and on the relationship of buffer value to the dissociation constant of the buffer and the concentration and the reaction of the buffer system. *Journal of Biological Chemistry* **52**, 525–70.

Wilkes, P.R.H., Walker, R.L., McDonald, D.G. & Wood, C.M. (1981). Respiratory, ventilatory, acid–base and ionoregulatory physiology of the white sucker *Catostomus commersoni*: The influence of hyperoxia. *Journal of Experimental Biology* **91**, 239–54.

Wood, C.M. & McDonald, D.G. (1982). Physiological mechanisms of acid toxicity to fish. In *Acid rain/fisheries*, ed. R.E. Johnson, pp. 197–226. Bethesda, MD: American Fisheries Society.

Wood, C.M. & McDonald, D.G. (1987). The physiology of acid aluminum stress in trout. *Annales de la Societe Royale Zoologique de Belgique* **117(S1)**, 399–410.

Wood, C.M. & Soivio, A. (1991). Environmental effects on gill function – an introduction. *Physiological Zoology* **64(N1)**, 1–3.

Woodbury, J.W. (1965). Regulation of pH. In *Physiology and biophysics*, ed. T.C. Ruch & H.D. Patton, pp. 899–934. Philadelphia: Saunders.

MICHAEL R. BLATT and
ALEXANDER GRABOV

H+-mediated control of ion channels in guard cells of higher plants

Introduction

Among biological systems for studying transport and signalling functions, stomatal guard cells are one of a very few highly successful cell models for higher plants. The dominance of the guard cell in this respect owes much to its unique situation and physiological function, features that have been used to considerable experimental advantage. Unlike the vast majority of cells in higher plants, guard cells lack functional plasmodesmata at maturity, and so are physically isolated from the neighbouring epidermal and mesophyll cells (Wille & Lucas, 1984; Willmer & Fricker, 1996). As a direct consequence, electrophysiological investigation of intact guard cells is possible without the complications of electrical coupling between cells. Thus, despite their comparatively small size, quantitative voltage clamp studies of guard cells have been carried out from several species, notably *Nicotiana*, *Arabidopsis* and *Vicia*. Furthermore, because this cell type can be isolated relatively easily, at least in the limited numbers required for such studies, guard cells have afforded virtually the only opportunity for comparisons of ion channel and other transport characteristics recorded in enzymatically isolated protoplasts and *in situ* (Lemtiri-Chlieh, 1996).

No less important, access into the network of regulatory processes that control guard cell membrane ion pumps and channels has been possible through the unique physiology of stomata. Guard cells are situated at the end of the transpiration stream within the plant. They surround small pores in the epidermis through which gas exchange for photosynthesis within the leaf mesophyll takes place. It is also through these pores that the bulk of water loss from the plant occurs. Thus, guard cell control of stomatal aperture is crucial to balancing these opposing constraints on the plant, minimising water loss from the leaf tissues while balancing the requirements for CO_2 in photosynthesis. It is no wonder, then, that the ability of these cells to integrate both environmental and internal signals has provided several experimental points of entry into signal cascades linking guard cell transport to stomatal control (Assmann, 1993; Blatt & Thiel, 1993).

Guard cells regulate stomatal aperture by swelling (to open) and shrinking (to close the pore) in volume, processes that are driven by the gain and loss of osmotically active solutes, respectively (Raschke, 1979; Willmer & Fricker, 1996) and, because the cells lack functional plasmodesmata, these fluxes must take place across the plasma membrane. The fluxes entailed – notably K^+, Cl^- and in some instances also organic acids such as malate (Willmer & Fricker, 1996) – are prodigious: guard cells of the broad bean *Vicia* have been estimated to take up and release 2–4 pmol of KCl between the open and closed states of the stoma. On a cell volume basis, these changes are equivalent to approximately 300 mosM in osmotically active solutes over periods of as little as 20–30 minutes (Willmer & Fricker, 1996). How are these fluxes realised? Remarkably, the transport pathways that mediate these solute fluxes across the plasma membrane are relatively few. They include a H^+-ATPase (Blatt, 1987), two dominant classes of K^+ channels that give rise to current rectifying inward ($I_{K,in}$) and outward ($I_{K,out}$), and one (or possibly two) anion channels that mediate anion efflux (Schroeder, Raschke & Neher, 1987; Hosoi, Iino & Shimazaki, 1988; Blatt, 1988; Hedrich, Busch & Raschke, 1990; Schroeder & Keller, 1992; Linder & Raschke, 1992; Blatt, 1992; Grabov *et al.*, 1997). Additional pathways for ion transport have been inferred from a background conductance (channel?) (Blatt, 1991; Thiel, MacRobbie & Blatt, 1992) and from evidence for energy-dependent K^+ and Cl^- uptake (MacRobbie, 1988; Clint & Blatt, 1989).

Co-ordinating these several transporters is essential if the guard cell is to achieve the requisite solute flux. Early proposals focused on its interaction with the H^+-ATPase (Raschke, 1987), suggesting that control of K^+ and other ion fluxes might be regulated fundamentally through the electrochemical 'poise' of the membrane, with the ion channels responding passively to the membrane voltage generated by the H^+-ATPase. It is now clear, however, that thermodynamic considerations alone do not account for guard cell transport behaviour. Rather, control of these fluxes is the result of a highly concerted process that conjoins the activities of all of the major ion currents across the membrane. Nonetheless, how this level of co-ordination comes about remains poorly understood. Only for abscisic acid (ABA) has much substance of the events in transport integration and control been consolidated to date. In vegetative tissues, ABA mediates adaptive responses to adverse environmental conditions, including drought and high salinity (Davies & Jones, 1991; Willmer & Fricker, 1996). During water stress conditions, it is synthesised in the roots and accumulates in leaf tissues to promote stomatal closure and reduce transpirational water loss.

It must be stressed that our present understanding of events – and, indeed, of the relevant transport pathways – is restricted almost entirely to the plasma membrane. Nonetheless, it implies an equivalent degree of transport co-ordination at the tonoplast and between the two membranes because the greater proportion of solute lost during stomatal closure, both in terms of total and relative amounts, originates from the vacuole (MacRobbie, 1995a, 1995b; Willmer & Fricker, 1996). Two different cation channels are known at the guard cell tonoplast that are capable of carrying K^+ (Ward & Schroeder, 1994; Schulzlessdorf & Hedrich, 1995). An anion channel was also recently identified with high selectivity for Cl^- over K^+ and dependent on protein phosphorylation for activity (Pei *et al.*, 1996). These channels are likely to prove important targets under ABA control, although their respective contributions still remain to be demonstrated. Equally, the vacuole constitutes an important source and sink for Ca^{2+} and H^+ generally, and probably contributes to ABA-mediated and other stimulus-mediated changes in cytosolic-free $[Ca^{2+}]$ ($[Ca^{2+}]_i$) (Allen & Sanders, 1995) and cytosolic pH (pH_i), but much detail is still missing.

With respect to ion flux at the plasma membrane, the sequence of events that follow ABA exposure can be ordered roughly as follows.

1.   ABA evokes the development of an inward-directed current, probably mediated by slow-activating anion channels (Grabov *et al.*, 1997), to depolarise the membrane and generate a driving force for K^+ efflux (Thiel *et al.*, 1992).
2.   It inactivates the inward-rectifying K^+ channels that normally mediate in K^+ uptake (Blatt, 1990).
3.   It may promote fast-activating anion (Cl^-) channels (Hedrich *et al.*, 1990; Schroeder & Keller, 1992; Thiel *et al.*, 1992).
4.   Finally, it activates current through outward-rectifying K^+ channels (Blatt, 1990; Lemtiri–Chlieh & MacRobbie, 1994) that, together with both slow-activating and fast-activating anion channels, facilitates the net loss of salt from the cells.

It is likely, too, that ABA reduces H^+-ATPase activity, either indirectly *via* secondary control or directly through substrate (H^+) withdrawal (Thiel *et al.*, 1992; Blatt & Armstrong, 1993; Goh, Oku & Shimazaki, 1995). However, the progression of these events is not linear. Alterations in all of the channel currents almost certainly occur in parallel, only with differing response halftimes, and not surprisingly, as it has long been evident that at least one signalling pathway additional to that mediated by $[Ca^{2+}]_i$ underpins K^+ and anion channel control. Overwhelming evidence now supports the idea of pH_i-mediated signalling in guard cells. It also highlights many questions that are still unanswered about ABA-evoked stomatal closure. The remainder of this chapter reviews several of these issues (and their background) that form the centre of current debate about guard cell signal transduction and ion channel control.

Shortfalls in the Ca^{2+} model

To appreciate the arguments concerning roles for pH_i in guard cells, a brief review of the evidence for Ca^{2+} as a second messenger is in order. In fact, it was recognised early on that the effects of increased $[Ca^{2+}]_i$ in several respects mimicked channel responses to ABA (Blatt, 1990; Thiel et al., 1992; Lemtiri–Chlieh & MacRobbie, 1994). Raising $[Ca^{2+}]_i$ to micromolar concentrations greatly reduced current through the inward-rectifying K^+ channels, $I_{K,in}$ (Schroeder & Hagiwara, 1989), an effect that is most evident as a displacement in the voltage sensitivity of $I_{K,in}$. Increased $[Ca^{2+}]_i$ was also found to promote anion currents within the physiological voltage range (Hedrich et al., 1990; Schroeder & Keller, 1992). Finally, initial measurements showed that $[Ca^{2+}]_i$ frequently rose from resting values near 100 nM to values often in excess of 1 µM during ABA exposures (McAinsh, Brownlee & Hetherington, 1990, 1992; Fricker et al., 1991). Thus, forceful arguments have made for ABA stimulus transduction that proceeds via a rise in $[Ca^{2+}]_i$ with the combined effect of preventing K^+ uptake and promoting anion efflux. Furthermore, because the anion currents would favour membrane depolarisation, it has been argued that increasing $[Ca^{2+}]_i$ should facilitate K^+ efflux by activating current through the outward-rectifying K^+ channels, $I_{K,out}$, at these voltages (MacRobbie, 1992; Assmann, 1993; Ward, Pei & Schroeder, 1995).

Even so, the $[Ca^{2+}]_i$ hypothesis has presented some difficulties for ABA-mediated signalling in guard cells almost from the start. As a result, attaching a physiological interpretation to $[Ca^{2+}]_i$ action has not been straightforward. One issue of continuing debate centres on the fact that the rise in $[Ca^{2+}]_i$ is not always observed with ABA treatments, or even during stomatal closure (Fricker et al., 1991; Gilroy et al., 1991; McAinsh et al., 1992). A plausible apologia can be made for this correlative shortfall, pointing to the technical problems inherent in fluorescent measurements of Ca^{2+} in plant cells (Fricker et al., 1991;

Willmer & Fricker, 1996): (1) that the intrinsic cytsolic volume is small, either sandwiched between plasma membrane and tonoplast or dominated by perinuclear cytoplasm, making it difficult to obtain reliable measurements with sufficient spatio-temporal resolution; and (2) that changes in $[Ca^{2+}]_i$ relevant to ion channel control almost certainly occur locally, adjacent to the plasma membrane.

Equally plausible, however, is the argument that $[Ca^{2+}]_i$ changes are secondary and serve another function in adapting ion channel response to the needs for solute flux (Blatt & Thiel, 1993). In this context, Allan *et al.* (1994) observed that environmental factors, notably growth temperature, can play a critical role in determining the scope and magnitude of $[Ca^{2+}]_i$ signals recorded in response to ABA. Furthermore, recent work from this laboratory (Grabov & Blatt, 1998) has indicated that transients and oscillations in $[Ca^{2+}]_i$, often associated with stomatal closure (McAinsh *et al.*, 1995; Webb *et al.*, 1996), are subject to membrane voltage. These studies have shown that $[Ca^{2+}]_i$ increases can be driven by membrane hyperpolarisation to activate Ca^{2+} influx and Ca^{2+}-induced Ca^{2+} release from intracellular stores. Significantly, these latter observations link $[Ca^{2+}]_i$ increases with membrane voltage and a feedback mechanism controlling $I_{K,in}$ (and, most probably, the anion channels). So, the results suggest that $[Ca^{2+}]_i$ may condition the responsiveness of the guard cells for the needs of closing stimuli to adjust channel activities.

These issues aside, by far the strongest argument in favour of a second, parallel signal cascade rests on the fact that increases in $[Ca^{2+}]_i$ do not mimic the effects of ABA in contolling at least one set of ion channels at the guard cell plasma membrane. Current through the outward-rectifying K^+ channels is activated in the presence of ABA (Blatt, 1990; Thiel *et al.*, 1992), in some instances by as much as 14-fold (1400 per cent) at any one voltage (Lemtiri–Chlieh & MacRobbie, 1994). This response does not depend on a rise in $[Ca^{2+}]_i$, nor has $I_{K,out}$ proven at all sensitive to $[Ca^{2+}]_i$ (Hosoi *et al.*, 1988; Schroeder & Hagiwara, 1989; Grabov & Blatt, 1997). Furthermore, the current is not affected by upstream intermediates of $[Ca^{2+}]_i$-related signal cascades, although in each case an influence on $I_{K,in}$ has been reported complementary to a rise in $[Ca^{2+}]_i$. Neither G-protein agonists and antagonists such as GTPγS, GDPβS, cholera and pertussis toxins (Fairley & Assmann, 1991) nor mas7, a serpentine receptor mimetic (Armstrong & Blatt, 1995) altered the characteristics of these channels. Equally, photolytic release of inositol-1,4,5,-trisphosphate was unable to trigger an increase in $I_{K,out}$, although it inactivated $I_{K,in}$ in a $[Ca^{2+}]_i$-dependent manner (Blatt, Thiel & Trentham, 1990). These latter observations are consistent with a G-protein-mediated cascade that controls $I_{K,in}$ (Assmann, 1993; Blatt & Thiel, 1993), but they also solidify the evidence against $[Ca^{2+}]_i$-related control of $I_{K,out}$. In short, 'the need to address the possible concurrent functioning of other signalling pathways in the guard cells' (Blatt *et al.*, 1990) has stood out in relief behind the arguments for $[Ca^{2+}]_i$ signalling.

The pH$_i$ signal

Despite these shortfalls in the $[Ca^{2+}]_i$ model, a role for pH$_i$ has surfaced only slowly. This delay is less surprising, perhaps, because the initial indications of stimulus-evoked changes in pH$_i$ were linked to control of the H^+-ATPase by auxin and fusicoccin (Bertl & Felle, 1985; Felle, 1988a) and, hence, to the role of cytosolic H^+ as a substrate for transport rather than as a second messenger in its own right. Two key developments have led to the proposal of a major, ABA-coupling signal cascade that passes through the pH$_i$ intermediate (Blatt & Thiel, 1993): (1) ABA was found to alkalinise the cytosol in guard cells (Irving, Gehring & Parish, 1992; Blatt & Armstrong, 1993) as well as in *Zea* and *Petroselenium* mesophyll cells (Gehring, Irving & Parish, 1990); and (2) experimentally imposed changes in pH$_i$ were observed to alter guard cell K^+ channels in a manner distinct from that of changes in extracellular pH (pH$_e$) (Blatt, 1992; Blatt & Armstrong, 1993). The first point is important because these observations implicitly separated the changes in pH$_i$ from the H^+-ATPase. Common dogma had long maintained that ABA reduces H^+-ATPase activity (Kasamo, 1979, 1981). However, the intuitive expectation was that decreasing H^+ extrusion *via* the pump should *decrease* pH$_i$ (raise $[H^+]_i$). So here were the first indications that hormonal control of pH$_i$ might reflect more than just changes in the pool of substrate for primary transport. The second point, furthermore, ascribed an action to changes in pH$_i$ that could not be related directly – albeit, perhaps indirectly (Blatt & Clint, 1989) – to H^+ flux across the plasma membrane and, instead, called into consideration its function as an internal signal to control ion channels.

Crucial to this proposal, 'pH clamp' experiments indicated that the alkaline shift in pH$_i$ is both necessary and sufficient to account for the activation of $I_{K,out}$ in ABA (Blatt & Armstrong, 1993). In these experiments, guard cells were loaded with pH buffers and with the weak acid butyrate to suppress changes in pH$_i$ with ABA and to raise and lower pH$_i$ either before or after adding ABA. In each case, the current response was fully accommodated as a simple function of $[H^+]_i$ without recourse to additional regulatory events (Blatt & Armstrong, 1993). Furthermore, subsequent work also implicated pH_i in channel control initiated by auxin (Thiel *et al.*, 1993; Blatt & Thiel, 1994), suggesting a central role in cellular stimulus–response coupling in guard cells. Indeed, Thiel *et al.* (1993) found that auxin action on $I_{K,out}$ could be mimicked in the absence of auxin using a peptide synthesised to the C-terminal sequence of a maize auxin-binding protein. Perhaps the most remarkable aspect of these studies was the change in pH_i – roughly 0.5 pH units within 20 seconds, at a conservative estimate – evoked by the peptide. Such a profound pH_i response rivals evoked changes of $[Ca^{2+}]_i$ in many plant cells, both in amplitude and kinetics, and remains difficult to explain in terms of H^+ transport across either plasma membrane or tonoplast.

It is worth noting that the action of pH_i on $I_{K,out}$ showed an unusual 'fingerprint'. Blatt and Armstrong (1993) found that increasing pH_i activated $I_{K,out}$ in a voltage-independent manner so that the current rose roughly in proportion at all voltages (Fig. 1, upper frame). Titration of the current as a function of the measured pH_i also suggested a co-operative binding of two H^+ to effect the response, and these results have been substantiated by combined voltage clamp and BCECF fluorescence ratio measurements of pH_i (Grabov & Blatt, 1997). Significantly, pH_i had little effect on the kinetics of $I_{K,out}$ activation. Instead, it appeared to mobilise K^+ channels into a pool that could be activated on membrane depolarisation (Blatt & Armstrong, 1993). Miedema and Assmann (1996) have since confirmed these observations, showing that the action of pH_i is mediated by H^+ interaction with binding sites closely associated with the membrane. Thus, it appears that cytosolic H^+ interacts directly with the channel or closely associated protein(s) at the cytosolic face of the membrane.

Nor is the action of pH_i restricted to control of $I_{K,out}$. Despite its association with $[Ca^{2+}]_i$ (above), regulation of $I_{K,in}$ – or one of the major signalling elements controlling these channels – is also sensitive to pH_i. Furthermore, the action of pH_i on $I_{K,in}$ shows a fingerprint that is fundamentally different to that of its action on $I_{K,out}$. Blatt and Armstrong (1993; see also Blatt, 1992) reported that lowering pH_i with weak acids activated $I_{K,in}$ and that the current was suppressed by alkaline pH_i loads (Fig. 1, upper frame). Titration of $I_{K,in}$ (Grabov & Blatt, 1997) has since demonstrated that the current behaves as a simple function of $[H^+]_i$. The data suggest H^+ binding at a site with a pK_a near 6.4, consistent with the titration of a single histidine residue and entirely consonant with several recent studies of cloned K^+ channels from yeast (Lesage *et al.*, 1996) and mammalian neuromuscular tissues (Coulter *et al.*, 1995).

One immediate question is whether pH_i might influence the K^+ chan-

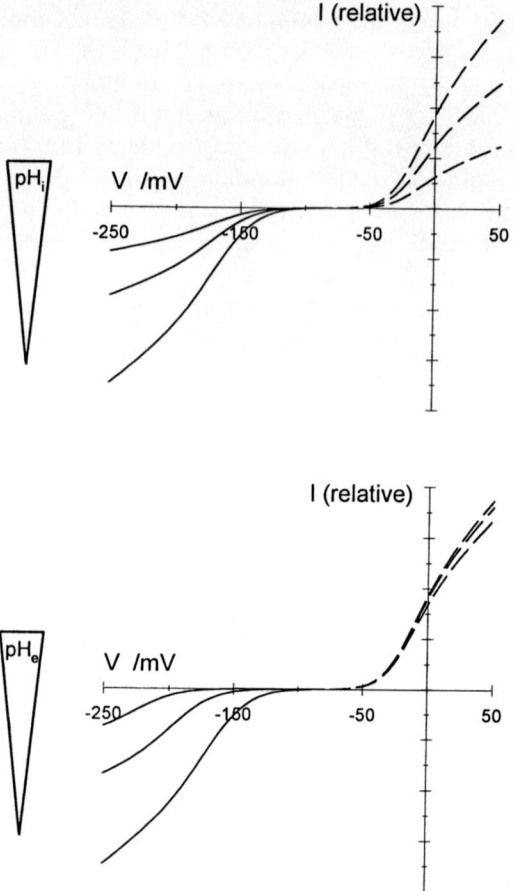

Fig. 1. K⁺ channel control by cytosolic pH and pH_e. Steady-state current-voltage relations for the inward-rectifying ($I_{K,in}$, solid lines) and outward-rectifying ($I_{K,out}$, dashed lines) K⁺ channels as observed in 10 mM external K⁺ ($E_K \approx -75$ mV). Responses to the pH parameter (magnitudes indicated by the triangular shapes, left) are shown in parallel by the families of current-voltage curves.

nels indirectly by interfering with Ca^{2+} signal transmission through a titration of Ca^{2+}-binding sites. To a first approximation, the answer is negative. The Ca^{2+} and H⁺ messengers converge on the K⁺ channels to control their activity *via* two independent mechanisms. In support of this conclusion, Grabov and Blatt (1997) found that pH_i-mediated control of $I_{K,in}$ could be observed without measurable changes in $[Ca^{2+}]_i$.

Furthermore, the effect of pH_i was evident predominantly as a voltage-independent change in $I_{K,in}$ conductance, whereas the action of $[Ca^{2+}]_i$ was profoundly voltage sensitive (analogous to the effects of pH_c; see Fig. 1, lower frame, also **Shortfalls in the Ca^{2+} model** above). These results clearly distinguished the actions of pH_i on $I_{K,in}$ from those of $[Ca^{2+}]_i$. Not surprisingly then, additional studies have indicated that changes in pH_i may contribute to regulating $I_{K,in}$ in lieu of $[Ca^{2+}]_i$. Work with synthetic peptides to the auxin-binding protein C-terminus (Thiel *et al.*, 1993) and with ABA (Blatt & Armstrong, 1993; Armstrong *et al.*, 1995), both of which inactivated $I_{K,in}$ and raised pH_i, failed to yield any consistent changes in $[Ca^{2+}]_i$, and a similar dependence on pH_i was implicated for the channel response to supramicromolar concentrations of auxin (Blatt & Thiel, 1994). In short, alkaline pH_i also contributes to $I_{K,in}$ control and may predominate over changes in $[Ca^{2+}]_i$ in some instances.

Universality of the pH_i signal

While the preceding arguments underline the importance of the pH_i signal, there exists at least one set of circumstances in which the ion channel characteristics appear divorced from regulation by pH_i. Stomata normally close in response to elevation in P_{CO_2}, much as they do in the presence of ABA. Like ABA, elevated P_{CO_2} promotes solute efflux from the guard cells (Willmer & Fricker, 1996) and, thus, might be anticipated to draw on a similar sequence of changes in K^+ and anion channel activities to facilitate these ion fluxes. The expectation has been borne out. Recent voltage clamp studies with *Vicia* guard cells (Brearley, Venis & Blatt, 1997) demonstrated that raising P_{CO_2} from ambient (350 µl/l) to 1000 µl/l evokes a rapid inactivation of $I_{K,in}$ and activation of $I_{K,out}$ as well as enhancing the anion channel current and altering its voltage-dependent kinetics. Significantly, however, parallel measurements with the H^+-sensitive fluorescent dye BCECF failed to uncover any measurable change in pH_i and, in 10 000 µl/l P_{CO_2} even a small (0.1–0.2 unit) decrease in pH_i was observed (Brearley *et al.*, 1997).

From a physiological standpoint, these observations do make sense. Raising P_{CO_2} leads to an increase in dissolved carbonic acid which, as a weak acid, will favour cytosolic acidification (Colman & Espie, 1985). Yet decreasing pH_i inactivates $I_{K,out}$ (above), an effect that must inevitably suppress K^+ salt loss from the guard cells and stomatal closure. So, providing an alternative control pathway is essential to circumventing the potential dilemma otherwise inherent when the response to a weak acid must be to facilitate K^+ efflux for stomatal closure. Because elevating P_{CO_2} actually promoted the activity of $I_{K,out}$, even in the face of mild decreases in pH_i, Brearley *et al.* (1997) concluded that the CO_2 signal must be transmitted *via* another, as yet unidentified, signal cascade. For the present, the identity of this third signalling pathway is a mystery.

Channel sensitivity to pH_e

A brief mention of channel sensitivity to pH_e is in order, especially given that pH_i in plants is, to a limited extent, affected by this parameter (Felle, 1989; Guern *et al.*, 1991). In fact, the inward-rectifying K^+ channels are subject to control by pH_e but, again, in a manner that is fundamentally distinct from that of pH_i, indicating that these channels possess two separate mechanisms (and two different H^+-binding sites) for sensing pH, one each for external and internal H^+. Unlike the effects of pH_i, Blatt (1992) found that reducing pH_e from 7.4 to 5.5 enhanced $I_{K,in}$ conductance five to seven-fold near -200 mV and displaced its voltage-dependence roughly $+45$ mV (Fig. 1, lower frame, shift in curve along the horizontal axis). Titration of the current against pH_e suggested H^+ binding to a single site, but in this case the pK_a was profoundly voltage sensitive. Increasing membrane voltage (inside negative) was paralleled by an increase in the apparent pK_a, as if H^+ binding were favoured by an electrical field driving H^+ to a site situated deep within the membrane. These characteristics pose a striking contrast to actions on K^+ and other channels in animal cells for which H^+ is often a potent blocker (Moody, 1984; Tytgat, Nilius & Carmeliet, 1990), but they make good 'physiological sense' in a H^+-coupling membrane: the sensitivity to pH_e may be seen to ensure that K^+ uptake cannot short-circuit other energetic demands on the H^+-ATPase and membrane voltage (Blatt, 1992). It is of interest that parallel sensitivities to pH_i and pH_e have recently been found for guard cell anion channels (Schulzlessdorf, Lohse & Hedrich, 1996). However, interpretation in this case is not straightforward, because neither pH dependence conforms to the expected requirements for charge balance or solute efflux during stomatal closure.

Origin of changes in pH_i

What is the origin of these changes in pH_i? Because of the high buffer capacity of guard cells (approximately100 mM/pH unit; Blatt & Armstrong, 1993; Grabov & Blatt, 1997), the pH_i rise of 0.2–0.3 units in ABA (Blatt & Armstrong, 1993) must be achieved either by eliminating approximately 30 mM H^+ from the cytosol or by a comparable increase in H^+ binding and sequestration within the cell. Of the possible sinks for H^+, transport by the H^+-ATPase across the plasma membrane can be ruled out unequivocally. Such a large flux of H^+ would entail substantial currents (approximately 10 $\mu A/cm^2$ over 5 minutes) and these currents have not been observed in response to ABA (Blatt, 1990; Thiel *et al.*, 1992; Blatt & Armstrong, 1993; Armstrong *et al.*, 1995; Grabov *et al.*, 1997). Two other potential sinks for H^+ are vacuolar H^+ transport, and metabolic H^+ consumption during organic acid breakdown and CO_2 release (and note that neither of these explanations excludes the other). Of these, the second possibility is attractive because it would tie the pH_i signal to the metabolic turnover of osmolites that is known to take place during stomatal closure. Guard cells of many higher plants, including *Vicia*, use malate as well as Cl^- to balance K^+ uptake during stomatal opening (Willmer & Fricker, 1996). Much of this malate is either decarboxylated or lost from the cells during closure (Davies and Jones, 1991; Willmer & Fricker, 1996). Because its pK_a and that of pyruvate are situated well below the normal cytosolic pH range, the acid anion itself is unlikely to affect pH_i. However, the associated release of OH^-, formation of HCO_3^- *via* carbonic anhydrase and its equilibration with carbonic acid could lead to a rise in pH_i.

Movement of H^+ into the vacuole is an equally likely explanation for the pH_i rise, in part because the H^+ flux could contribute to charge balance during K^+ efflux from the vacuole. Roughly 50–70 per cent of the vacuolar K^+ salt content passes across the tonoplast before exiting the guard cell during stomatal closure (MacRobbie, 1987, 1991). A rough calculation shows that in accounting for a 0.3 pH_i unit rise, only 2–4 per cent of this flux need be balanced by H^+. In fact, vacuolar acidification on this scale is known to take place during closure (Penny & Bowling, 1975). Recent studies also lend support to this argument (Frohnmeyer, Grabov & Blatt, 1997), demonstrating that evacuolation of guard cell as well as mesophyll protoplasts prevents pH_i changes in auxin without affecting H^+ secretion.

Is the pH_i signal G-protein coupled?

Less obvious at present are the events that trigger these changes in pH_i in response to ABA or other stimuli. There is certainly some evidence to support a role for heterotrimeric G-proteins in regulating $I_{K,in}$. However, none of these data can be tied to pH_i regulation of the current, nor are they wholly consistent with coupling, even through the $[Ca^{2+}]_i$ intermediate. In their original studies, Fairley and Assmann (1991) showed that $I_{K,in}$ was sensitive to the G-protein (Gα) antagonists cholera and pertussis toxins in *Vicia* guard cell protoplasts; the current was inactivated in the presence of the non-hydrolysable GTP analogues GTP-γ-S; and inactivation by GTP-γ-S was overcome by buffering changes in $[Ca^{2+}]_i$. These studies placed guard cells firmly within the common themes of animal cell signalling (Berridge, 1993; Bootman & Berridge, 1995), indicating that control of $I_{K,in}$ could pass from Gα activation and, plausibly, stimulation of inositol-1,4,5-trisphosphate production by phospholipase C to a rise in $[Ca^{2+}]_i$. Subsequently, Wu and Assmann (1994) have added evidence for coupling *via* a membrane-delimited Gα, and Kelly, Esser and Schroeder (1995) have suggested that $[Ca^{2+}]_i$ may exert a dual effect with G-proteins on $I_{K,in}$.

G-proteins may also control $I_{K,in}$ activity independent of $[Ca^{2+}]_i$. Armstrong and Blatt (1995) found that mas7, that mimics G-protein-coupled (7-TMS) receptors and activates Gα, was effective in inactivating $I_{K,in}$. Mas7 action was blocked by GDP-β-S, consistent with Gα control of $I_{K,in}$, but the effect could not be overcome by cytosolic Ca^{2+} buffering or by neomycin sulphate, an antagonist of phospholipase C and inositol-1,4,5-trisphosphate-mediated Ca^{2+} release. Nonetheless, these studies yielded no evidence for mas7 action passing through the pH_i intermediate. Unlike the effects on auxin-evoked and ABA-evoked responses of the K^+ channels, treatments with weak acid to clamp pH_i near 7.0 failed to ameliorate the effects of mas7 on $I_{K,in}$. Furthermore, mas7 had no influence on $I_{K,out}$. In fact, there remains an overwhelming lack of evidence to support a function for G-proteins in hormonal responses of guard cells *per se*, whether coupled through $[Ca^{2+}]_i$, pH_i or otherwise. Thus, it appears ever more probable that relevant signalling events upstream in these cells differ from the classic G-protein-receptor model.

pH_i interaction with $[Ca^{2+}]_i$

Because both pH_i and $[Ca^{2+}]_i$ contribute to the regulation of guard cell ion channels, interactions between these second messengers carry a particular interest in the context of modulating and adapting the signals themselves. Clearly, the action of one messenger may substitute for the other under defined experimental conditions. Inactivation of $I_{K,in}$ does occur, mediated by $[Ca^{2+}]_i$, even when pH_i is buffered (Blatt & Armstrong, 1993; Lemtiri–Chlieh & MacRobbie, 1994). Conversely, changes in stomatal aperture (Gilroy *et al.*, 1991; McAinsh *et al.*, 1992; Allan *et al.*, 1994), K^+ channel gating and pH_i (Thiel *et al.* 1993; Armstrong *et al.*, 1995; Grabov & Blatt, 1997) may be observed without measurable changes in $[Ca^{2+}]_i$. These extremes aside, however, it is probable that stimulus-evoked pH_i and $[Ca^{2+}]_i$ signals in many circumstances interact to control $I_{K,in}$ and the anion channels.

What are the probable focal points for such interactions? Apart from binding events associated with the channels themselves, cytosolic H^+ could also affect $[Ca^{2+}]_i$ signal transmission indirectly by titrating Ca^{2+}-binding domains of targets upstream of the ion channels (Moody, 1984; Busa, 1986) or by affecting events of cytosolic Ca^{2+} release (Busa, 1986). Both the binding of inositol-1,4,5-trisphosphate to its receptor as well as the size of internal Ca^{2+} stores show a marked pH_i sensitivity, for example in mammalian secretory tissues (Taylor & Richardson, 1991; for plants, see Blackford, Rea & Sanders, 1990). The effect of alkaline pH_i is to promote Ca^{2+} release at resting inositol-1,4,5-trisphosphate concentrations (Danthuluri, Kim & Brock, 1990). In guard cells, the gating of vacuolar SV-type ion channels is sensitive to pH_i (Schulzlessdorf & Hedrich, 1995) and favours vacuolar Ca^{2+} release at acid pH_i. Note that the action of pH_i in each of these instances affects events upstream of the $[Ca^{2+}]_i$ messenger itself, although the associated mechanisms are different. Equally, it is conceivable that both the $[Ca^{2+}]_i$ and pH_i signals may converge on a downstream target. Similarly, interactions of this kind are well known and affect the efficacy of action rather than the second messenger itself. Such is the case in animals for which the gating of many Ca^{2+}-dependent ion channels is potentiated by alkaline pH_i which promotes Ca^{2+} binding (Christensen & Zeuthen, 1987; Laurido *et al.*, 1991).

In fact, pH_i interactions with $[Ca^{2+}]_i$ of both kinds are demonstrable in *Vicia* guard cells. Grabov and Blatt (1997) found that $[Ca^{2+}]_i$ generally was unaffected over pH_i values ranging from about 7.2 to 7.8 when guard cells were challenged with weak acids. In this circumstance, the cumulative effects of pH_i and $[Ca^{2+}]_i$ on the current were complementary, so that decreasing $[Ca^{2+}]_i$ and pH_i independently favoured $I_{K,in}$ at physiological voltages. For pH_i, the action was to augment $I_{K,in}$ amplitude in a scalar fashion, while for decreasing $[Ca^{2+}]_i$, it was to displace the voltage sensitivity of the current to more positive voltages. Below pH_i values of 7.0, however, $[Ca^{2+}]_i$ was frequently observed to increase, with the net effect that $I_{K,in}$ amplitude was reduced at intermediate clamp voltages but was largely unaffected or increased at more negative voltages. Again, the response reflects additive but, in this case, opposing actions of $[Ca^{2+}]_i$ and pH_i. It is interesting, too, that these $[Ca^{2+}]_i$ increases did not correlate with the timecourse of pH_i loads as might be expected for simple, bulk titration of Ca^{2+}-binding sites. Instead, it appears that pH_i may 'prime' signalling elements that mediate in $[Ca^{2+}]_i$ control and indirectly trigger a rise in $[Ca^{2+}]_i$ (Blatt & Grabov, 1997b).

Because local $[Ca^{2+}]_i$ 'hot spots' are thought to underpin Ca^{2+}-induced Ca^{2+} release in many cell types (Bootman & Berridge, 1995; Berridge, 1996), it is worth considering whether similar 'pH_i hot spots' might also occur, either superimposed on or associated with local changes in $[Ca^{2+}]_i$. This premise certainly deserves attention now in guard cells.

Factors affecting pH_i signal transmission

The efficacy of the pH_i signal itself is subject to downstream modulaton by protein phosphorylation. Indeed, its link to the *ABI1* gene of *Arabidopsis* provides one of the strongest arguments favouring a primary role for pH_i as a second messenger. The *abi1* gene is one of a family of *ab*scisic acid-*in*sensitive mutants originally characterised by its ability to germinate and grow on high concentrations of ABA (Koornneef, Reuling & Karssen, 1984) and displays ABA-related phenotypes, including an inability to control stomatal aperture. Sequencing and biochemical analyses have shown that the protein is a type 2C protein (serine/threonine) phosphatase, dependent on Mg^{2+} and insensitive to Ca^{2+} (Leung *et al.*, 1994; Meyer, Leube & Grill, 1994; Bertauche, Leung & Giraudat, 1996), and *in vivo* the mutant *abi1* (dominant negative) gene is known to confer an equivalent subset of phenotypes in transgenic *Nicotiana*, including the strong tendency to wilt (Armstrong *et al.*, 1995).

Armstrong *et al.* (1995) linked the mutant phenotype to the guard cell K^+ channels in *abi1*-transgenic tobacco. These studies showed that the *abi1* gene not only reduced $I_{K,out}$ in the absence of ABA, but eliminated the response of the current in its presence. The transgene was also found to interfere with ABA-evoked control of $I_{K,in}$ and to block stomatal closure. Furthermore, the background of $I_{K,out}$ activity, as well as $I_{K,in}$ and $I_{K,out}$ responses to ABA and stomatal closure, could be 'rescued' by treatments with protein kinase antagonists. So, how can *abi1*-transgene action be understood? The simplest explanation is that the (dominant) mutant protein phosphatase interferes with a wild-type homologue in tobacco preventing protein dephosphorylation and that the kinase antagonists are able to redress the resulting imbalance in phosphorylation to recover the ion channel sensitivities to ABA.

Armstrong *et al.* (1995) suggested that the abi1 phosphatase interfered with transmission of the pH_i signal, because guard cells from both the wild-type and *abi1*-transformed tobacco showed the characteristic alkalinisation of pH_i in ABA. Subsequent work has confirmed this prediction, demonstrating that experimentally imposed changes in pH_i, like the changes in pH_i in ABA, are largely ineffective in altering the K^+

currents of the *abi*-transgenic but not of wild-type plants (Blatt & Grabov, 1997a). Thus, it appears that the phosphatase either targets a common signalling intermediate with the pH_i signal to modify its efficacy in K^+ channel regulation or it is, itself, the target for pH_i regulation upstream from the K^+ channels.

One last point bears mention in relation to the specificity of the abi1 protein phosphatase. Subsequent work (Grabov *et al.*, 1997) has indicated that the anion channels, too, are subject to control by protein (de-)phosphorylation in their response to ABA. The anion channels, however, have proven insensitive to the presence of the *abi1* gene, with the current showing a similar degree of stimulation by ABA in both wild-type and *abi1*-transgenic tobacco and an equivalent sensitivity to the protein phosphatase antagonist calyculin A. Thus, while the anion channels are subject to regulation by protein phosphorylation, the effect appears to be specific to 1/2A-type protein phosphatases (Schmidt *et al.*, 1995; Grabov *et al.*, 1997) and distinct from the action of the 2C-type protein phosphatase.

Conclusion

There can be little doubt now about the signalling role of pH_i or its independence from $[Ca^{2+}]_i$-mediated regulation in stomatal guard cells. Nonethless, its emergence as a second messenger also serves to highlight our relative ignorance about the origin(s) of the pH_i changes as well as the nature of their interactions with other signalling circuits in the guard cells. In a broader context, too, the role of the pH_i signal is likely to be of particular importance to plants generally, and not simply because of their dependence on the H^+ electrochemical gradient in powering transmembrane solute transport. Like ABA-evoked changes in $[Ca^{2+}]_i$, the pH_i signal may be central to a stimulus–response coupling in guard cells, but the pH_i cascade is clearly only one of several signalling pathways with overlapping functions. Indeed, any current perspective of this, now expanding network of controls will almost certainly appear naive before the close of this decade. While the present task remains to characterise these and related signalling events, both upstream and downstream from the ionic second messengers, equally it must also be to understand how these events are integrated *in vivo*.

Acknowledgements

This work was supported by the Gatsby Charitable Foundation, the Royal Society, Human Frontiers Science Program grant RG303/95M and EC Biotech grant CT960062. AG is currently funded on BBSRC grant 32/C098-1.

References

Allan, A.C., Fricker, M.D., Ward, J.L., Beale, M.H. & Trewavas, A.J. (1994). Two transduction pathways mediate rapid effects of abscisic acid in *Commelina* guard cells. *Plant Cell* **6**, 1319–28.

Allen, G.J. & Sanders, D. (1995). Calcineurin, a type 2B protein phosphatase, modulates the Ca^{2+}–permeable slow vacuolar ion channel of stomatal guard cells. *Plant Cell* **7**, 1473–83.

Armstrong, F. & Blatt, M.R. (1995). Evidence for K^+ channel control in *Vicia* guard cells coupled by G–proteins to a 7TMS receptor. *Plant Journal* **8**, 187–98.

Armstrong, F., Leung, J., Grabov, A., Brearley, J., Giraudat, J. & Blatt, M.R. (1995). Sensitivity to abscisic acid of guard cell K^+ channels is suppressed by *abi1–1*, a mutant *Arabidopsis* gene encoding a putative protein phosphatase. *Proceedings of the National Academy of Sciences USA* **92**, 9520–4.

Assmann, S.M. (1993). Signal transduction in guard cells. *Annual Review of Cell Biology* **9**, 345–75.

Berridge, M.J. (1993). Inositol trisphosphate and calcium signalling. *Nature* **361**, 315–25.

Berridge, M.J. (1996). Microdomains and elemental events in calcium signaling. *Cell Calcium* **20**, 95–6.

Bertauche, N., Leung, J. & Giraudat, J. (1996). Protein phosphatase activity of the ABI1 (Abscisic acid–Insensitive 1) protein from *Arabidopsis thaliana*. *European Journal of Biochemistry* **241**, 193–200.

Bertl, A. & Felle, H. (1985). Cytoplasmic pH of root hair cells of *Sinapis alba* recorded by a pH–sensitive microelectrode. Does fusicoccin stimulate the proton pump by cytoplasmic acidification? *Journal of Experimental Botany* **36**, 1142–9.

Blackford, S., Rea, P.A. & Sanders, D. (1990). Voltage sensitivity of H^+/Ca^{2+} antiport in higher plant tonoplast suggests a role in vacuolar calcium accumulation. *Journal of Biological Chemistry* **265**, 9617–20.

Blatt, M.R. (1987). Electrical characteristics of stomatal guard cells: the contribution of ATP–dependent, 'electrogenic' transport revealed by current–voltage and difference–current–voltage analysis. *Journal of Membrane Biology* **98**, 257–74.

Blatt, M.R. (1988). Potassium–dependent bipolar gating of potassium channels in guard cells. *Journal of Membrane Biology* **102**, 235–46.

Blatt, M.R. (1990). Potassium channel currents in intact stomatal guard cells: rapid enhancement by abscisic acid. *Planta* **180**, 445–55.

Blatt, M.R. (1991). Ion channel gating in plants: physiological implications and integration for stomatal function. *Journal of Membrane Biology* **124**, 95–112.

Blatt, M.R. (1992). K$^+$ channels of stomatal guard cells: characteristics of the inward rectifier and its control by pH. *Journal of General Physiology* **99**, 615–44.

Blatt, M.R. & Armstrong, F. (1993). K$^+$ channels of stomatal guard cells: abscisic acid–evoked control of the outward rectifier mediated by cytoplasmic pH. *Planta* **191**, 330–41.

Blatt, M.R. & Clint, G.M. (1989). Mechanisms of fusicoccin action kinetic modification and inactivation of potassium channels in guard cells. *Planta* **178**, 509–23.

Blatt, M.R. & Grabov, A. (1997a). Signalling networks and guard cell function. *Physiologia Plantarum* **100**, 481–90.

Blatt, M.R. & Grabov, A. (1997b). Signal redundancy, gates and integration in the control of ion channels for stomatal movement. *Journal of Experimental Botany* **48**, 529–37.

Blatt, M.R. & Thiel, G. (1993). Hormonal control of ion channel gating. *Annual Review of Plant Physiology Molecular Biology* **44**, 543–67.

Blatt, M.R. & Thiel, G. (1994). K$^+$ channels of stomatal guard cells: bimodal control of the K$^+$ inward–rectifier evoked by auxin. *Plant Journal* **5**, 55–68.

Blatt, M.R., Thiel, G. & Trentham, D.R. (1990). Reversible inactivation of K$^+$ channels of *Vicia* stomatal guard cells following the photolysis of caged inositol 1,4,5–trisphosphate. *Nature* **346**, 766–9.

Bootman, M.D. & Berridge, M.J. (1995). The elemental principles of calcium signaling. *Cell* **83**, 675–8.

Brearley, J., Venis, M.A. & Blatt, M.R. (1997). CO$_2$–mediated control of K$^+$ and anion channels in guard cells of *Vicia*. *Planta* **203**, 145–54.

Busa, W.B. (1986). Mechanisms and consequences of pH–mediated cell regulation. *Annual Review of Physiology* **48**, 389–402.

Christensen, O. & Zeuthen, T. (1987). Maxi K$^+$ channels in leaky epithelia are regulated by intracellular Ca^{2+}, pH and membrane potential. *Pflügers Archiv, European Journal of Physiology* **408**, 249–59.

Clint, G.M. & Blatt, M.R. (1989). Mechanisms of fusicoccin action: evidence for concerted modulations of secondary K$^+$ transport in a higher–plant cell. *Planta* **178**, 495–508.

Colman, B. & Espie, G.S. (1985). CO$_2$ uptake and transport in leaf mesophyll cells. *Plant Cell and Environment* **8**, 449–57.

Coulter, K.L., Perier, F., Radeke, C.M. & Vandenberg, C.A. (1995). Identification and molecular localization of a pH–sensing domain for the inward rectifier potassium channel HIR. *Neuron* **15**, 1157–68.

Danthuluri, N.R., Kim, D. & Brock, T.A. (1990) Intracellular alkalinization leads to Ca^{2+} mobilization from agonist–sensitive pools in bovine aortic endothelial cells. *Journal of Biological Chemistry* **265**, 19071–6.

Davies, W.J. & Jones, H.G. (1991). *Abscisic acid: physiology and biochemistry*, pp. 1–266. Oxford: Bios Scientific.

Fairley, G.K. & Assmann, S.M. (1991). Evidence for G–protein regulation of inward potassium ion channel current in guard cells of fava bean. *Plant Cell* **3**, 1037–44.

Felle, H. (1988a). Cytoplasmic free calcium in *Riccia fluitans* L. and *Zea mays* L.: interaction of Ca^{2+} and pH? *Planta* **176**, 248–55.

Felle, H. (1988b). Auxin causes oscillations of cytosolic free calcium and pH in *Zea mays* coleoptiles. *Planta* **174**, 495–9.

Felle, H. (1989). pH as a second messenger in plants. In *Second messengers in plant growth and development*, ed. W.F. Boss & D.J. Morrè, pp. 145–66. New York: Alan R. Liss.

Fricker, M.D., Gilroy, S., Read, N.D. & Trewavas, A.J. (1991). Visualisation and measurement of the calcium message in guard cells. In *Molecular biology of plant development*, ed. W. Schuch & G. Jenkins, pp. 177–190. Cambridge: Cambridge University Press.

Frohnmeyer, H., Grabov, A. & Blatt, M.R. (1997). A role for the vacuole in auxin-mediated control of cytoplasmic pH by *Vicia* mesophyll and guard cell protoplasts. *Plant Journal* **13**, 109–16.

Gehring, C.A., Irving, H.R. & Parish, R.W. (1990). Effects of auxin and abscisic acid on cytosolic calcium and pH in plant cells. *Proceedings of the National Academy of Sciences USA* **87**, 9645–9.

Gilroy, S., Fricker, M.D., Read, N.D. & Trewavas, A.J. (1991). Role of calcium in signal transduction of *Commelina* guard cells. *Plant Cell* **3**, 333–44.

Goh, C.H., Oku, T. & Shimazaki, K. (1995). Properties of proton–pumping in response to blue–light and fusicoccin in guard–cell protoplasts isolated from adaxial epidermis of *Vicia* leaves. *Plant Physiology* **109**, 187–94.

Grabov, A. & Blatt, M.R. (1997). Parallel control of the inward–rectifier K^+ channel by cytosolic–free Ca^{2+} and pH in *Vicia* guard cells. *Planta* **201**, 84–95.

Grabov, A. & Blatt, M.R. (1998). Membrane voltage initiates Ca^{2+} waves and potentiates Ca^{2+} increases with abscisic acid in stomatal guard cells. *Proceedings of the National Academy of Sciences USA* **95**, 4778–83.

Grabov, A., Leung, J., Giraudat, J. & Blatt, M.R. (1997). Alteration of anion channel kinetics in wild type and *abi1–1* transgenic *Nicotiana benthamiana* guard cells by abscisic acid. *Plant Journal* **12**, 203–13.

Guern, J., Felle, H., Mathieu, Y. & Kurkdjian, A. (1991). Regulation of intracellular pH in plant–cells. *International Review of Cytology* **127**, 111–73.

Hedrich, R., Busch, H. & Raschke, K. (1990). Ca^{2+} and nucleotide dependent regulation of voltage dependent anion channels in the plasma membrane of guard cells. *EMBO Journal* **9**, 3889–92.

Hosoi, S., Iino, M. & Shimazaki, K. (1988). Outward–rectifying K^+ channels in stomatal guard cell protoplasts. *Plant Cell Physiology* **29**, 907–11.

Irving, H.R., Gehring, C.A. & Parish, R.W. (1992). Changes in cytosolic pH and calcium of guard cells precede stomatal movements. *Proceedings of the National Academy of Sciences USA* **89**, 1790–4.

Kasamo, K. (1979). Effect of abscisic acid on membrane–bound epidermal ATPase from tobacco leaves. *Plant Cell Physiology* **20**, 293–300.

Kasamo, K. (1981). Effect of abscisic acid on the K^+ efflux and membrane potential of *Nicotiana tabacum* L. leaf cells. *Plant Cell Physiology* **22**, 1257–67.

Kelly, W.B., Esser, J.E. & Schroeder, J.I. (1995). Effects of cytosolic calcium and limited, possible dual, effects of G–protein modulators on guard cell inward potassium channels. *Plant Journal* **8**, 479–89.

Koornneef, M., Reuling, G. & Karssen, C.M. (1984). The isolation and characterization of abscisic acid–insensitive mutants of *Arabidopsis thaliana*. *Physiologia Plantarum* **61**, 377–83.

Laurido, C., Candia, S., Wolff, D. & Latorre, R. (1991). Proton modulation of Ca^{2+}–activated K^+ channel from rat skeletal muscle incorporated into planar bilayers. *Journal of General Physiology* **98**, 1025–43.

Lemtiri-Chlieh, F. (1996). Effects of internal K^+ and ABA on the voltage–dependence and time–dependence of the outward K^+–rectifier in Vicia guard cells. *Journal of Membrane Biology* **153**, 105–16.

Lemtiri–Chlieh, F. & MacRobbie, E.A.C. (1994). Role of calcium in the modulation of Vicia guard cell potassium channels by abscisic acid: a patch–clamp study. *Journal of Membrane Biology* **137**, 99–107.

Lesage, F., Guillemare, E., Fink, M. *et al.* (1996). A pH–sensitive yeast outward rectifier K^+ channel with 2 pore domains and novel gating properties. *Journal of Biological Chemistry* **271**, 4183–7.

Leung, J., Bouvier–Durand, M., Morris, P.–C., Guerrier, D., Chefdor, F. & Giraudat, J. (1994). *Arabidopsis* ABA response gene *ABI1*: features of a calcium–modulated protein phosphatase. *Science* **264**, 1448–52.

Linder, B. & Raschke, K. (1992). A slow anion channel in guard cells, activating at large hyperpolarization, may be principal for stomatal closing. *FEBS Letters* **313**, 27–30.

MacRobbie, E.A.C. (1987). Ionic relations of guard cells. In *Stomatal function*, ed. E. Zeiger, G.D. Farquhar & I.R. Cowan, pp. 125–62. Stanford: Stanford University Press.

MacRobbie, E.A.C. (1988). Stomatal guard cells. In *Solute transport in plant cells and tissues*, ed. D.A. Baker & J.L. Hall, pp. 453–97. Harlow: Longman Press.

MacRobbie, E.A.C. (1991). Effect of ABA on ion transport and stomatal regulation. In *Abscisic acid physiology and biochemistry*, ed. W.J. Davies & H.G. Jones, pp. 153–68. Oxford: Bios Scientific.

MacRobbie, E.A.C. (1992). Calcium and ABA–induced stomatal closure. *Proceedings of the Royal Society of London, B, Biological Sciences* **338**, 5–18.

MacRobbie, E.A.C. (1995a). Effects of ABA on $^{86}Rb^+$ fluxes at plasmalemma and tonoplast of stomatal guard cells. *Plant Journal* **7**, 835–43.

MacRobbie, E.A.C. (1995b). ABA–induced ion efflux in stomatal guard–cells – multiple actions of aba inside and outside the cell. *Plant Journal* **7**, 565–76.

McAinsh, M.R., Brownlee, C. & Hetherington, A.M. (1990). Abscisic acid–induced elevation of guard cell cytosolic Ca^{2+} precedes stomatal closure. *Nature* **343**, 186–8.

McAinsh, M.R., Brownlee, C. & Hetherington, A.M. (1992). Visualizing changes in cytosolic–free Ca^{2+} during the response of stomatal guard cells to abscisic acid. *Plant Cell* **4**, 1113–22.

McAinsh, M.R., Webb, A.A.R., Taylor, J.E. & Hetherington, A.M. (1995). Stimulus–induced oscillations in guard cell cytosolic–free calcium. *Plant Cell* **7**, 1207–19.

Meyer, K., Leube, M.P. & Grill, E. (1994). A protein phosphatase 2C involved in ABA signal transduction in *Arabidopsis thaliana*. *Science* **264**, 1452–5.

Miedema, H. & Assmann, S.M. (1996). A membrane–delimited effect of internal pH on the K^+ outward rectifier of *Vicia faba* guard cells. *Journal of Membrane Biology* **154**, 227–37.

Moody, W. Jr (1984). Effects of intracellular H^+ on the electrical properties of excitable cells. *Annual Review of Neuroscience* **7**, 257–78.

Pei, Z.M., Ward, J.M., Harper, J.F. & Schroeder, J.I. (1996). A novel chloride channel in *Vicia faba* guard cell vacuoles activated by the serine/threonine kinase, CDPK. *EMBO Journal* **15**, 6564–74.

Penny, M.G. & Bowling, D.J.F. (1975). Direct determination of pH in the stomatal complex of *Commelina*. *Planta* **122**, 209–12.

Raschke, K. (1979). Movements of stomata. In *Encyclopedia of plant physiology, New Series*, Vol. 7, ed. W. Haupt & M.E. Feinleib, pp. 373–441. Berlin: Springer.

Raschke, K. (1987) Action of abscisic acid on guard cells. In *Stomatal function*, ed. E. Zeiger, G.D. Farhquar & I.R. Cowan, pp. 253–279. Stanford: Stanford University Press.

Schmidt, C., Schelle, I., Liao, Y.J. & Schroeder, J.I. (1995). Strong regulation of slow anion channels and abscisic acid signaling in guard cells by phosphorylation and dephosphorylation events. *Proceedings of the National Academy of Sciences USA* **92**, 9535–9.

Schroeder, J.I. & Hagiwara, S. (1989) Cytosolic calcium regulates ion channels in the plasma membrane of *Vicia faba* guard cells. *Nature* **338**, 427–30.

Schroeder, J.I. & Keller, B.U. (1992). Two types of anion channel currents in guard cells with distinct voltage regulation. *Proceedings of the National Academy of Sciences USA* **89**, 5025–9.

Schroeder, J.I., Raschke, K. & Neher, E. (1987). Voltage dependence of K$^+$ channels in guard–cell protoplasts. *Proceedings of the National Academy of Sciences USA* **84**, 4108–12.

Schulzlessdorf, B. & Hedrich, R. (1995). Protons and calcium modulate SV–type channels in the vacuolar lysosomal compartment – channel interaction with calmodulin inhibitors. *Planta* **197**, 655–71.

Schulzlessdorf, B., Lohse, G. & Hedrich, R. (1996). GCAC1 recognizes the pH gradient across the plasma membrane: a pH–sensitive and ATP–dependent anion channel links guard cell membrane potential to acid and energy metabolism. *Plant Journal* **10**, 993–1004.

Taylor, C.W. & Richardson, A. (1991). Structure and function of inositol trisphosphate receptors. *Pharmacology and Therapeutics* **51**, 97–137.

Thiel, G., Blatt, M.R., Fricker, M.D., White, I.R. & Millner, P.A. (1993). Modulation of K$^+$ channels in Vicia stomatal guard cells by peptide homologs to the auxin–binding protein C–terminus. *Proceedings of the National Academy of Sciences USA* **90**, 11493–7.

Thiel, G., MacRobbie, E.A.C. & Blatt, M.R. (1992). Membrane transport in stomatal guard cells: the importance of voltage control. *Journal of Membrane Biology* **126**, 1–18.

Tytgat, J., Nilius, B. & Carmeliet, E. (1990). Modulation of the T–type cardiac Ca^{2+} channel by changes in proton concentration. *Journal of Physiology* **96**, 973–90.

Ward, J.M., Pei, Z.M. & Schroeder, J.I. (1995). Roles of ion channels in initiation of signal transduction in higher plants. *Plant Cell* **7**, 833–44.

Ward, J.M. & Schroeder, J.I. (1994). Calcium–activated K$^+$ channels and calcium–induced calcium release by slow vacuolar ion channels in guard–cell vacuoles implicated in the control of stomatal closure. *Plant Cell* **6**, 669–83.

Webb, A.A.R., McAinsh, M.R., Mansfield, T.A. & Hetherington, A.M. (1996). Carbon dioxide induces increases in guard cell cytosolic free calcium. *Plant Journal* **9**, 297–304.

Wille, A. & Lucas, W. (1984). Ultrastructural and histochemical studies on guard cells. *Planta* **160**, 129–42.

Willmer, C. & Fricker, M.D. (1996). *Stomata*, pp. 1–375. London: Chapman and Hall.

Wu, W.H. & Assmann, S.M. (1994). A membrane–delimited pathway of G protein regulation of the guard cell inward K^+ channel. *Proceedings of the National Academy of Sciences USA* **91**, 6310–14.

J.A. RAVEN

pH regulation of plants with CO_2-concentrating mechanisms

Introduction

Inorganic carbon influx and O_2 efflux are generally the largest *net* solute fluxes across the plasmalemma of photosynthetic organisms, reaching values of several $\mu mol/m^2/s$ (Raven, 1988, 1995). The efflux of the O_2 produced by photosystem II of photosynthesis is thought to occur by diffusion through the lipid phase of the plasmalemma, and of other membranes between the site of O_2 production and the plasmalemma. The lipid solution mechanism also accounts for the entry (as CO_2) of inorganic carbon in the majority of species of photosynthetic organisms, and for the majority of global photosynthetic carbon assimilation (Raven, 1991, 1997a).

Diffusive entry of CO_2 and acid–base regulation

Diffusive entry of CO_2, followed by fixation in photosynthesis by the carboxylase activity of RUBISCO (ribulose bisphosphate carboxylase-oxygenase; 'the most abundant protein in the world': Ellis, 1979) and the photosynthetic carbon reduction cycle to form carbohydrate, causes no net H^+ uptake or production in the steady state, i.e. after the intracellular CO_2 level is decreased when illumination replaces CO_2 production by net CO_2 consumption. Even in the steady state of photosynthesis, there are H^+ fluxes and pH gradients associated with the CO_2 flux because carbonic anhydrase in the cytosol and the stroma enlists HCO_3^- as a transported inorganic carbon species which can move in parallel with CO_2, with, of course, the necessity for (buffered) H^+ to move in parallel with HCO_3^- (Raven & Glidewell, 1981; Cowan, 1986; Majeau, Arnoldo & Coleman, 1994; Price *et al.*, 1994). Downstream reactions of the immediate (carbohydrate) products of photosynthesis to produce new plant material involve net H^+ production with NH_4^+ as N source (approximately 1.3 H^+ per N assimilated) or (via symbiosis) N_2 as N source (greater than 0.3 H^+ per N assimilated), or net OH^- production with NO_3^- as N source (less than 0.70 OH^- per N) when SO_4^{2-} is the S source (Raven & Wollenweber, 1992). These H^+/OH^- fluxes across the plasmalemma are superimposed on any H^+ or OH^- fluxes at the plasmalemma associated with CO_2-concentrating mechanisms.

Evidence for the occurrence of CO_2-concentrating mechanisms and for the inorganic carbon species entering cells during the operation of CO_2-concentrating mechanisms

Gas exchange data

The diffusive entry of CO_2, followed by fixation by RUBISCO, has predictable characteristics in terms of gas exchange, provided the kinetics of the RUBISCO of the organism concerned are known, as well as the relation of the RUBISCO activity in the organism to the CO_2-saturated and light-saturated photosynthetic rate, and the pathway of phosphoglycolate metabolism (Raven, 1984, 1997a, 1997c). These characteristics include the $K_{1/2(CO_2)}$, and its O_2 dependence, and the CO_2 compensation concentration (the external CO_2 level reached when the organism is illuminated in a confined space) and its O_2 dependence. The competitive interaction of CO_2 and O_2 means that the $K_{1/2(CO_2)}$ is lower, and the CO_2 compensation concentration is higher, at higher O_2 concentrations when RUBISCO *in vivo* is supplied with CO_2 diffusion (Raven, 1984, 1997a, 1997c). Furthermore, the $K_{1/2(CO_2)}$ in vivo is greater than that *in vitro* when CO_2 entry is by diffusion, provided that the *in vitro* CO_2-saturated rate of carboxylation by RUBISCO does not exceed the *in vivo* light-saturated and CO_2-saturated rate of photosynthesis (Raven, 1984, 1997a, 1997c).

In some cases, the $K_{1/2(CO_2)}$ and the CO_2 compensation concentration, and their O_2 dependences, are all smaller than predicted. When the discrepancies are well outside the range of errors in measurements, then the assumption is that intracellular CO_2 is higher than the external concentration (or, for a plant in air, the equilibrium value in solution), i.e. a CO_2-concentrating mechanisms is operating. Such data are available for many free-living algae, algae endosymbiotic in invertebrates, many aquatic vascular plants, some terrestrial lichens and hornworts, and C_4 and crassulacean acid metabolism (CAM) plants (Raven, 1995, 1997a, 1997c; Al-Moghrabi *et al.*, 1996).

Estimation of intracellular inorganic carbon concentrations

For illuminated CAM plants, with their closed stomata, the intercellular (and hence intracellular) CO_2 and O_2 levels can be estimated by extracting gas samples from intercellular gas spaces (Raven *et al.*, 1994). The intercellular and intracellular CO_2 concentrations in CAM plants in the light – when they are refixing *via* RUBISCO the CO_2 produced by decarboxylation of the malic acid produced at night by dark CO_2 fixation – are high enough to saturate RUBISCO, even with the high intercellular and intracellular O_2 concentrations found in such plants in the light when stomata are closed.

For C$_4$ plants, the CO_2 contents of leaves have been obtained by rapid freezing of leaf samples in liquid N_2, with subsequent analysis of CO_2 after unfreezing in a sealed vessel. Sufficient CO_2 to saturate RUBISCO (even with the high O_2 levels likely in many C$_4$ plants: Raven *et al.*, 1994) can be calculated from this total inorganic carbon measurement if the inorganic pool of the CO_2 is assumed to be confined to the bundle sheath cells, and if it is further assumed that this CO_2 does not equilibrate with HCO_3^- at the pH of bundle sheath cell cytosol/stroma so that the CO_2/HCO_3^- ratio is much higher than the equilibrium value at pH 7.4–7.8 (see Raven, 1995).

This technique has only been used for a small number of the C$_4$ plants identified as such on structural, enzymic, $^{14}CO_2$ fixation product and/or natural abundance $^{13}C/^{12}C$ evidence, but always showed the predicted high internal CO_2 levels.

A number of methods (Raven, 1997a) have been used for 'lower plants' in which gas exchange data suggest the occurrence of a CO_2-concentrating mechanism. A widely used technique is silicone oil centrifugation or filtration of microalgae or isolated cells at short time intervals (seconds to tens of seconds) after addition of ^{14}C-inorganic carbon to the culture, with estimation of intracellular ^{14}C-inorganic carbon after correction for extracellular inorganic carbon. For aquatic macrophytes, the quantity of O_2 evolved after removal of *external* inorganic carbon has been used to estimate the *internal* inorganic carbon pool, assuming that one O_2 is evolved per inorganic carbon assimilated. Net CO_2 uptake upon illumination when photosynthetic CO_2 assimilation is blocked (e.g. with glycolaldehyde) has been used to detect CO_2-concentrating mechanisms in terrestrial lichens and hornworts. Analysis of the extent of CO_2 accumulation from these total internal inorganic carbon estimates requires assumptions about compartmentation, measurements (or assumptions) as to intracellular pH (pH$_i$), and assumptions about the extent of equilibration of CO_2 with other dissolved inorganic carbon species. In many cases, such analysis shows an internal accumulation of CO_2 which is sufficient to account for the gas exchange characteristics (see under 'Gas exchange data' above) in terms of the kinetics of RUBISCO and of NADPH and ATP production without the need for 'special pleading' in terms of compartmentation or lack of CO_2-HCO_3^-

equilibration. However, in some cases, and especially for marine algae, the intracellular accumulation of CO_2 assuming equilibrium between CO_2 and other inorganic carbon species is apparently inadequate to account for the gas exchange data in terms of a CO_2-concentrating mechanism under natural conditions (Raven, 1997a).

Inorganic carbon species entering the cell when CO_2-concentrating mechanisms are operating

There is little experimental support for CO_3^{2-} entry or H_2CO_3 entry, so the two contenders are CO_2 and HCO_3^- (Raven, 1997a). With a CO_2-concentrating mechanism, the CO_2 concentration gradient is such as to cause a net passive CO_2 efflux out of the compartment in which the high steady-state CO_2 level is maintained (Raven, 1997a; Al-Moghrabi et al., 1996). In eukaryotes, this does not immediately rule out passive CO_2 entry at the plasmalemma because active inorganic carbon influx at the plastid level (some algae?) and symplasmically into certain cell types (C_4 plants) can account for certain CO_2-concentrating mechanisms. CO_2 entry can be directly shown by real-time sampling of the solution for CO_2 by mass spectrometry and by isotope disequilibrium studies. HCO_3^- entry can also be addressed by isotope disequilibrium studies, and by demonstrating that the rate of inorganic carbon entry exceeds the (uncatalysed) rate at which the medium can supply CO_2 from HCO_3^-. All but the first of these (net CO_2 decrease) is vitiated by any external carbonic anhydrase activity. Clearly, HCO_3^- 'use' by extracellular conversion to CO_2 followed by CO_2 uptake will have a very different impact on intracellular acid–base balance from direct uptake of HCO_3^-, bearing in mind that CO_2 is the inorganic carbon species fixed by RUBISCO.

The conclusions from such measurements are that many aquatic organisms with CO_2-concentrating mechanisms have entry of both CO_2 and HCO_3^- across the plasmalemma. No relevant data seem to be available for terrestrial organisms with CO_2-concentrating mechanisms, although some of the data on aquatic organisms are relevant to atmospheric CO_2 uptake, e.g. for intertidal algae which are exposed to air on an approximately twice-daily basis. It is likely that diffusive entry of CO_2 by lipid solution occurs in C_4 and CAM plants. Mediated CO_2 uptake by passive uniport, or by primary or secondary active transport at the plasmalemma, is now (perhaps grudgingly) accepted as a means by which CO_2 enters cells of aquatic organisms with CO_2-concentrating mechanisms (Raven, 1970). Clearly, the consequences for whole-cell or tissue pH regulation of CO_2 entry followed by CO_2 fixation to produce (initially) a neutral reduced product are very different from those related to HCO_3^- entry followed by CO_2 fixation to produce an uncharged reduced product. In the case of CO_2 entry, no transplasmalemma net H^+ fluxes are involved in the photosynthetic production of a neutral reduced product. However, when HCO_3^- crosses the plasmalemma, one H^+ must enter (or one OH^- leave) per CO_2 fixed into a neutral reduced product *in the steady state* and one equivalent of positive charge must enter (or one equivalent of negative charge leave) per equivalent of *net* HCO_3^- accumulation during the initiation of the operation of a CO_2-concentrating mechanism (e.g. when the light is turned on).

Mechanisms of CO₂-concentrating mechanisms and their implications for acid–base regulation

Crassulacean acid metabolism

Crassulacean acid metabolism involves the storage in the vacuole of malic acid produced by fixation at night of exogenous CO_2 (entering the cells as CO_2) using neutral solutes (starch or vacuolar sugars) to provide the C_3 variety for phosphoenolpyruvate carboxylase (PEPC) activity. This involves a large (2–3 pH unit) decrease in vacuolar pH between end of day and end of night. De-acidification during the day removes malic acid from the vacuole (with an increase in vacuolar pH) with decarboxylation of malic acid, and the ultimate conversion of all four carbon atoms to carbohydrate, the three-carbon moiety by gluconeogenesis, and the CO_2 by RUBISCO and the carbon reduction cycle. This sequence involves no necessary H^+ exchange between cells and their aqueous medium (cell wall in terrestrial plants) (Raven, 1995).

C₄ pathway

Here, external CO_2 is fixed by PEPC in mesophyll cells in a sequence converting a C_3 monocarboxylate anion into a C_4 dicarboxylate anion. The C_4 anion diffuses through plasmodesmata to bundle sheath cells where it is decarboxylated; the C_3 anion product diffuses back into mesophyll cells to initiate another turn of the C_3–C_4 cycle, while the CO_2 is fixed by RUBISCO and the carbon reduction cycle in the mesophyll cells. Each turn of the C_3–C_4 cycle generates one H^+ in the mesophyll cell and consumes one H^+ in the bundle sheath cell (Osmond & Smith, 1976; Raven, 1995). Thus, for every CO_2 fixed (net) by RUBISCO, *at least* one H^+ is consumed in the bundle sheath cells, because some of the CO_2 delivered to the bundle sheath cells leaks out rather than being refixed by RUBISCO (Raven, 1995). One means by which inorganic carbon leakage is minimised is the absence of carbonic anhydrase in bundle sheath cells; this prevents rapid conversion of CO_2 (released by C_4 anion decarboxylation in bundle sheath cells) to HCO_3^- which could diffuse to mesophyll cells where carbonic anhydrase activity (needed to convert CO_2 derived from the atmosphere into HCO_3^-, the substrate for PEPC) occurs (Raven, 1995). This necessary absence of carbonic anhydrase (see Hewett-Emmett & Tashian, 1996) from bundle sheath cells means that HCO_3^- cannot be used as a quantitatively significant means of transport of OH- from bundle sheath to mesophyll cells during net CO_2 fixation by RUBISCO in bundle sheath cells (Raven, 1995). Other mobile buffers with appropriate pK_a values (phosphates, amino acids) are presumably involved. The pH difference between the cytosol of the bundle sheath (higher pH) and of the mesophyll (lower pH) cells of up to 1 pH unit has been computed as necessary to drive the flux of (buffered) H^+/OH^- of possibly up to 18 µmol H^+/OH^- per m^2 wall area per second through the plasmodesmata (Raven, 1995), but they have not been measured directly, e.g. by fluorescent indicators (Yin *et al.*, 1996). It is also not clear if perturbations of leaf cell pH (e.g. by gaseous NH_3) specifically influence C_4 metabolism *via* effects on OH^- movement from bundle sheath to mesophyll (Raven, 1995; Yin *et al.*, 1996).

CO_2-concentrating mechanisms based on active transport across membranes

CO_2-concentrating mechanisms based on active transport of inorganic carbon across membranes

Active influx of CO_2 at the plasmalemma, followed by fixation of CO_2 by RUBISCO and the carbon reduction cycle into a neutral reduced compound, has no effect on whole-cell acid–base balance if it involves the primary active transport (ATP-driven) mechanism of Rotatore, Lew and Colman (1992). Secondary active transport with H^+ as the driving ion would tend to acidify the cytosol. However, considerations of plasmalemma electrical capacitance, and of cytosol volume per unit plasmalemma area and cytosol H^+ buffer capacity, show that negligible cytosol pH change would occur before complete membrane depolarisation occurred (see Raven & Wollenweber, 1992). The plasmalemma H^+ pump driven by ATP (or Na^+/H^+ antiport driven by Na^+ primary active transport) would restore the membrane potential without a detectable change in cytosol pH.

Active influx of HCO_3^- at the plasmalemma followed by fixation of CO_2 by RUBISCO and the carbon reduction cycle into a neutral reduced compound produces one excess OH⁻ per carbon fixed. A 'simple' explanation would be HCO_3^-/OH^- antiport (1:1) at the plasmalemma. However, such a mechanism could not lead to an accumulation of CO_2 in the cell to a higher concentration than that in the medium, as can be seen from the analysis of Beardall and Raven (1981) using the Nernst and Henderson–Hasselbalch equations. Accumulation of CO_2 by secondary active transport (symport) of HCO_3^- with H^+ as driving ion would need an H^+/HCO_3^- ratio in excess of 1, leading to the requirement for active H^+ efflux which would maintain charge balance and, incidentally, acid–base balance. Data showing sulphonated stilbene inhibition of HCO_3^- use in some marine algae, allied in some cases with Cl^--dependence of HCO_3^- use, are consistent with some direct $HCO_3^-:OH^-$ or indirect ($HCO_3^-:Cl^-$ antiport; $Cl:H^+$ symport) exchange of $1HCO_3^-$ for 1OH- (see Raven, 1997a, 1997c; Larsson *et al.*, 1997). Ritchie, Nadolny and Larkum (1996) have ruled out, on kinetic and energetic grounds, H^+ and Na^+ (and K^+) as driving ions for HCO_3^- symport in the cyanobacterium *Synechococcus*, while Raven (1984) and Poole and Raven (1997) show that co-transport of HCO_3^- with H^+, Na^+ or Cl^- is very unlikely in the red macroalga *Porphyra* or the marine green macroalgae *Enteromorpha* or *Ulva*. In cyanobacteria, measurements of transmembrane electrical potential differences in the light with and without external HCO_3^- are consistent with primary active electrogenic HCO_3^- influx (Kaplan *et al.*, 1982; Ritchie *et al.*, 1996). Such electrogenic HCO_3^- influx increases the inwardly directed H^+ electrochemical potential difference, so that regulated passive H^+ influx *via* a channel or transporter could account for acid–base regulation. However, there is no direct evidence on this point.

There is some evidence for accumulation of inorganic carbon in the chloroplasts by mechanisms other than CO_2 entry and ionisation in response to an inside-alkaline pH difference, although the mechanism is unclear (Raven, 1997a). An attractive hypothesis as to the operation of CO_2-concentrating mechanisms based on active transport of inorganic carbon across membranes is that HCO_3^- is delivered to the RUBISCO-containing compartment regardless of the species (CO_2 or HCO_3^-) taken up at the outside of the membrane (plasmalemma of cyanobacteria; plastid envelope of chloroplasts), so that OH^- is taken up from the cytosol or stroma during CO_2 transport to these compartments across the membrane. CO_2 supply to RUBISCO involves closely regulated carbonic anhydrase activity in carboxysomes (cyanobacteria) or pyrenoids (in plastids of some eukaryotes with active inorganic carbon transport-based CO_2-concentrating mechanisms), both of which are major sites of occurrence of RUBISCO. CO_2 production from

HCO_3^- at the site of RUBISCO activity generates one OH^- per CO_2. The OH^- must return to the membrane where it is consumed in the CO_2 to HCO_3^- conversion in the CO_2 uptake/HCO_3^- release transporter, or is effluxed (or neutralised by entering H^+) in the case of HCO_3^- uptake/ HCO_3^- release. OH^- cannot move from RUBISCO to the membrane as HCO_3^- because the net HCO_3^- flux is from the membrane to RUBISCO, so (as for C_4 plants) phosphates and amino acids must be invoked as the mobile buffers (Raven, 1995). Computed (but not yet measured, e.g. using fluorescent indicators) pH differences required to drive these fluxes of up to 2.5 μmol OH^-/m^2 plasmalemma per second are less than 0.5 pH units (Raven, 1995; see also Smith, 1985).

Even if the hypothesis outlined above for a CO_2-concentrating mechanism which restricts CO_2 leakage (see also Fridlyand, Kaplan & Reinhold, 1996) proves to be correct for those organisms with overt inorganic carbon accumulation, it is clear that there is still a significant amount of work to be done on the acid–base regulation aspects of the mechanism, both in aqueous phases, as in C_4 plants, and at membranes.

CO_2-concentrating mechanisms based on HCO_3^- to CO_2 conversion in compartments whose pH is lower than that of compartments containing RUBISCO

In these mechanisms, HCO_3^- from the medium enters a compartment maintained at a low pH by active H^+ transport. Such a compartment has a higher *equilibrium* $CO_2 : HCO_3^-$ ratio than does the medium, and a higher *rate* of uncatalysed conversion of HCO_3^- (plus H^+) to CO_2 than in the medium. The CO_2 generated in this compartment moves (passively) into the compartment containing RUBISCO. Such a mechanism was first proposed and quantified by Walker, Smith and Cathers (1980) for extracellular acidic zones of some characeans and freshwater vascular plants (Prins *et al.*, 1982), with the plasmalemma as the H^+-pumping (P-type ATPase) membrane, and CO_2 diffusing to RUBISCO across the plasmalemma, cytosol and plastid membranes (and, of course, leaking to the medium). Pronina and Semenenko (1984, 1992) and Pronina *et al.* (1981) – cf. Raven (1997a, 1997b, 1997c) – suggested a similar mechanism for the thylakoid lumen (maintained acid in the light due to light-driven H^+ pumps), with passive HCO_3^- fluxes all the way from the medium into the thylakoid lumen, and CO_2 diffusing out of the thylakoid to RUBISCO. Raven (1997a, 1997c) suggested a similar role for vacuoles, with the acidic pH maintained by V-type H^+-ATPase. The volume of the thylakoid lumen is relatively so small that, despite its low pH, carbonic anhydrase is necessary to account for

CO_2 production at the rate at which it is consumed by photosynthesis; this may also be the case for some surface acid zones (Price *et al.*, 1985). The surface acid zone mechanism is well authenticated although it may act in series with some other CO_2-concentrating mechanism acting at the plasmalemma (Brechignac & Lucas, 1987) in the plastid (CO_2 release in a C_4-like pathway: Reiskind *et al.*, 1987) level. Some (but by no means all) data on carbonic anhydrase compartmentation are consistent with the thylakoid lumen hypothesis (Badger & Price, 1994; Karlsson *et al.*, 1996, 1998; Raven, 1997a, 1997b; Stemler, 1997). The vacuole hypothesis of CO_2 production (Raven, 1997a, 1997c) remains untested. All three mechanisms require very low carbonic anhydrase activity in the compartment which contains RUBISCO, so that CO_2 remains out of equilibrium with HCO_3^- (i.e. at a higher than equilibrium concentration) before it is fixed by RUBISCO. This also permits a CO_2-concentrating mechanism (in the strict sense of a CO_2-concentrating mechanism) to operate with a *lower total* inorganic carbon concentration averaged over all compartments inside than outside the cells (Raven, 1997a, 1997b).

The consequences for acid–base regulation of such mechanisms of HCO_3^- use are that OH- has to be disposed of (as for other means of HCO_3^- use). For the extracellular acid zones, OH^- generated in the cytosol must be (passively) excreted to some other zones on the surface; these often correspond to sites of $CaCO_3$ precipitation. For the thylakoid and vacuolar mechanisms, OH^- is generated in the cytosol (when CO_2 is generated in the vacuole) or stroma (when CO_2 is generated in the thylakoid lumen); OH^- efflux would (except at exceptionally high external pH values) be passive across the plasmalemma (and plastid envelope).

An internal means of OH^- disposal following HCO_3^- uptake and internal conversion to CO_2 is that of $CaCO_3$ precipitation. This occurs in coccolithophorids (members of the algal division Haptophyta), where it is sometimes considered to be a means of generating CO_2 (Brownlee *et al.*, 1994) according to the overall equation:

$$Ca^{2+} + 2HCO_3^- \rightarrow CaCO_3 + H_2O + CO_2$$

Brownlee *et al.* (1994) point out that the uncatalysed rate of production of CO_2 from HCO_3^- in coccolithophorid cells, granted the measured pH_i and HCO_3^-, is most probably inadequate to supply all of the CO_2 consumed by RUBISCO. It is very unlikely that the HCO_3^- concentration could be high enough, and the pH low enough, at the site of coccolith formation and also permit uncatalysed production of CO_2

from HCO_3^- at a rate sufficient to satisfy the demands of photosynthesis and also permit $CaCO_3$ precipitation (Brownlee *et al.*, 1994; Anning *et al.*, 1996), especially in view of the small fraction of the cell volume occupied by the Golgi/coccolith vesicle. Because the present evidence is that carbonic anhydrase in coccolithophorids is confined to the chloroplasts (Quiroga & González, 1993; Nimer, Guan & Merrett, 1994), we cannot look to carbonic anhydrase to catalyse CO_2 production in the Golgi/coccolith vesicle. These arguments strengthen the conclusion (Brownlee *et al.*, 1994) that the reaction in the Golgi/coccolith vesicle can be approximated by:

$$HCO_3^- + Ca^{2+} \rightarrow CaCO_3 + H^+$$

and that calcification *per se* does not generate CO_2, but rather functions to neutralise OH^- generated when HCO_3^- enters the cells and CO_2 is fixed. The energetics and stoichiometry of one Ca^{2+} entering the vesicle for each H^+ leaving can be most readily accommodated by a primary active $Ca^{2+} : H^+$ antiport driven by ATP (see McConnaughey, 1994) using a P-type ATPase such as occurs in endomembranes (Mintz & Guillain, 1997). A Ca^{2+}-stimulated ATPase has been demonstrated in coccolithophorid coccolith-forming vesicles (Wainwright, Kwon & González, 1992; Araki & González, 1994); however, this V-type ATPase does not catalyse a Ca^{2+} flux, but rather a Ca^{2+}-stimulated H^+ flux (Araki & González, 1994). Because the pH_i (determined with fluorescent indicators) and HCO_3^- concentration of coccolithophorids during exponential growth (Brownlee *et al.*, 1994; Anning *et al.*, 1996; Nimer, Merrett & Brownlee, 1996) are not such as to yield a significant *equilibrium* CO_2 concentration in excess of that in the medium, it is tempting to suggest a thylakoid lumen location for carbonic anhydrase in the coccolithophore plastids, permitting the mechanism to increase the CO_2 concentration around RUBISCO proposed by Pronina and Semenenko (1984, 1992), Pronina *et al.*, (1981), Fridlyand and Kaler (1987), Raven (1997a, 1997b, 1997c), and Karlsson *et al.* (1998). This is shown in Figure 1. This mechanism is consistent with most of the available data, but is *not* the most energy-efficient possibility for coccolith formation and photosynthesis in coccolithophorids (Brownlee *et al.*, 1994; Anning *et al.*, 1996). It requires that functions other than that of a minimum-energy means of acid–base regulation in HCO_3^--using photosynthesis (Anning *et al.*, 1996) be sought to explain the selective significance of coccolith production (e.g. light scattering; density regulation; avoidance of very high external pH (pH_e) values in dense blooms). As with OH^- flux from RUBISCO to the plasmalemma during

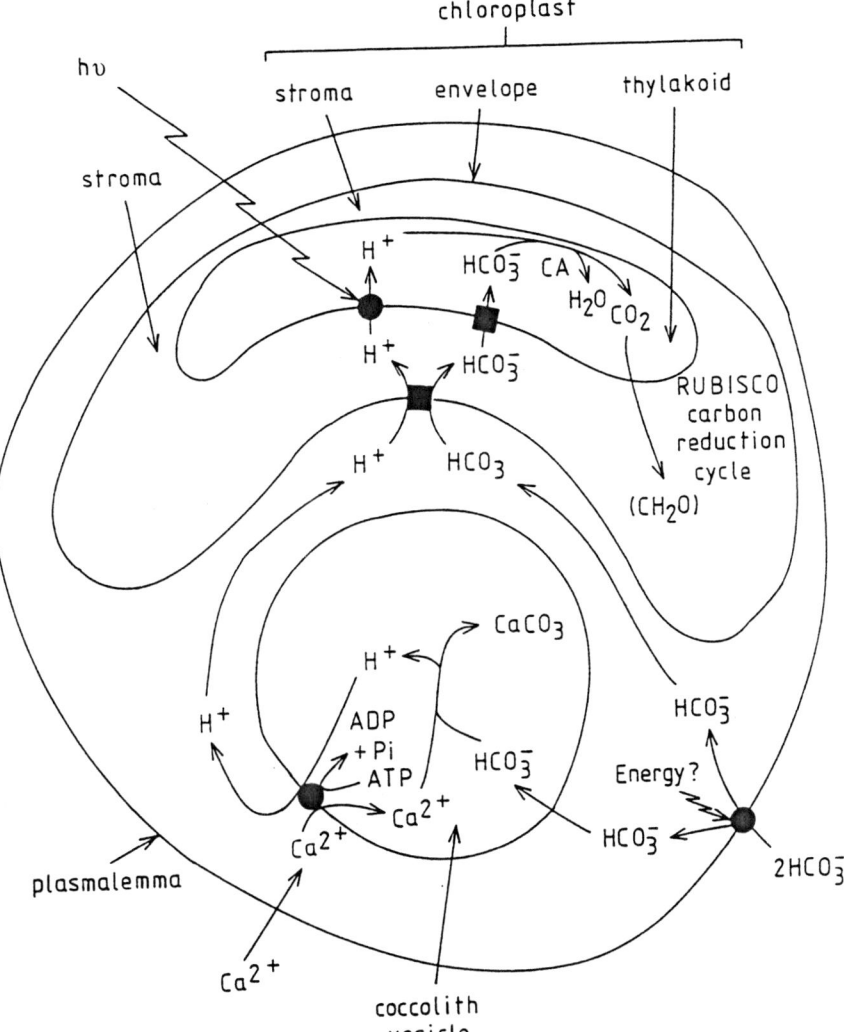

Fig. 1. Possible scheme for HCO₃⁻ use in photosynthesis and CaCO₃ precipitation in coccolithophores involving carbonic anhydrase (CA) in the thylakoid lumen and a 1Ca²⁺ : 1H⁺ antiporter ATPase in the coccolith vesicle membrane. Filled circles = primary active transport; filled squares = channel for passive uniport, or symport.

HCO$_3^-$ use, so with OH$^-$ flux from RUBISCO to the Golgi/coccolith vesicle, the flux of OH$^-$ through the cytosol cannot involve transport as HCO$_3^-$. There is also the problem of the necessary high fluxes of Ca^{2+} through the cytosol during coccolithogenesis, despite the low mean concentration in the cytosol of free (and chelated?) Ca^{2+} (Raven, 1980; Brownlee et al., 1994).

Conclusions

The need for transplasmalemma H$^+$/OH$^-$ fluxes in aquatic organisms with CO$_2$-concentrating mechanisms is less than had previously been thought because there is increasing evidence that inorganic carbon can enter cells as CO$_2$ as well as HCO$_3^-$ in plants with CO$_2$-concentrating mechanisms based on active transport of inorganic carbon at the plasmalemma.

All organisms with CO$_2$-concentrating mechanisms need significant OH$^-$/H$^+$ fluxes through the cytosol (and/or stroma); in all but CAM plants the H$^+$/OH$^-$ fluxes cannot be aided by HCO$_3^-$ fluxes (in the same direction as OH$^-$ flux) as the net HCO$_3^-$ flux is in the same direction as that of H$^+$, i.e. the opposite direction to that of OH$^-$.

The various proposed mechanisms underlying CO$_2$-concentrating mechanisms need more rigorous testing, especially with regard to the location of carbonic anhydrase activities, the mechanism of transmembrane active and passive inorganic carbon and H$^+$/OH$^-$ fluxes, and the magnitude of pH gradients in aqueous phases in steady-state photosynthesis.

Acknowledgements

Funding from BBSRC to study acid–base regulation and from NERC to study photosynthetic acquisition of inorganic carbon in aquatic plants is gratefully acknowledged.

References

Al-Moghrabi, S., Goiran, C., Allemand, D., Speziale, N. & Jaubert, J. (1996). Inorganic carbon uptake for photosynthesis by the symbiotic coral-dinoflagellate association. II. Mechanisms for bicarbonate uptake. *Journal of Experimental Marine Biology and Ecology* **199**, 227–48.

Anning, T., Nimer, N., Merrett, M.J. & Brownlee, C. (1996). Costs and benefits of calcification in coccolithophorids. *Journal of Marine Systems* **9**, 45–56.

Araki, Y. & Gonzàlez, E.L. (1998). V-type and P-type Ca^{2+}-stimulated

ATPase in a calcifying strain of *Pleurochrysis* sp. (Haptophyceae). *Journal of Phycology* **34**, 79–88.

Badger, M.R. & Price, G.D. (1994). The role of carbonic anhydrase in photosynthesis. *Annual Review of Plant Physiology and Plant Molecular Biology* **45**, 369–92.

Beardall, J. & Raven, J.A. (1981). Transport of inorganic carbon and the 'CO_2 concentrating mechanism' in *Chlorella emersonii* (Chlorophyceae). *Journal of Phycology* **92**, 1–20.

Brechignac, F. & Lucas, W.J. (1987). Photorespiration and internal CO_2 accumulation in *Chara corallina* as inferred from the influence of DIC and CO_2 on photosynthesis. *Plant Physiology* **83**, 63–9.

Brownlee, C., Nimer, N., Dong, L.F. & Merrett, M.J. (1994). Cellular regulation during calcification in *Emiliania huxleyi*. In *The haptophyte algae*, ed. J.C. Green & B.S.C. Leadbeater, pp. 133–48. Systematics Association Special Publication 51. Oxford: Clarendon Press.

Cowan, I.R. (1986). Economics of carbon fixation in higher plants. In *On the economy of plant form and function*, ed. T.J. Givnish, pp. 133–70. Cambridge: Cambridge University Press.

Ellis, R.J. (1979). The most abundant protein in the world. *Trends in Biochemical Science* **4**, 241–4.

Fridlyand, L.E. & Kaler, V.L. (1987). Possible CO_2 concentrating mechanisms in chloroplasts of C_3 plants: role of carbonic anhydrase. *General Physiology and Biophysics* **6**, 617–37.

Fridlyand, L.E., Kaplan, E. & Reinhold, L. (1996). Quantitative evolution of the role of a putative CO_2-scavenging entity on the cyanobacterial CO_2-concentrating mechanism. *Biosystems* **37**, 229–38.

Hewett-Emmett, D. & Tashian, R.E. (1996). Functional diversity, conservation and convergence in the evolution of the α-, β- and δ-carbonic anhydrase gene families. *Molecular Phylogeny and Evolution* **5**, 50–77.

Kaplan, A., Zenvirth, D., Reinhold, L. & Berry, J.A. (1982). Involvement of a primary electrogenic pump in the mechanism for HCO_3^- uptake by the cyanobacterium *Anabaena variabilis*. *Plant Physiology* **69**, 978–82.

Karlsson, J., Clarke, A., Chen, Z., Mason, C.B., Moroney, J.V. & Samuelsson, G. (1996). Chloroplastic carbonic anhydrase of *Chlamydomonas reinhardtii*: molecular, cloning, sequencing and western blot analysis of high-C_i-requiring mutants. In *Abstracts of the First European Phycological Congress, Cologne, August 1996*, Abstract 248, p. 36. Cambridge: Cambridge University Press.

Karlsson, J., Clarke, A.C., Chen, Z-Y. *et al.* (1998). A novel α-type carbonic anhydrase associated with the thylakoid membrane in *Chlamydomonas reinhardtii*, is required for growth at ambient CO_2. *EMBO Journal* **17**, 1208–16.

Larsson, C., Axelsson, L., Ryberg, H. & Beer, S. (1997). Photosynthetic carbon utilization by *Enteromorpha intestinalis* (Chlorophyta) from a Swedish rockpool. *European Journal of Phycology* **32**, 49–54.

Majeau, N., Arnoldo, M. & Coleman, J.R. (1994). Modification of carbonic anhydrase activity by antisense and overexpression constructs in transgenic tobacco. *Plant Molecular Biology* **25**, 377–85.

McConnaughey, T.A. (1994). Calcification, photosynthesis and global carbon cycles. In *Past and present biomineralization processes: consideration about the carbon cycle*, ed. F. Doumerge, pp. 137–58. Monaco: Monaco Musée Oceanographique.

Mintz, E. & Guillain, F. (1997). Ca^{2+} transport by the sarcoplasmic reticulum ATPase. *Biochimica et Biophysica Acta* **1318**, 52–70.

Nimer, N.A., Guan, Q. & Merrett, M.J. (1994). Extra- and intracellular carbonic anhydrase in relation to culture age in a high-calcifying strain of *Emiliania huxleyi* Lohman. *New Phytologist* **126**, 611–17.

Nimer, N.A., Merrett, M.J. & Brownlee, C. (1996). Inorganic carbon transport in relation to culture age and inorganic carbon concentration in a high-calcifying strain of *Emiliania huxleyi* (Prymnesiophyceae). *Journal of Phycology* **32**, 813–18.

Osmond, C.B. & Smith, F.A. (1976). Symplasmic transport of metabolites during C_4-photosynthesis. In *Intercellular communication in plants: studies on plasmodesmata*, ed. B.E.S. Gunning & A.W. Polards, pp. 229–41. Berlin: Springer Verlag.

Poole, L.J. & Raven, J.A. (1997). The biology of *Enteromorpha*. *Advances in Botanical Research* **12**, 1–148.

Price, G.D., Badger, M.R., Bassett, M.E. & Whitecross, M.I. (1985). Involvement of plasmalemmasomes and carbonic anhydrase in photosynthetic utilization of bicarbonate in *Chara corallina*. *Australian Journal of Plant Physiology* **12**, 241–56.

Price, G.D., von Caemmerer, S., Evans, J.R. *et al.* (1994). Specific reduction of chloroplast carbonic anhydrase activity by antisense RNA in transgenic tobacco plants has a minor effect on photosynthetic CO_2 assimilation. *Planta* **193**, 331–40.

Prins, H.B.A., Snel, J.F.H., Zanstra, P.E. & Helder, R.J. (1982). The mechanism of bicarbonate assimilation by the polar leaves of *Potamogeton* and *Elodea*. CO_2 concentration at the leaf surface. *Plant, Cell and Environment* **5**, 207–14.

Pronina, N.A., Avramova, S., Georgiev, D. & Semenenko, V.E. (1981). A pattern of carbonic anhydrase activity in *Chlorella* and *Scenedesmus* on cell adaptation to high light intensity and low CO_2 concentration. *Fiziologia Rastenii* **28**, 43–52.

Pronina, N.A. & Semenenko, V.E. (1984). Localization of membrane bound and soluble carbonic anhydrase in the *Chlorella* cell. *Fiziologia Rastenii* **31**, 241–51.

Pronina, N.A. & Semenenko, V.E. (1992). Role of the pyrenoid in concentration, generation and fixation of CO_2 in the chloroplast of microalgae. *Soviet Plant Physiology* **39**, 470–6.

Quiroga, O. & Gonzàlez, E.L. (1993). Carbonic anhydrase in the chloroplast of a coccolithophorid (Prymnesiophyceae). *Journal of Phycology* **29**, 321–4.

Raven, J.A. (1970). Exogenous inorganic carbon sources in plant photosynthesis. *Biological Reviews* **45**, 167–221.

Raven, J.A. (1980). Nutrient transport in microalgae. *Advances in Microbial Physiology* **21**, 47–226.

Raven, J.A. (1984). *Energetics and transport in aquatic plants.* New York: A.R. Liss.

Raven, J.A. (1988). Algae. In *Solute transport in plant cells and tissues*, ed. D.A. Baker & J.L. Hall, pp. 166–219. London: Longmans.

Raven, J.A. (1991). Implications of inorganic C utilization: ecology, evolution and geochemistry. *Canadian Journal of Botany* **69**, 908–24.

Raven, J.A. (1995). Symplasmic proton fluxes in photosynthesizing and developing plant tissues. *Biological Reviews* **70**, 189–224.

Raven, J.A. (1997a). Inorganic carbon acquisition by marine autotrophs. *Advances in Botanical Research* **27**, 85–209.

Raven, J.A. (1997b). CO_2 concentrating mechanisms: a direct role for thylakoid lumen acidification? *Plant, Cell and Environment* **20**, 147–54.

Raven, J.A. (1997c). Putting the C in phycology. *European Journal of Phycology* **32**, 319–33.

Raven, J.A. & Glidewell, S.M. (1981). Processes limiting photosynthetic conductance. In *Physiological processes limiting plant productivity*, ed. C.B. Johnson, pp. 109–36. London: Butterworths.

Raven, J.A., Johnston, A.M., Kübler, J. & Parsons, R. (1994). The influence of natural and experimental high O_2 concentrations on O_2-evolving photolithotrophs. *Biological Reviews* **69**, 271–86.

Raven, J.A. & Wollenweber, B. (1992). Temporal and spatial aspects of acid–base regulation. In *Current topics in plant biochemistry and physiology* **11**, ed. D.D. Randall, R.E. Sharp, A.J. Novacky & D.G. Blevins, pp. 220–98. University of Missouri-Columbia: Interdisciplinary Plant Group.

Reiskind, J.B., Madsen, T.V., van Ginkel, L.C. & Bowes, G. (1997). Evidence that inducible C_4-type photosynthesis is a chloroplastic CO_2-concentrating mechanism in *Hydrilla*, a submersed monocot. *Plant, Cell and Environment* **20**, 211–20.

Ritchie, R.J., Nadolny, C. & Larkum, A.W.D. (1996). Driving forces for bicarbonate transport in the cyanobacterium *Synechococcus* R-2 (PCC 7942). *Plant Physiology* **112**, 1573–84.

Rotatore, C., Lew, R.R. & Colman, B. (1992). Active uptake of CO_2 during photosynthesis in the green alga *Eremosphaera viridis* is mediated by a CO_2-ATPase. *Planta* **188**, 539–45.

Smith, F.A. (1985). Historical perspective on HCO_3^- assimilation. In *Inorganic carbon uptake by aquatic photosynthetic organisms*, ed. W.J. Lucas & J.A. Berry, pp. 1–15. Rockville, MD: American Society of Plant Physiologists.

Stemler, A.J. (1997). The case for chloroplast thylakoid carbonic anhydrase. *Physiologia Plantarum* **99**, 348–53.

Wainwright, I.M., Kwon, D.-K. & Gonzàlez, E. (1992). Isolation and characterization of golgi from *Coccolithus pelagicus* (Prymnesiophyceae). *Journal of Phycology* **28**, 643–8.

Walker, N.A., Smith, F.A. & Cathers, I.R. (1980). Bicarbonate assimilation by freshwater charophytes and higher plants. I. Membrane transport of bicarbonate ions is not proven. *Journal of Membrane Biology* **57**, 51–8.

Yin, Z-H., Kaiser, W.M., Heber, U. & Raven. J.A. (1996). Acquisition and assimilation of gaseous ammonia as revealed by intracellular pH changes in leaves of higher plants. *Planta* **200**, 380–7.

R.G. RATCLIFFE

Intracellular pH regulation in plants under anoxia

Introduction

The pH of an intracellular compartment (pH_i) depends on its buffering capacity, i.e. the concentration of weak acids, and on its ionic composition (Stewart, 1983). These properties are determined by metabolism and ion transport and since many biochemical and biophysical processes are sensitive to pH, it is reasonable to assume that any tendency for the pH to drift away from some notionally optimal or normal value will be corrected by altering the balance between the proton-consuming and proton-generating processes within the cell. In fact, Raven (1986) showed that pH regulation is an essential requirement for the growth of all plant cells; and, more generally, the ability to maintain electrochemical potential differences for the proton across membranes ($\Delta\bar{\mu}_H{}^+$), with the underlying contribution from any difference in pH, is a fundamental requirement of cellular energetics (Nicholls & Ferguson, 1992).

Intracellular pH values in plant cells can be measured by several methods (Guern et al., 1991), and most of the recent work has been done using microelectrodes, fluorescent probes and nuclear magnetic resonance (NMR) spectroscopy. Microelectrodes allow fast, real-time monitoring of single cells, as well as the simultaneous measurement of the functionally related membrane potential (Felle & Bertl, 1986; Felle, 1987, 1993); fluorescent probes allow non-invasive measurements of subcellular pH values (Kosegarten et al., 1997) and, in conjunction with laser scanning confocal microscopy, allow the construction of pH maps within cells and tissues (Gibbon & Kropf, 1994); and NMR provides a range of methods for measuring cytoplasmic and vacuolar pH values, while simultaneously recording other metabolically important information (Ratcliffe, 1994). The various methods all have their strengths and weaknesses (Guern et al., 1991; Felle, 1993; Ratcliffe, 1994), but there is generally reasonable agreement between the pH values that they generate (for example, see Gibbon & Kropf, 1994). The same strengths and weaknesses affect the versatility with which the three methods can be used to detect pH changes in response to physiological perturbations, but again the different measurements seem to be consistent when direct comparisons are possible.

While pH_i regulation is important, and while pH measurements can be made relatively easily, a full understanding of the phenomenon in plants remains elusive. The difficulty arises because of the multiplicity of the processes that contribute to the proton balance: any metabolic event or ion transport process that alters the intracellular buffering capacity or the ionic composition can influence the pH. The problem is analogous to the situation faced in the analysis of metabolic pathways (ap Rees & Hill, 1994; Fell, 1997). Here, it has been established that the control of the flux through a pathway is distributed throughout the pathway, and that the flux control coefficients assigned to particular steps are likely to be sensitive to the prevailing physiological conditions. As a result, the concepts of regulatory enzyme and rate-determining step have had to be abandoned in the analysis of metabolic regulation. Similarly with pH regulation: many biochemical and biophysical events influence the buffering capacity and ionic composition of an intracellular compartment, and the relative importance of these processes is likely to vary with the conditions experienced by the cell. Thus, while it is tempting to assert that pH regulation is largely dependent on some process, for example the operation of the plasma membrane H^+-ATPase (Michelet & Boutry, 1995), such statements are unlikely to be universally true.

To assess the significance of specific biochemical and biophysical processes for pH regulation, it is necessary to design experiments that highlight the pH-perturbing or pH-stabilising effects of the process of interest. A good example, and one which happens to demonstrate a significant contribution from the plasma membrane H^+-ATPase, can be found in an NMR investigation of cytoplasmic pH regulation in an *Acer pseudoplatanus* cell suspension (Gout, Bligny & Douce, 1992). The cytoplasmic pH of a plant cell is usually observed to be tightly regulated over a range of external pH values (Fox & Ratcliffe, 1990) and in *A. pseudoplatanus* the cytoplasmic pH was independent of the external pH (pH_e) between pH 4.5 and 7.5. However, the cytoplasmic pH decreased at pH_e values below 4.5, and this was associated with an increase in the rate of O_2 consumption and a decrease in the ATP concentration. The loss of pH control occurred when the O_2 consumption rate reached the uncoupled O_2 uptake rate, and it was concluded that the plasma membrane H^+-ATPase was no longer able to combat the influx of H^+ at low pH because of a shortage of ATP. In critical support of this conclusion, it was shown that resupplying inorganic phosphate to phosphorus-starved cells allowed the cytoplasmic pH to recover to its normal value of 7.5 at pH_e values of 4.5 and 6.0 only after the ATP level had itself recovered to its usual value (Fig. 1). In related experiments, it was shown that citrate and malate were synthesised in response to an alkalinisation of the cytoplasm, as would be predicted by a biochemical pH-stat model (Davies, 1986), but that the kinetics of the process were too slow to account for the response of the cytoplasmic pH to changes in the pH_e. Thus, these experiments provided strong evidence for the importance of the plasma membrane H^+-ATPase – and good evidence

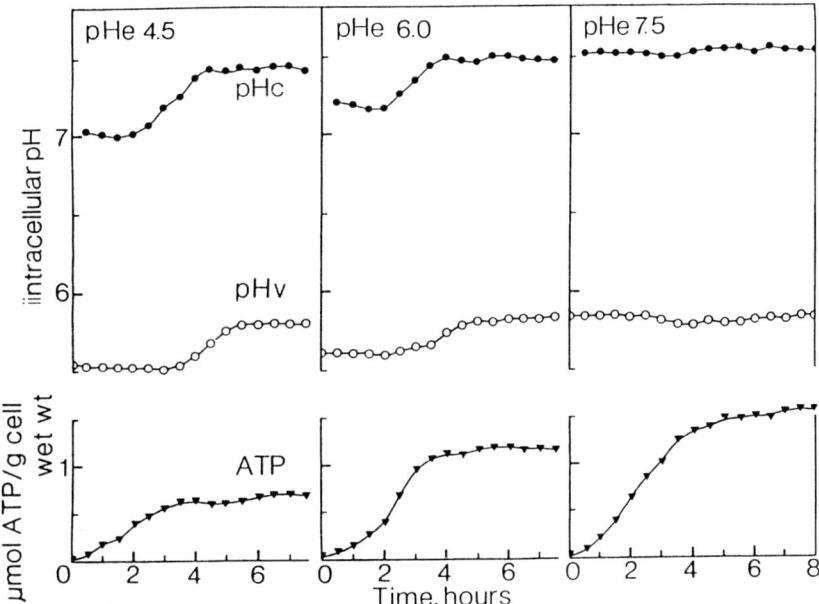

Fig. 1. Recovery of the ATP level, the cytoplasmic pH (pHc) and the vacuolar pH (pHv) following the resupply of inorganic phosphate to a phosphorus-starved *Acer pseudoplatanus* cell suspension at different pHe values. The phosphorus-starved cells have a very low ATP level and they are unable to maintain a normal pHc at acidic pHe values. Adding inorganic phosphate to the suspending medium increased the ATP level and this allowed pHc to recover to 7.5. These *in vivo* ^{31}P-NMR results show the importance of the plasma membrane H^+-ATPase in preventing an acidification of the cytoplasm at low pH_e. Reprinted from Gout *et al.* (1992) by permission of the American Society for Biochemistry and Molecular Biology.

against a significant role for a biochemical pH-stat involving the synthesis of organic acids – in the regulation of pH_i in *A. pseudoplatanus* cells in response to changes in pH_e (Gout *et al.*, 1992).

It is unclear how far one can extrapolate the conclusions on *A. pseudoplatanus* to other plants and other situations, and the aim of this chapter is to review recent work on pH_i regulation in anoxic plant tissues.

Cytoplasmic pH changes under anoxia

An acidification of the cytoplasm is commonly observed when plant cells are deprived of an adequate supply of O_2 (Fig. 2A). For example, in flooding intolerant maize root tips, the imposition of anoxia typically leads to an immediate decrease of 0.5 to 0.6 pH units in the cytoplasmic pH (Saint-Ges *et al.*, 1991; Roberts *et al.*, 1992; Fox, McCallan & Ratcliffe, 1995b). This pH response is readily observed experimentally, e.g. by NMR methods (Fig. 2A; Ratcliffe, 1995, 1997; Roberts & Xia, 1996) or by microelectrodes (Felle, 1996), and it is associated with important changes in metabolism (Ratcliffe, 1995, 1997; Roberts & Xia, 1996) and gene expression (Sachs, Subbaiah & Saab, 1996). The initial fall in the cytoplasmic pH of the maize root tip, which occurs within the first 15 to 20 minutes of O_2 deprivation, is followed by a period of stabilisation or partial recovery, but this eventually gives way to a further slow acidification of the cytoplasm that leads to cell death (Roberts *et al.*, 1984a).

While the pH response sketched out in the previous paragraph is typical, several factors can influence the extent of the cytoplasmic acidification following the onset of anoxia. First, the size of the initial fall in pH is species and tissue dependent. Thus, rice shoots showed a

---

Fig. 2. Graphs showing the effect of a switch from aerobic to anaerobic conditions on (**A**) the cytoplasmic pH and (**B**) the rate of ethanol production for maize root tips incubated at pH 10 in the presence (●) and absence (o) of a permeant weak base. The cytoplasmic pH, measured by *in vivo* ^{31}P-NMR, fell rapidly following the switch to anoxia at time zero, but the presence of methylamine caused the pH to recover to an aerobic value. This resulted in a decrease in the synthesis of ethanol, showing that pH is a regulator of pyruvate decarboxylase activity *in vivo*. Reprinted from Fox *et al.* (1995b) by permission of Springer Verlag.

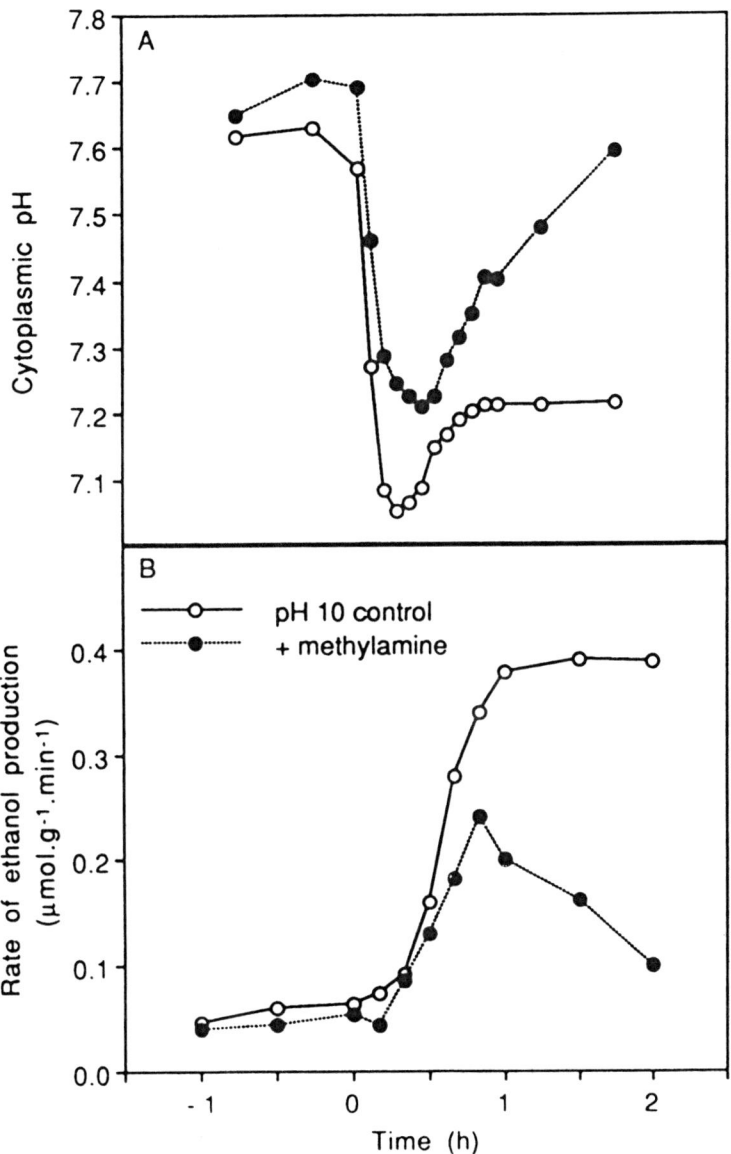

Fig. 2.

smaller initial cytoplasmic acidification (0.4 pH units) than wheat roots (0.8 pH units) (Menegus *et al.*, 1991); and tissues from flooding tolerant species often show only a very slight acidification, for example less than 0.2 pH units in rice coleoptiles (Fan, Lane & Higashi, 1992) and the stem tissue of *Potamogeton pectinatus* (M.B. Jackson, R.G. Ratcliffe & J.E. Summers, unpublished results). Secondly, the fall in the cytoplasmic pH observed in flooding intolerant species can be reduced by using a hypoxic pretreatment to acclimatise the plant to a low O_2 level. This strategy, which mimics the gradual depletion of the O_2 supply that is likely to occur during the onset of anaerobiosis in the field, has a profound effect on the response of maize roots (Saglio, Drew & Pradet, 1988; Johnson, Cobb & Drew, 1989; Hole *et al.*, 1992; Xia & Saglio, 1992) and wheat roots (Waters *et al.*, 1991b) to anoxia, and leads to a higher cytoplasmic pH in the roots of anoxic acclimated maize (Xia & Roberts, 1994). Thirdly, the pH response can be affected by other environmental factors. For example, the stabilisation of the cytoplasmic pH in maize root tips under anoxia is affected by pH_e: at pH 10, the cytoplasmic pH passed through a minimum before recovering to pH 7.2 (Fig. 2A); whereas at pH 4, the cytoplasmic pH fell to 7.0 and showed no sign of recovery (Fox *et al.*, 1995b). This result, and the observations of Xia and Roberts (1996), who reported that low pH_e also promotes cytoplasmic acidosis in acclimated maize root tips, is consistent with the general conclusion that the adverse effects of anoxia are likely to be influenced by a range of environmental factors as well as by the shortage of O_2 (Waters *et al.*, 1991a).

The changes in cytoplasmic pH under anoxia can be expected to have an impact on the activities of a whole range of enzymes with non-zero elasticity coefficients for H^+ in the physiological pH range; and they are also important because cytoplasmic acidosis correlates inversely with survival under anoxia (Roberts *et al.*, 1984a; Xia & Roberts, 1994, 1996). Thus, avoiding the uncontrolled onset of cytoplasmic acidosis is likely to be an important strategy for survival in the absence of O_2, and this focuses attention on the origins of the pH change and the mechanisms that allow the pH of the cytoplasm to be maintained within acceptable limits. It has already been argued that pH regulation under anoxia necessarily depends on all the biochemical and biophysical events that contribute to the overall proton balance in the cytoplasm (Ratcliffe, 1995), and recent progress towards highlighting the contributions of a number of these processes is described in the following sections. It is convenient to retain the distinction between biochemical mechanisms involving metabolism and biophysical processes involving ion transport for this purpose, although ultimately this distinction has no fundamental significance in relation to the factors that determine the pH of an intracellular compartment.

Biochemical pH regulation under anoxia

Many enzyme activities are pH dependent and this leads to the possibility of self-compensating metabolic networks in which metabolism shifts towards H^+ consumption as pH falls, and H^+ production as pH increases (Davies, 1986; Raven, 1986). Davies, Grego and Kenworthy (1974) proposed many years ago that a biochemical pH-stat of this kind played an important role in pH regulation in plants under anoxia, and it remains a dominant idea to this day. The model envisaged that acidification of the cytoplasm would simultaneously inhibit lactate dehydrogenase, with its alkaline pH optimum and its pH-dependent sensitivity to ATP, and activate pyruvate decarboxylase, with its acidic pH optimum, leading to a switch from the production of lactate to the production of ethanol and a concomitant stabilisation of the cytoplasmic pH (Fig. 3). *In vivo* NMR methods have been invaluable in establishing the validity of several aspects of this model (Ratcliffe, 1995, 1997; Roberts & Xia, 1996), and in particular in demonstrating the role of pH as an *in vivo* regulator of pyruvate decarboxylase (see Fig. 2B; Roberts *et al.*, 1984b; Fox *et al.*, 1995b). Thus, there is little doubt that a biochemical pH-stat does operate, and that a fall in cytoplasmic pH under anoxia can trigger the switch from one fermentation pathway to another.

However, it would be incorrect to assume that the Davies model is

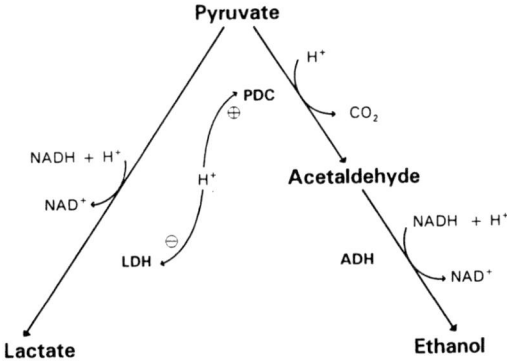

Fig. 3. The biochemical pH-stat model for the involvement of H^+ in the switch from lactate to ethanol as a fermentation end-product. ADH, alcohol dehyrogenase; LDH, lactate dehydrogenase; PDC, pyruvate decarboxylase.

able to provide a complete understanding of pH_i regulation under anoxia, and several difficulties have emerged in the light of more detailed analyses of the metabolic response to anoxia (Ratcliffe, 1995). First, the model assumes that the acidification of the cytoplasm is entirely due to the fermentation of glucose to lactate, which, with the associated hydrolysis of the ATP formed by glycolysis, leads to the production of two H^+ per glucose molecule (Hochachka & Mommsen, 1983). In fact, three independent analyses of the metabolic response of anoxic maize root tips indicated that lactate production was insufficient, just right or too great to explain the initial fall in pH (for a review, see Ratcliffe, 1995), and while fermentation to lactate clearly introduces a new source of cytoplasmic protons, the evidence does not suggest that the acidification of the cytoplasm is the result of this process alone. Secondly, the model assumes that lactate production is a transient process. While this is true in some species, lactate production can also be negligible, as in rice shoots (Menegus *et al.*, 1991), or prolonged, as in *Limonium* spp. (Rivoal & Hanson, 1993) and in acclimated maize root tips (Xia & Saglio, 1992; Xia & Roberts, 1994). Finally, the model assumes that cytoplasmic pH regulation during the onset of anoxia can be completely understood in terms of the biochemical pH-stat involving lactate dehydrogenase and pyruvate decarboxylase; whereas more recent evidence indicates that other mechanisms, including lactate efflux from lactate-accumulating tissues (Xia & Saglio, 1992; Rivoal & Hanson, 1993; Xia & Roberts, 1994), need to be considered.

Despite these reservations about the generality of the Davies model, and irrespective of the origin of the cytoplasmic acidification that occurs under anoxia, it is clear that the activation of proton-consuming pathways could be an effective mechanism for combating a fall in pH. Recent work in this area has been aimed at identifying pathways that can be activated by a fall in pH, and by continuing efforts to establish their quantitative significance. Roberts *et al.* (1992) used NMR methods to assess the contribution to pH regulation in anoxic maize root tips of several pathways leading to ethanol, lactate and alanine, and concluded: (i) that the synthesis of lactate and alanine accounted for the cytoplasmic acidification following the onset of anoxia; (ii) that the fall in cytoplasmic pH was limited by the activation of both pyruvate decarboxylase and malic enzyme; and (iii) that the proton-consuming synthesis of γ-aminobutyrate (GABA) from glutamate was quantitatively insignificant during the initial response to O_2 deprivation.

In related work, NMR has been used to demonstrate the pH-dependent activation of the pathway to GABA in hypoxic carrot cells (Carroll *et al.*, 1994) and in *Datura stramonium* root cultures (Ford,

Ratcliffe & Robins, 1996). Evidence for the direct involvement of pH in the regulation of glutamate decarboxylase was obtained by manipulating the cytoplasmic pH during the anoxic response, and it is clear from these and other experiments (Crawford *et al.*, 1994) that activation of the decarboxylation of glutamate has the potential to contribute to pH regulation under conditions that lead to cytoplasmic acidification. The fact that this mechanism is quantitatively insignificant during the *initial* stabilisation of the cytoplasmic pH in maize root tips (Roberts *et al.*, 1992; Roberts & Xia, 1996) does not alter the fact that the synthesis of GABA over longer periods of anoxia will necessarily assist in the maintenance of a relatively high cytoplasmic pH.

Nitrate reduction and assimilation constitute another potentially important mechanism for the biochemical consumption of protons during anaerobiosis. An increased nitrate supply prolongs the survival of anoxic maize root tips (Roberts, Andrade & Anderson, 1985), and nitrate is certainly utilised during the anaerobic germination of rice seeds (Reggiani *et al.*, 1993). Both of these observations suggest that nitrate assimilation may be a significant factor in pH regulation under anoxia, and evidence in favour of this proposal has recently been obtained in rice coleoptiles (Fan, Higashi & Lane, 1995; Fan *et al.*, 1997). A detailed analysis of [^{15}N]nitrate labelling experiments in this tissue showed that anoxia increased the incorporation of label into alanine and GABA, and it was concluded that anaerobic nitrate assimilation did indeed have a significant role in regulating cytoplasmic pH and in maintaining energy production. The role of nitrate assimilation in controlling pH is likely to extend beyond the simple stoichiometric consumption of protons because the pH-dependent properties of nitrate reductase suggest that nitrate assimilation may be able to act as a biochemical pH-stat (Kaiser & Brendle-Behnisch, 1995). Nitrate reductase is reversibly activated by a decrease in pH (Kaiser & Brendle-Behnisch, 1995), in a process that is mediated by dephosphorylation of the protein, and nitrate reduction is also known to be stimulated in anoxic barley roots (Botrel, Magné & Kaiser, 1996). Thus, although it has yet to be proved *in vivo*, it seems likely that cytoplasmic acidification under anoxia could trigger the activation of nitrate reductase and thus counteract any drift towards cytoplasmic acidosis.

To date, evidence for biochemical pH regulation during anaerobiosis has been largely based on metabolic analyses, both *in vivo* and *in vitro*, coupled with demonstrations *in vivo* that pH is an effective regulator of an enzyme in the proposed pathway and the occasional use of specific inhibitors where available. However, by analogy with much of the current activity in the field of metabolic regulation, further progress in delineating the extent of the biochemical mechanisms of pH$_i$ regulation might be expected from the analysis of mutants with unusual anaerobic properties and transgenic plants with altered levels of potentially important enzymes. Analysis of maize lines differing in ADH activity showed that the normal aerobic level of the enzyme is far in excess of the level required for energy production *via* fermentation (Roberts *et al.*, 1989), and so experiments with transgenic material have focused on plants with altered levels of lactate dehydrogenase and pyruvate decarboxylase. In fact, while the obvious targets for these molecular genetics approaches are the pathways to lactate and ethanol from pyruvate, the survey below shows that little new information on the role of these pathways in controlling pH has emerged so far.

Rivoal and Hanson (1994) showed that overexpression of lactate dehydrogenase, at a level up to 50 times higher than in control plants, had no effect on the balance between lactate and ethanol production in tomato roots, and they speculated that this arose either because of a difference in the response of the cytoplasmic pH to anoxia in the transgenic plants or because of a difference in the *in vivo* pH sensitivity of the lactate dehydrogenase transgene product. Unfortunately, no pH measurements have been reported for this transgenic material, and so the explanation for the minimal effect of the overexpression of lactate dehydrogenase on metabolism, as well as its significance for the biochemical pH-stat model of pH regulation under anoxia, is still unclear. There has also been some progress in the construction of transgenic plants deficient in lactate dehydrogenase (Germain, Saglio & Ricard, 1995; Dunford, Kruger & Ratcliffe, 1997), but these plants are still being characterised and no measurements of intracellular pH have been reported.

Overexpression of pyruvate decarboxylase has been achieved by transforming tobacco plants with a gene from the obligate anaerobe *Zymomonas mobilis* (Bucher, Brändle & Kuhlemeier, 1994). The leaves of a transgenic line with constitutively high levels of pyruvate decarboxylase were found to accumulate 10–35 times more acetaldehyde, and 8–20 times more ethanol, during the first two to four hours of anoxia than their wild-type counterparts, and it was concluded that pyruvate decarboxylase activity could be a significant factor in the control of ethanol synthesis during short-term anoxia. In principle, this could lead to improved pH regulation under anoxia and this question has recently been examined using *in vivo* ^{31}P-NMR (D.L. Couldwell, R. Dunford, N.J. Kruger, C. Kuhlemeier & R.G. Ratcliffe, unpublished results). Tobacco leaves respond to anoxia with a cytoplasmic acidification in much the same way as other plant tissues (Fig. 4), but overexpression

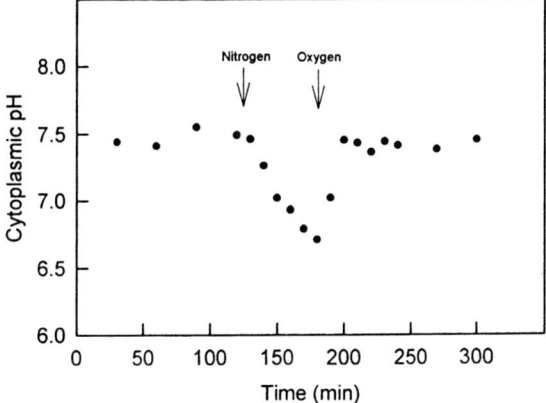

Fig. 4. The effect of anoxia on the cytoplasmic pH of tobacco leaves. The definition of the timecourse is less good than that for the maize root tips in Fig. 2A – reflecting the technical problems associated with applying the *in vivo* ^{31}P-NMR technique to leaf material – but it is sufficient to show the overall similarity. Comparison of wild-type and transgenic plants showed that overexpression of pyruvate decarboxylase had no effect on the pH response (data not shown). (Observations of D.L. Couldwell, R. Dunford, N.J. Kruger, C. Kuhlemeier and R.G. Ratcliffe.)

of pyruvate decarboxylase leads to no difference between the transgenic material and the wild type. This indicates that ethanol production in the transgenic leaves is not replacing acidifying fermentation processes, and that the cytoplasmic pH behaviour in Figure 4 must have its origin in other H^+-generating mechanisms such as ion transport. More generally, these measurements emphasise that fermentation to ethanol, as a pH-neutral process, is only able to stabilise pH when it can replace a H^+-generating process such as fermentation to lactate, and in other circumstances it has no pH-regulatory effect at all. Thus, ethanol production is unable to combat an acidification arising, for example, from H^+ leakage from the vacuole and this highlights an essential limitation of an exclusively metabolic approach to pH regulation.

Biophysical pH regulation under anoxia

The contribution of ion transport to pH regulation in plants during anoxia was overshadowed for many years by the attractiveness of the biochemical pH-stat model, but more recently attention has begun to focus on the role played by biophysical processes, such as lactate transport and proton pumping by the plasma membrane and tonoplast H⁺-ATPases.

Lactate transport is an essential feature of intracellular pH regulation in the anaerobic tissues of animals (Poole & Halestrap, 1993; Wang, Heigenhauser & Wood, 1996), but it is only recently that it has become apparent that it may be important in some plant tissues. As reviewed elsewhere (Ratcliffe, 1995), lactate production can occur over extended periods in some species and this tends to be associated with the release of substantial quantities of lactate to the surrounding medium (Xia & Saglio, 1992; Rivoal & Hanson, 1993; Xia & Roberts, 1994). Xia and Saglio (1992) showed that a hypoxic pretreatment increased the release of lactate from anoxic maize root tips, and subsequently it was shown by ^{31}P-NMR that this resulted in improved pH regulation in the cytoplasm (Xia & Roberts, 1994). The hypoxic treatment of maize root tips apparently induced the synthesis of a lactate carrier (Xia & Saglio, 1992), although the precise mechanism of the carrier has yet to be elucidated. It would seem reasonable to assume that lactate efflux occurs by a H⁺/lactate⁻ co-transport mechanism analogous to that found in the plasma membrane of mammalian cells, and that this provides the basis for the improved pH regulation and survival of the acclimated roots. However, Xia and Roberts (1996) recently reported that the greater lactate efflux from acclimated maize root tips was associated with a lower H⁺ efflux rate than in the non-acclimated tissue, and this is difficult to reconcile with the tighter control of the cytoplasmic pH. Blocking lactate export from acclimated roots with an inhibitor could help to resolve this question, and in any case it should not be forgotten that the proton efflux measurements are the net result of all the processes that influence the pH of the external medium, which makes it difficult to draw firm conclusions about the contribution of H⁺/lactate⁻ co-transport.

In contrast to lactate efflux, which appears to be a rather specialised mechanism of pH regulation in anaerobic plant cells, the ubiquity of the plasma membrane and tonoplast proton pumps suggests that these transport processes might play a critical role in pH regulation under anoxia. Strictly speaking, these electrogenic pumps hyperpolarise the membrane, and it is the subsequent redistribution of the so-called strong ions that leads to pH changes in the compartments separated by the membrane (Stewart, 1983). While this means that the extent of proton pumping is not necessarily directly related to changes in pH, many experiments are at least qualitatively consistent with the idea that an activation of these pumps can lead to an alkalinisation of the cytoplasm, and a simultaneous acidification of either the suspending medium or the vacuole. The significance of this process for pH regulation in anaerobic plant tissues is uncertain, but several recent papers have addressed this question using NMR, fluorescence and microelectrode methods. This work is reviewed here, and leads to the conclusion that the contribution of the proton pumps is variable, and not necessarily dominant.

Extrapolating from aerobic conditions – where there is good evidence that the plasma membrane H^+-ATPase can sometimes make a significant contribution to cytoplasmic pH regulation (Gout *et al.*, 1992) – to the situation under anoxia is complicated by the metabolic response of the tissue to O_2 deprivation. Typically, the cytoplasmic pH falls, the ATP concentration decreases and the ADP level rises under anoxia, and in principle these changes could have a direct effect on the activity of the H^+-ATPases and their potential to contribute to pH regulation. Thus, the acidification of the cytoplasm would be expected to activate the plasma membrane H^+-ATPase (Guern *et al.*, 1991), while the decrease in the ATP level could lead to substrate limitation of both ATPases, and the increase in ADP could lead to inhibition of the tonoplast H^+-ATPase (Brauer & Tu, 1994). The K_m values for MgATP for the ATPases are quite high, typically 0.3–1.4 mM for the plasma membrane pump (Sze, 1985; Michelet & Boutry, 1995) and around 0.2 mM for the tonoplast pump (Sze, 1985; Brauer & Tu, 1994), and the possible significance of this for pH_i regulation under anoxia was first noticed by Saint-Ges *et al.* (1991). These authors observed that the cytoplasmic pH and ATP level followed similar time courses in anoxic maize root tips, and they speculated that the underlying cause of the similarity could be substrate limitation of the ATPases.

This hypothesis has been tested by manipulating the ATP level independently of the anoxic response using metabolic inhibitors (Fox *et al.*, 1995a; Xia, Saglio & Roberts, 1995; A. Hene & R.G. Ratcliffe, unpublished results). NMR experiments on non-acclimated maize root tips showed: (i) that increasing the ATP level by up to 100 per cent had no effect on the cytoplasmic pH under anoxia; and (ii) that reducing the ATP level below its normal anaerobic value prevented the modest recovery in cytoplasmic pH that would normally have been expected under the anoxic conditions used (Fox *et al.*, 1995a; A. Hene & R.G. Ratcliffe, unpublished results). These experiments indicate that substrate limitation of the H^+-ATPases is only likely to become a significant factor in cytoplasmic pH regulation in anoxic tissues with unusually low ATP levels. Similar experiments on intact, acclimated maize seedlings showed that reducing the ATP level under anoxia by more than 50 per cent had no effect on survival, or on the time course of the cytoplasmic pH (Xia *et al.*, 1995); and again it was concluded that the contribution of ATP-dependent processes to cytoplasmic pH regulation was unaffected by changes in the ATP level within the physiological range. Thus, the higher ATP level in acclimated maize root tips under anoxia is not a critical feature of the acclimation process (Xia *et al.*, 1995); and substrate limitation of the H^+-ATPases does not appear to be a significant factor in cytoplasmic pH regulation in anoxic maize roots.

The effect of anoxia on the operation of the plasma membrane H^+-ATPase has been investigated by microelectrode measurements (Felle, 1996) and by monitoring the changes in pH_c caused by H^+ extrusion (Xia & Roberts, 1996). Microelectrodes showed that the root hairs of *Medicago sativa* responded to the removal of O_2 with an immediate fall of 0.5 pH units in the cytoplasmic pH and a decrease in the membrane potential to its diffusion value (Felle, 1996). The depolarisation of the membrane indicates that the proton pump was deactivated, and at first sight it might appear that this was the cause of the cytoplasmic acidification. However, Felle argued that cytoplasmic acidosis under anoxia was not the result of pump deactivation, because depolarisation and repolarisation of the membrane occurred without changes in cytoplasmic pH in aerobic experiments with oligomycin, antimycin A, cyanide and fusicoccin (Felle, 1996). The strength of this argument is difficult to judge in the absence of data for treatments that might have stimulated the proton pump under anoxia, but because the changes in cytoplasmic pH during the onset and recovery from anoxia appeared to occur more quickly than the changes in membrane potential, it seems more likely that the data should be interpreted in terms of a failure to maintain the membrane potential during the imposition of an acid load generated in response to the onset of anoxia. This, then, raises the important question of why the pump is not activated by the cytoplasmic acidification that occurs under anoxia when it is so readily activated under aerobic conditions by the uptake of a permeable weak acid. The answer to this question, on the basis of Felle's data, would be that anoxia turns off the pump, as indicated by the depolarisation of the root hairs, and so the pump is unable to respond to H^+ in the usual way. This would also provide a rationalisation for much of the data from the NMR experiments in which ATP levels were manipulated under anoxia (Fox *et al.*, 1995a; Xia *et al.*, 1995; A. Hene & R.G. Ratcliffe, unpublished results), but it would not be consistent with the measurements on proton extrusion from maize roots (Xia & Roberts, 1996) described in the following paragraph.

Thus, in contrast to the microelectrode study, Xia and Roberts (1996) showed that proton extrusion was stimulated by fusicoccin in both acclimated and non-acclimated root tips under anoxia, indicating that the plasma membrane H^+-ATPase could operate under these conditions. It was also shown that net proton extrusion increased under conditions that increased cytoplasmic acidosis: acclimated roots, which maintained a higher cytoplasmic pH value under anoxia, showed lower extrusion rates than non-acclimated roots; and net proton influx observed under aerobic conditions at a pH_e of 4.5 decreased following the switch to anoxia (Xia & Roberts, 1996). These observations appear to be consistent with a stimulation of the plasma membrane H^+-ATPase by reduced cytoplasmic pH under anoxia, although in the absence of a specific inhibitor for the pump it is not possible to be certain that it is the change in H^+-ATPase activity that is entirely responsible for the observed changes in proton extrusion.

Perhaps more importantly, the data on the maize root tips show that proton extrusion makes little contribution to cytoplasmic pH regulation in anoxic maize root tips at a pH_e around 6 (Xia & Roberts, 1996). Moreover, non-acclimated root tips show a higher net rate of proton efflux than acclimated tissues, indicating that proton pumping to the external medium does not provide an explanation for the improved cytoplasmic pH regulation observed in acclimated tissues. At a more acidic pH_e of 4.5, net proton influx is reduced under anoxia and it seems likely that proton efflux across the plasma membrane is activated under these conditions and that proton transport now makes a significant contribution to the regulation of the cytoplasmic pH. In fact, the capacity for pH regulation under these acidic conditions is insufficient to prevent a downward drift in cytoplasmic pH and a consequent reduction in survival, but the increased activity of the proton pump at least delays the eventual loss of pH control.

While investigations of pH_i regulation generally focus on the cytoplasm, the regulation of the vacuolar pH is also relevant because of the importance of the pH difference between the cytoplasm and the vacuole in energising the tonoplast membrane, and because of the potential significance of the vacuole as both a sink and a source for protons during anaerobiosis. Furthermore, the tonoplast H^+-pyrophosphatase has been shown to be induced by anoxia in rice seedlings (Carystinos *et al.*, 1995), and this immediately raises questions about the potential contribution of the transporter to pH regulation in anoxic tissues (Carystinos *et al.*, 1995; Xia *et al.*, 1995). In a recent study, Brauer *et al.* (1997) used fluorescence and NMR methods to investigate the relative importance of the tonoplast H^+-ATPase and the tonoplast H^+-pyrophosphatase for the maintenance of the vacuolar pH in maize root hair cells under aerobic and anaerobic conditions. Bafilomycin A_1, a specific inhibitor of the tonoplast H^+-ATPase, caused an increase in vacuolar pH under aerobic conditions, suggesting a role for the H^+-ATPase in determining vacuolar pH. Attempts to limit this contribution by using metabolic inhibitors to reduce the ATP level were unsuccessful, most probably because the treatments did not reduce the cytoplasmic ATP concentration below the K_m value for the tonoplast H^+-ATPase. However, it was also observed that the sensitivity of the vacuolar pH to bafilomycin A_1 was abolished by treatments, including anoxia, that increased the ADP level. This result was rationalised in terms of the *in vitro* sensitivity of the tonoplast H^+-ATPase to ADP, and it was concluded that the results were consistent with the view that the vacuolar pH is maintained by the tonoplast H^+-ATPase under aerobic conditions and by the tonoplast H^+-pyrophosphatase under anoxia (Brauer *et al.*, 1997).

Conclusions

Simple generalisations about pH_i regulation in anaerobic plant tissues seem increasingly implausible as more information becomes available about the origin of the pH changes that occur under anoxia and the regulatory mechanisms that oppose the drift towards cytoplasmic acidosis. Plants can draw on a range of biochemical and biophysical mechanisms to control pH, and the work that has been reviewed here emphasises the variable contribution of the different mechanisms in different situations. Experiments that highlight the operation of specific mechanisms *in vivo* have established that the long-running biochemical pH-stat model of anaerobic pH regulation, in which there is a pH-induced switch from lactate to ethanol production, is part of the story; but a complete understanding of the considerable variation in the effectiveness of pH regulation across the whole spectrum of anoxia tolerance in plants remains elusive.

A final point concerns the extent to which the pH changes that occur under anoxia are solely responsible for triggering adjustments in the pH-regulatory mechanisms. It has been argued elsewhere (Ratcliffe, 1997) that the evidence for pH-induced changes in metabolism under anoxia is largely qualitative – in the sense that it has not been proved that the changes in pH are quantitatively sufficient to alter metabolic fluxes to the observed extent – and the possibility that the signalling function of the pH changes is influenced by other factors seems increasingly likely. In this context, it is perhaps particularly significant that changes in cytoplasmic calcium have been shown to occur over the same timescale as the acidification of the cytoplasm under anoxia (Subbaiah, Bush & Sachs, 1994; Sedbrook *et al.*, 1996), and the relation between these two events is now under investigation (Sachs *et al.*,

1996). It is too early to draw any conclusions on this point, not least because of the shortage of information on the anaerobic calcium response and the discrepancies between the two published papers (Subbaiah *et al.*, 1994; Sedbrook *et al.*, 1996), but the complexity of the interaction between the calcium and pH-signalling pathways in guard cells (see Blatt & Grabov, this volume) suggests that the situation in anoxic tissues will be far from simple.

Acknowledgements

The author acknowledges the financial support of the Biotechnology and Biological Sciences Research Council. He is also grateful to Dr. M.R. Knight for a critical reading of the manuscript.

References

ap Rees, T. & Hill, S.A. (1994). Metabolic control analysis of plant metabolism. *Plant, Cell and Environment* **17**, 587–99.

Botrel, A., Magné, C. & Kaiser, W.M. (1996). Nitrate reduction, nitrite reduction and ammonium assimilation in barley roots in response to anoxia. *Plant Physiology and Biochemistry* **34**, 645–52.

Brauer, D. & Tu, S. (1994). Effects of ATP analogs on the proton pumping by the vacuolar H^+-ATPase from maize roots. *Physiologia Plantarum* **91**, 442–8.

Brauer, D., Uknalis, J., Triana, R., Shachar-Hill, Y. & Tu, S. (1997). Effects of bafilomycin A_1 and metabolic inhibitors on the maintenance of vacuolar acidity in maize root hair cells. *Plant Physiology* **113**, 809–16.

Bucher, M., Brändle, R. & Kuhlemeier, C. (1994). Ethanolic fermentation in transgenic tobacco expressing *Zymomonas mobilis* pyruvate decarboxylase. *EMBO Journal* **13**, 2755–63.

Carroll, A.D., Fox, G.G., Laurie, S., Phillips, R., Ratcliffe, R.G. & Stewart, G.R. (1994). Ammonium assimilation and the role of γ-aminobutyric acid in pH homeostasis in carrot cell suspensions. *Plant Physiology* **106**, 513–20.

Carystinos, G.D., MacDonald, H.R., Monroy, A.F., Dhindsa, R.S. & Poole, R.J. (1995). Vacuolar H^+-translocating pyrophosphatase is induced by anoxia or chilling in seedlings of rice. *Plant Physiology* **108**, 641–9.

Crawford, L.A., Bown, A.W., Breitkreuz, K.E. & Guinel, F.C. (1994). The synthesis of γ-aminobutyric acid in response to treatments reducing cytosolic pH. *Plant Physiology* **104**, 865–71.

Davies, D.D. (1986). The fine control of cytosolic pH. *Physiologia Plantarum* **67**, 702–6.

Davies, D.D., Grego, S. & Kenworthy, P. (1974). The control of the production of lactate and ethanol by higher plants. *Planta* **118**, 297–310.

Dunford, R., Kruger, N.J. & Ratcliffe, R.G. (1997). Comparison of two cDNA clones for lactate dehydrogenase from potato. *Journal of Experimental Botany* **48**, Supplement 82.

Fan, T.W-M., Higashi, R.M., Frenkiel, T.A. & Lane, A.N. (1997). Anaerobic nitrate and ammonium metabolism in flood-tolerant rice coleoptiles. *Journal of Experimental Botany* **48**, 1655–66.

Fan, T.W-M., Higashi, R.M. & Lane, A.N. (1995). Use of ^{15}N and ^{13}C isotope labelling and multinuclear NMR for exploring nitrate metabolism in anaerobic rice coleoptiles. *Plant Physiology* **108**, S12.

Fan, T.W-M., Lane, A.N. & Higashi, R.M. (1992). Hypoxia does not affect rate of ATP synthesis and energy metabolism in rice shoot tips as measured by ^{31}P NMR *in vivo*. *Archives of Biochemistry and Biophysics* **294**, 314–18.

Fell, D. (1997). *Understanding the control of metabolism*. London: Portland Press.

Felle, H. (1987). Proton transport and pH control in *Sinapis alba* root hairs. A study carried out with double-barrelled pH microelectrodes. *Journal of Experimental Botany* **38**, 340–54.

Felle, H.H. (1993). Ion-selective microelectrodes: their use and importance in modern plant cell biology. *Botanica Acta* **106**, 5–12.

Felle, H.H. (1996). Control of cytoplasmic pH under anoxic conditions and its implication for plasma membrane proton transport in *Medicago sativa* root hairs. *Journal of Experimental Botany* **47**, 967–73.

Felle, H. & Bertl, A. (1986). The fabrication of H$^+$-selective liquid-membrane microelectrodes for use in plant cells. *Journal of Experimental Botany* **37**, 1416–28.

Ford, Y.Y., Ratcliffe, R.G. & Robins, R.J. (1996). Phytohormone-induced GABA production in transformed root cultures of *Datura stramonium*: an *in vivo* ^{15}N NMR study. *Journal of Experimental Botany* **47**, 811–18.

Fox, G.G., Hene, A., McCallan, N.R. & Ratcliffe, R.G. (1995a). The role of cytoplasmic pH and ATP in the anoxic response: non-invasive experiments with *in vivo* NMR spectroscopy. *Plant Physiology* **108**, S43.

Fox, G.G., McCallan, N.R. & Ratcliffe, R.G. (1995b). Manipulating cytoplasmic pH under anoxia: a critical test of the role of pH in the switch from aerobic to anaerobic metabolism. *Planta* **195**, 324–30.

Fox, G.G. & Ratcliffe, R.G. (1990). ^{31}P NMR observations on the effect of the external pH on the intracellular pH values in plant cell suspension cultures. *Plant Physiology* **93**, 512–21.

Germain, V., Saglio, P.H. & Ricard, B. (1995). Molecular and physiological analyses of tomato roots transformed with antisense constructs targeted to L-lactate dehydrogenase. *Plant Physiology* **108**, S65.

Gibbon, B.C. & Kropf, D.L. (1994). Cytosolic pH gradients associated with tip growth. *Science* **263**, 1419–21.

Gout, E., Bligny, R. & Douce, R. (1992). Regulation of intracellular pH values in higher plant cells. Carbon-13 and phosphorus-31 nuclear magnetic resonance studies. *Journal of Biological Chemistry* **267**, 13903–9.

Guern, J., Felle, H., Mathieu, Y. & Kurkdjian, A. (1991). Regulation of intracellular pH in plant cells. *International Review of Cytology* **127**, 111–73.

Hochachka, P.W. & Mommsen, T.P. (1983). Protons and anaerobiosis. *Science* **219**, 1391–7.

Hole, D.J., Cobb, B.G., Hole, P.S. & Drew, M.C. (1992). Enhancement of anaerobic respiration in root tips of *Zea mays* following low oxygen (hypoxic) acclimation. *Plant Physiology* **99**, 213–18.

Johnson, J., Cobb, B.G. & Drew, M.C. (1989). Hypoxic induction of anoxia tolerance in root tips of *Zea mays*. *Plant Physiology* **91**, 837–41.

Kaiser, W.M. & Brendle-Behnisch, E. (1995). Acid–base modulation of nitrate reductase in leaf tissues. *Planta* **196**, 1–6.

Kosegarten, H., Grolig, F., Wieneke, J., Wilson, G. & Hoffmann, B. (1997). Differential ammonia-elicited changes of cytosolic pH in root hair cells of rice and maize as monitored by 2′,7′-bis-(2-carboxyethyl)-5 (and -6)-carboxyfluorescein fluorescence ratio. *Plant Physiology* **113**, 451–61.

Menegus, F., Cattaruzza, L., Mattana, M., Beffagna, N. & Ragg, E. (1991). Response to anoxia in rice and wheat seedlings. Changes in the pH of intracellular compartments, glucose-6-phosphate level and metabolic rate. *Plant Physiology* **95**, 760–7.

Michelet, B. & Boutry, M. (1995). The plasma membrane H^+-ATPase. A highly regulated enzyme with multiple physiological functions. *Plant Physiology* **108**, 1–6.

Nicholls, D.G. & Ferguson, S.J. (1992). *Bioenergetics 2*. London: Academic Press.

Poole, R.C. & Halestrap, A.P. (1993). Transport of lactate and other monocarboxylates across mammalian plasma membranes. *American Journal of Physiology* **264**, C761–82.

Ratcliffe, R.G. (1994). *In vivo* nuclear magnetic resonance studies of higher plants and algae. *Advances in Botanical Research* **20**, 43–123.

Ratcliffe, R.G. (1995). Metabolic aspects of the anoxic response in plant tissue. In *Environment and plant metabolism: flexibility and acclimation*, ed. N. Smirnoff, pp. 111–27. Oxford: BIOS Scientific Publishers.

Ratcliffe, R.G. (1997). *In vivo* NMR studies of the metabolic response of plant tissues to anoxia. *Annals of Botany* **79** (Supplement A), 39–48.

Raven, J.A. (1986). Biochemical disposal of excess H^+ in growing plants? *New Phytologist* **104**, 175–206.

Reggiani, R., Mattana, M., Aurisano, N. & Bertani, A. (1993). Utilization of stored nitrate during the anaerobic germination of rice seeds. *Plant and Cell Physiology* **34**, 379–83.

Rivoal, J. & Hanson, A.D. (1993). Evidence for a large and sustained glycolytic flux to lactate in anoxic roots of some members of the halophytic genus *Limonium*. *Plant Physiology* **101**, 553–60.

Rivoal, J. & Hanson, A.D. (1994). Metabolic control of anaerobic glycolysis. Overexpression of lactate dehydrogenase in transgenic tomato roots supports the Davies–Roberts hypothesis and points to a critical role for lactate secretion. *Plant Physiology* **106**, 1179–85.

Roberts, J.K.M., Andrade, F.H. & Anderson, I.C. (1985). Further evidence that cytoplasmic acidosis is a determinant of flooding intolerance in plants. *Plant Physiology* **77**, 492–4.

Roberts, J.K.M., Callis, J., Jardetzky, O., Walbot, V. & Freeling, M. (1984a). Cytoplasmic acidosis as a determinant of flooding intolerance in plants. *Proceedings of the National Academy of Sciences USA* **81**, 6029–33.

Roberts, J.K.M., Callis, J., Wemmer, D., Walbot, V. & Jardetzky, O. (1984b). Mechanism of cytoplasmic pH regulation in hypoxic maize root tips and its role in survival under hypoxia. *Proceedings of the National Academy of Sciences USA* **81**, 3379–83.

Roberts, J.K.M., Chang, K., Webster, C., Callis, J. & Walbot, V. (1989). Dependence of ethanolic fermentation, cytoplasmic pH regulation and viability on the activity of alcohol dehydrogenase in hypoxic maize root tips. *Plant Physiology* **89**, 1275–8.

Roberts, J.K.M., Hooks, M.A., Miaullis, A.P., Edwards, S. & Webster, C. (1992). Contribution of malate and amino acid metabolism to cytoplasmic pH regulation in hypoxic maize root tips studied using nuclear magnetic resonance spectroscopy. *Plant Physiology* **98**, 480–7.

Roberts, J.K.M. & Xia, J.H. (1996). NMR contributions to understanding of plant responses to low oxygen stress. In *Nuclear magnetic resonance in plant biology*, ed. Y. Shachar-Hill & P.E. Pfeffer, pp. 155–80. Rockville, MD: American Society of Plant Physiologists.

Sachs, M.M., Subbaiah, C.C. & Saab, I.N. (1996). Anaerobic gene expression and flooding tolerance. *Journal of Experimental Botany* **47**, 1–15.

Saglio, P.H., Drew, M.C. & Pradet, A. (1988). Metabolic acclimation to anoxia induced by low (2–4 kPa partial pressure) oxygen pretreatment (hypoxia) in root tips of *Zea mays*. *Plant Physiology* **86**, 61–6.

Saint-Ges, V., Roby, C., Bligny, R., Pradet, A. & Douce, R. (1991). Kinetic studies of the variations of cytoplasmic pH, nucleotide triphosphates (^{31}P NMR) and lactate during normoxic and anoxic transitions in maize root tips. *European Journal of Biochemistry* **200**, 477–82.

Sedbrook, J.C., Kronebusch, P.J., Borisy, G.G., Trewavas, A.J. & Masson, P.H. (1996). Transgenic *AEQUORIN* reveals organ-specific cytosolic Ca^{2+} responses to anoxia in *Arabidopsis thaliana* seedlings. *Plant Physiology* **111**, 243–57.

Stewart, P.A. (1983). Modern quantitative acid–base chemistry. *Canadian Journal of Physiology and Pharmacology* **61**, 1444–61.

Subbaiah, C.C., Bush, D.S. & Sachs, M.M. (1994). Elevation of cytosolic calcium precedes anoxic gene expression in maize suspension cultured cells. *The Plant Cell* **6**, 1747–62.

Sze, H. (1985). H^+-translocating ATPases: advances using membrane vesicles. *Annual Review of Plant Physiology* **36**, 175–208.

Wang, Y., Heigenhauser, G.J.F. & Wood, C.M. (1996). Lactate and metabolic H^+ transport and distribution after exercise in rainbow trout white muscle. *American Journal of Physiology* **271**, R1239–50.

Waters, I., Kuiper, P.J.C., Watkin, E. & Greenway, H. (1991a). Effects of anoxia on wheat seedlings. 1. Interaction between anoxia and other environmental factors. *Journal of Experimental Botany* **42**, 1427–35.

Waters, I., Morrell, S., Greenway, H. & Colmer, T.D. (1991b). Effects of anoxia on wheat seedlings. 2. Influence of O_2 supply prior to anoxia on tolerance to anoxia, alcoholic fermentation and sugar levels. *Journal of Experimental Botany* **42**, 1437–47.

Xia, J.H. & Roberts, J.K.M. (1994). Improved cytoplasmic pH regulation, increased lactate efflux, and reduced cytoplasmic lactate levels are biochemical traits expressed in root tips of whole maize seedlings acclimated to a low oxygen environment. *Plant Physiology* **105**, 651–7.

Xia, J.H. & Roberts, J.K.M. (1996). Regulation of H^+ extrusion and cytoplasmic pH in maize root tips acclimated to a low oxygen environment. *Plant Physiology* **111**, 227–33.

Xia, J.H. & Saglio, P.H. (1992). Lactic acid efflux as a mechanism of hypoxic acclimation of maize root tips to anoxia. *Plant Physiology* **100**, 40–6.

Xia, J.H., Saglio, P.H. & Roberts, J.K.M. (1995). Nucleotide levels do not critically determine survival of maize root tips acclimated to a low oxygen environment. *Plant Physiology* **108**, 589–95.

DONALD C. JACKSON

The role of turtle shell in acid–base buffering

Introduction

The first line of defense of an organism against an acid load is the available endogenous buffering of its body fluids. These buffers bind most of the protons and greatly minimize the fall in pH. Conventional descriptions in vertebrates generally define two major buffering compartments, the extracellular fluid and the intracellular fluid. Extracellular fluid buffering is dominated by bicarbonate, with a minor contribution from plasma proteins; whereas in the intracellular compartment, non-bicarbonate buffers such as proteins and phosphates predominate.

Often neglected in a consideration of body buffers is the skeleton, even though this structure, with its rich reservoir of mineral phosphates and carbonates, is potentially the major source of buffering power in the body. Skeletal contribution to routine buffering, however, is often less important because of slow exchange kinetics with the extracellular fluid and because of inaccessibility of much of the bone mineral. Under certain circumstances, though, bone *can* play a central role in acid–base balance. In clinical cases of chronic non-respiratory acidosis, skeletal buffers are heavily recruited and this response has led to extensive investigations of the role of bone in acid–base function in humans and other mammals. Normal stresses of animals can also produce acidotic states that involve skeletal buffers as, for example, in crustaceans in which hypoxic states can lead to lactate retention and dissolution of $CaCO_3$ from their exoskeletons.

A dramatic and ecologically relevant example of bone involvement in acid buffering is the contribution of the shell of the freshwater turtle in neutralizing lactic acid produced during prolonged anoxic submergence by these animals. Before considering this topic in detail, it is useful first to review principles of bone acid–base function revealed by mammalian studies.

Acid buffering by the mammalian skeleton

Although the mineral phase of bone is principally the calcium phosphate crystal, hydroxy apatite, the buffering role of the skeleton has largely been ascribed to skeletal carbonates. Early work by Irving and Chute (1932) described loss of bone CO_2 over several days of acid loading, and this observation has been confirmed in subsequent work (Burnell, 1971; Bettice, 1984). Associated with the loss of bone CO_2 is an efflux from bone of both calcium (Bushinsky et al., 1983) and sodium (Forbes, 1960; Burnell, 1971). Phosphate is not generally implicated in the skeletal buffering mechanism.

Skeletal buffering can be highly effective and can balance acid production under conditions of renal insufficiency when excretion of acid is impaired. In this way, a stable but acidotic blood acid–base state can be sustained for long periods (Schwartz et al., 1959). However, the cost of skeletal recruitment may be considerable because significant demineralization can occur and lead to bone pathology (Eiamong & Kurtzman, 1994). In addition, increased urinary excretion of calcium can deplete citrate levels and potentially lead to precipitation of calcium salts.

How bone buffers are recruited is still imperfectly understood, but considerable information is available, particularly from the work of Bushinsky and co-workers (Bushinsky, 1994) on in vitro neonatal mouse calvariae. Their work indicates that acidic incubation solutions, initially unsaturated with respect to $CaCO_3$, induce dissolution of bone $CaCO_3$ until chemical equilibrium is achieved (Bushinsky & Lechleider, 1987). Although this suggests that acid recruitment of bone buffering is a passive phenomenon, earlier work from the same laboratory (Bushinsky, Goldring & Coe, 1985) showed that calcium-mobilizing hormones (PTH, vitamin D) increase the rate of calcium efflux at each incubation pH, although the slope of the pH dependence was unchanged. Similarly, metabolic poisons or cell disruption shifted the calcium efflux – pH relationship to the left, but again with no change in slope. These results suggest that the buffer release involves both a passive chemical equilibrium component that is pH dependent and an active cell-mediated component that is pH independent. The cell-mediated component is thought to act via local release of acid from osteoclasts that dissolves bone mineral (Schlesinger, Mattsson & Blair, 1994). Other studies on neonatal mouse calvariae have revealed the difficulty in establishing the stoichiometry of the exchange process (Bushinsky et al., 1983). In acute acid incubations, calculated proton influx into bone exceeded calcium efflux by 16–21 times. The difference is thought to be accounted for by Na^+ and K^+ dissociation from binding sites on the bone surface and efflux into the extracellular fluid in exchange for protons, although this would relegate calcium to a surprisingly minor role. Ion microprobe analysis of bone surface revealed a greater decrease of Na^+ and K^+ than of calcium during acid incubation (Bushinsky, Levi-Setti & Coe, 1986).

The site within bone where the mobile ions reside and the precise exchange processes between extracellular fluid and bone are also problematic. Currently, it is thought that CO_2 exists in two forms in bone: first, as carbonate substituted for OH^- or phosphate in the apatite crystal, and second, as bicarbonate in the hydration layer at the surface of the apatite (Bushinsky, 1994). The superficial location provides ready access to bone extracellular fluid and may correspond to the labile bicarbonate component described by Poyart, Burseaux and Fréminet (1975).

The turtle shell

The role of bone in acid–base balance in the freshwater turtle is exaggerated compared to other vertebrates for both structural and functional reasons. Studies on the painted turtle, *Chrysemys picta bellii*, have characterized the features of this animal that account for its special properties. Structurally, the large calcified shell together with a conventional skeleton give the turtle an unusually large mass of bone compared to other vertebrates. Functionally, the significant trait is the turtle's capacity to sustain long periods of non-respiratory acidosis, specifically lactic acidosis, when submerged in O_2-poor water. Because renal function is minimal in this state, the usual pathway for acid excretion is ineffective and the turtle must rely exclusively on endogenous buffering and intercompartmental exchanges to defend its acid–base state.

Structural features of the shell

The turtle shell is a combination of a horny epithelial covering over a dermal armor to which elements of the axial skeleton are fused (Zangerl, 1969). The mass of the shell is large compared to the skeletal mass of vertebrates generally; in *Chrysemys*, for example, the shell is 32 percent of the body mass of the animal (Jackson, Toney & Okamoto, 1996). The remainder of the skeleton not attached to the shell is an additional 5.5 percent of body mass (unpublished observations), so that all the bony tissue accounts for over 37 percent of body mass. Estimated skeletal mass of a similarly sized mammal based on allometric analysis is only about 11 percent of body mass (Calder, 1984). Shell density is high, about 1.6 (Warburton & Jackson, 1995), so the shell's contribution to the weight of the turtle in water is proportionately even greater than in air, amounting in *Trachemys scripta* to some 75 percent of the total (Jackson, 1969). Because of the substantial sinking tendency of the shell, turtles require unusually large lung volumes (approximately 140 ml/kg) in order to achieve neutral buoyancy (Jackson, 1971). The large lung volume pays an additional dividend to a semi-aquatic diving turtle by providing it with its major on-board O_2 reserve (Fuster, Pagés & Palacios, 1997).

The composition of the shell is generally similar to vertebrate bone, consisting principally of calcium phosphate in a molar ratio of about 10 : 6 (Warburton & Jackson, 1995), the predicted ratio for apatite crystal. Also present are sizable deposits of magnesium, sodium, and CO_2. Analyses reveal that most of the body's inventory of each of these substances resides in the shell (Table 1), the site of well over 90 percent of the calcium, phosphate, magnesium, and CO_2, as well as over half of the body's sodium.

Functional features of the shell

The obvious and principal functions of the turtle's shell are to provide protection against the hazards of the outside world and to serve as an anchor point for skeletal muscles. The rigid shell necessarily impacts many other functions of the turtle, including its locomotion, respiration, and copulation. Body volume is severely constrained so that changes in one bodily component (eggs, fat, fluid, air) must be balanced by opposite changes in another (Jackson, 1971).

The shell serves other less obvious functions for the turtle (as does the skeleton in other animals) that relate to its large reserve of minerals. Extracellular calcium and phosphorus homeostasis (Clark, 1970) is thought to depend on exchanges between extracellular fluid and shell,

Table 1. *The importance of the shell as a storage site for minerals and CO_2*

	Shell	Rest of body	Shell (% total)
Calcium (mmol/g)	5.7	0.003	99.9
Phosphate (mmol/g)	3.3	0.018	98.9
Magnesium (mmol/g)	0.13	0.003	95.7
Sodium (mmol/g)	0.15	0.05	60.7
CO_2 (mmol/g)	1.34	0.023	96.8

and bone and possibly shell also serve as sources of calcium and phosphate for egg production (Edgren, 1960). Finally, the shell acts as a principal organ in the regulation of acid–base state during prolonged non-respiratory acidosis. It appears to function in two distinct ways in this regard, both of which exploit the shell's carbonates as buffers of lactic acid. First, buffering involves demineralization of the shell and movement of strong cations, calcium, magnesium, and possibly sodium, from the shell into the extracellular fluid, in exchange for protons. Second, lactate is sequestered within the shell and the associated proton is probably buffered by shell carbonate. These two mechanisms are discussed in detail below, but first the physiological circumstances that necessitate as well as permit the recruitment of shell buffering are discussed.

Anoxic tolerance of turtles

Freshwater turtles, particularly *Emydidae* such as *Chrysemys* and *Trachemys*, have a remarkable capacity for anoxia tolerance and anaerobic glycolysis. In nature, hibernating individuals in the temperate zone may spend several months submerged beneath ice, often buried in mud without access to O_2 (Ultsch, 1989). In the laboratory, experimental submergences in O_2-free water have demonstrated that life without O_2 is possible for turtles for extended periods throughout their normal temperature range. Durations from which they can fully recover vary inversely with temperature, ranging from at least 12 hours at 20 °C to three months or more at 3 °C (Herbert & Jackson, 1985a). A key adaptation is their capacity to depress metabolism when anoxic (Jackson, 1968; Herbert & Jackson, 1985b).

Depression of metabolism contributes to anoxia tolerance in two important ways: (1) it slows the rate at which glycolytic substrates (chiefly glycogen) are utilized, and (2) it reduces the rate at which acid end-products (chiefly lactate) are produced. These two processes are stoichiometrically linked, but they can be viewed as distinct problems for the anoxic animal, and either or both could represent limiting factors to anoxic survival. A downregulation of metabolism slows the rate at which either of these could approach a critical state.

Glycogen reserves in the turtle are large (Daw, Wenger & Berne, 1967) and probably do not become depleted even during the longest submergences at low temperature (Hochachka, 1982), but substrate supply could be a serious issue if the turtle began the winter without an adequate glycogen reserve.

Acid–base consequences of prolonged anoxia

Lactate accumulation may represent a more serious threat to the anoxic turtle than glycogen depletion. Lactate is the terminal product of anaerobic glycolysis and, as pointed out by Hochachka and Mommsen (1983), the production of lactate is associated with equimolar generation of H^+ when glycolytic ATP production and hydrolysis are matched, a condition that is approximated during long-term anoxia. Lactic acid is a relatively strong acid that is almost fully dissociated at physiological pH, so that its rate of production, the anaerobic metabolic rate, is also the rate at which the animal is subjected to an acid load. Suppression of this rate has clear advantages for the maintenance of acid–base balance.

During very long anoxic periods such as have been utilized in experimental submergences, lactate can nonetheless build up to very high concentrations in the plasma, despite the slow rate of production. For example, during 12 weeks of anoxic submergence at 3 °C, plasma lactate rose from 1 mmol/l to 145 mmol/l (Jackson & Heisler, 1982). The final concentration reached is far more than the buffering capacity available in the extracellular and intracellular compartments. Extracellular buffering, though unusually large in turtles because of high bicarbonate levels (Smith, 1929), is still only in the range of 40–50 mmol/l per pH unit. Lactic acid would, therefore, fully titrate available buffers at a concentration of about 50 mmol/l, and further addition of acid would produce a precipitous fall in pH. *In vivo*, however, blood pH is reduced only to 7.16 at a plasma lactate concentration of 145 mmol/l (Jackson & Heisler, 1983). Acid must have been transported out of the blood compartment or supplemental buffer transported in to explain this result. Both of these possible mechanisms are involved and the shell plays the central role in each.

Shell mechanism 1: export of buffer to the extracellular fluid

A characteristic pattern of ionic changes in plasma has been observed in several studies of experimental submergence anoxia (Ultsch & Jackson, 1982; Jackson & Heisler, 1982; Herbert & Jackson, 1985a) that include increases in the cations calcium and magnesium that parallel the rise in lactate (Fig. 1). Available evidence, both direct and circumstantial, indicates that the calcium and magnesium originate in the shell of the turtle. First, as noted earlier (Table 1), more than 95 percent of the body's supply of these elements is in the shell. Second, the other possible source (the soft tissues) has a limited supply of these ions, and skeletal muscle calcium and magnesium actually increase somewhat

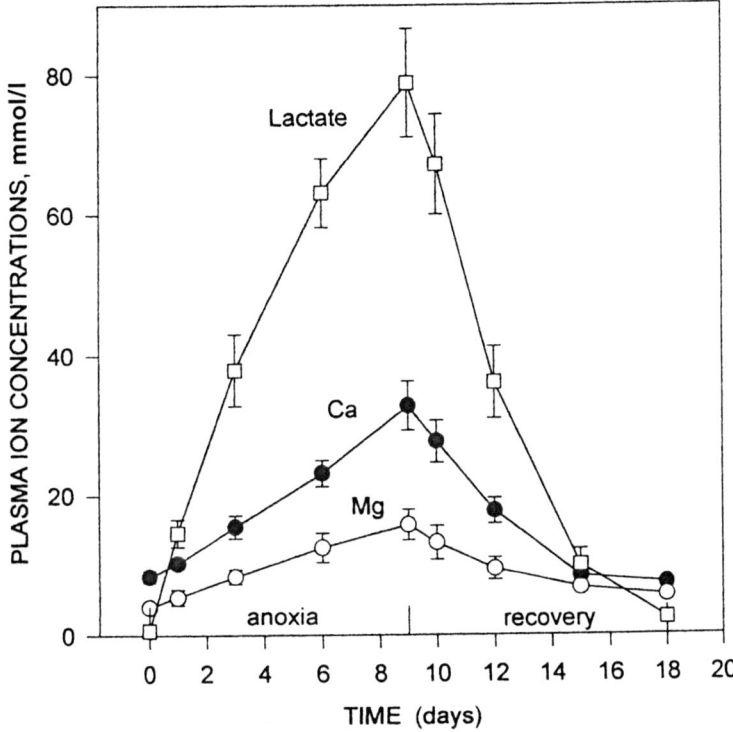

Fig. 1. Plasma concentrations of Lac⁻, Ca²⁺, and Mg²⁺ in the turtle, *Chrysemys picta bellii*, at 10 °C during nine days of anoxia and nine days of recovery.

during anoxia at 3 °C (Jackson & Heisler, 1983). In addition, shell magnesium and total CO_2 concentrations decreased significantly during prolonged anoxia (Warburton & Jackson, 1995), consistent with the release of $MgCO_3$ from the shell. No change in shell calcium concentration was observed, but the expected decrease, although large in terms of the plasma effect, is very small in relation to the total calcium present in the shell (less than 0.5 percent). The expected changes in shell magnesium and CO_2, however, are comparatively large as a percentage of the total present, and hence were detectable (Warburton & Jackson, 1995). However, no change was observed in either shell or plasma phosphate, leading to the conclusion that in turtle shell, as in mammalian bone, non-respiratory acidosis causes a dissolution of shell carbonates, but not of the phosphate-containing apatite crystal.

In-vitro measurements on powdered turtle shell (S. Visuri & D.C.

Jackson, unpublished observations) also support this conclusion. Incubation with acid solutions over a range of pH values caused release of calcium, magnesium, and CO_2, but not of phosphate except at very low, non-physiological, pH (Fig. 2). This *in vitro* incubation approach can be used to determine the stoichiometry of the exchange process, but preliminary results suggest that the system is complex. Not only do calcium, magnesium, and CO_2 efflux from the shell in acid solutions, but so also does sodium. Likewise, preliminary results on anoxic turtles *in vivo* indicate a significant fall in shell sodium concentration, even though the plasma sodium concentration does not change. This apparent paradox may be explained by a dilutional rise in plasma volume, but this interpretation is difficult to reconcile with other plasma solute and osmolality data.

Based on mammalian bone studies, shell demineralization may occur

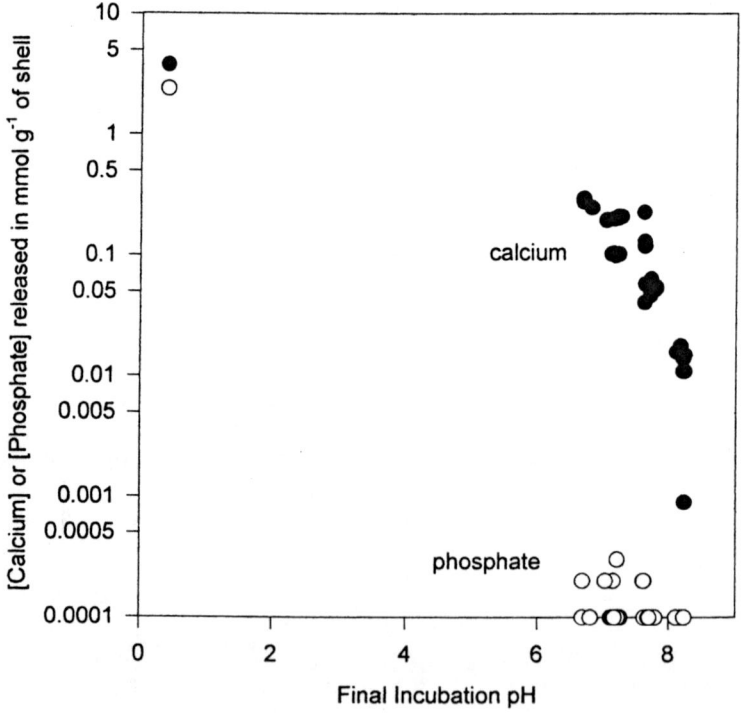

Fig. 2. Loss of Ca^{2+} from turtle shell powder incubated in buffered solutions at different pH. Note that phosphate, in contrast to Ca^{2+}, is released from the shell only at very low, unphysiological pH.

at the surfaces of the bone crystals or from the hydration layers that
surround the crystals. The pool of mineral buffer at these sites is labile
and accessible to the extracellular fluid of the shell. A consistent observation that indirectly supports this hypothesis is that the release of magnesium is far greater in proportion to calcium release than the total shell
elemental ratio would predict. This suggests that the ratio of
magnesium : calcium in the labile pool is higher than in the total shell,
and that the bulk of the calcium resides in the non-labile apatite crystal
structure. The similarity between the *in vivo* buffer release and the
release observed in powdered shell *in vitro* suggests that the dissolution
of carbonate buffer is largely passive, similar to the short-term response
of mammalian bone *in vitro* (Bushinsky & Lechleider, 1987), although
changes in plasma acid and alkaline phosphatases have been observed
during anoxia *in vivo* that are consistent with some active bone resorption (Warburton & Jackson, 1995). The lack of phosphate release supports the notion that the bulk of the bone mineral, the apatite, is largely
unaffected by chronic acidosis.

Shell mechanism 2: storage of lactate within the shell

Recently, somewhat serendipidously, another and perhaps even more
significant role for the turtle shell in buffering a lactic acid load was
found. During the course of a study in which the distribution within the
body of injected ^{14}C-lactate was determined during anoxia at 20 °C, ^{14}C
activity was tested on powdered shell, generated in the process of
accessing the soft tissues. Surprisingly, relatively high ^{14}C activities
were present in the shell samples at the end of anoxia which were as
high, in disintegrations/min per g, as in the soft tissues of the body.
When shell uptake was studied systematically, activity was found to be
distributed nearly uniformly throughout the shell (both plastron and
carapace) and the same result was obtained when the lactate was measured by direct chemical analysis using conventional enzymatic methods.
When these results were related to other measurements, taking into
account the sizes of the various compartments, it was determined that
about 30 percent of the total body lactate was located in the shell at the
end of six hours of anoxia at 20 °C (Jackson *et al.*, 1996).

Subsequent work has extended this observation by looking at shell
lactate uptake during anoxic submergences at 3 °C and 10 °C (Jackson,
1997). At 3 °C, turtles were submerged in N_2-equilibrated water for 12
weeks, at the end of which time shell and plasma samples were tested
for lactate concentrations. The results revealed that shell lactate (in
mmol/kg) was elevated to nearly the same value as plasma lactate (in

mmol/l). Furthermore, the shell lactate was distributed uniformly in the representative shell regions tested, just as was the case at 20 °C. This indicates that a single shell sample can reveal the behavior of the whole shell mass.

A second set of experiments was performed at 10 °C. Catheterized turtles were submerged for nine days and recovered with access to air for an additional nine days. Samples of plasma and shell (collected by a hole punch from the margin of the carapace) were taken periodically during and following the anoxic period. Once again, shell lactate was found to increase in concentration during anoxia by a similar magnitude to plasma, and both the rise during anoxia and the fall during recovery closely paralleled the plasma changes.

The concentrations of shell lactate at each temperature shown in Figure 3 are expressed as mmol/kg shell mass. But the shell is mostly

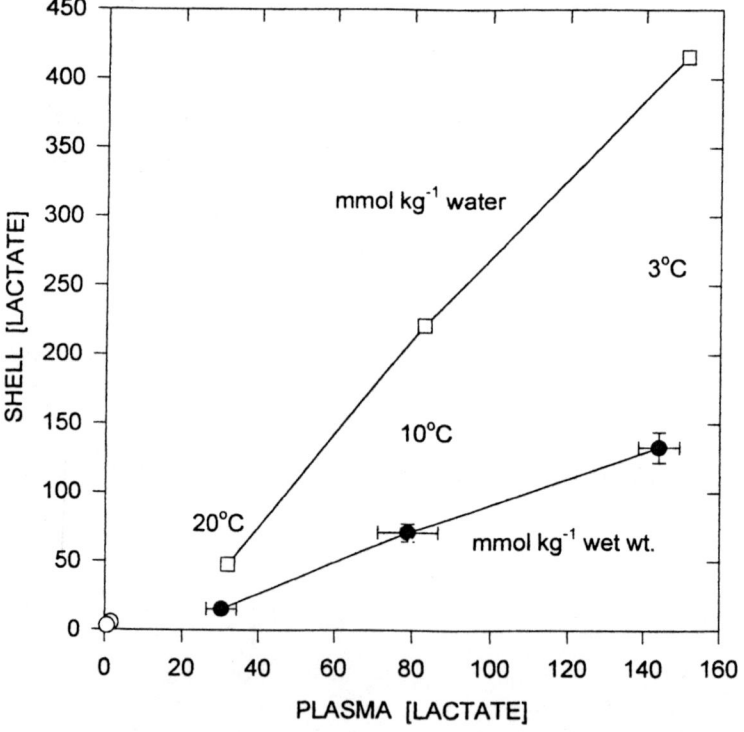

Fig. 3. Shell [Lac⁻] after prolonged anoxia at 20 °C (6 hours), 10 °C (9 days), and 3 °C (12 weeks), expressed as mmol/kg wet weight of shell (solid circles) and as mmol/kg shell water (open squares).

mineral and organic matrix and only 32 percent water, so it is instructive to express the lactate concentration as well per kg of shell water (Fig. 3). Note that very high concentrations exceeding 0.4 M are obtained.

Calculations based on previously determined body fluid volumes and tissue lactate distribution reveal that over 40 percent of the total body lactate was present in the shell at the end of the anoxic periods at both 3 °C and 10 °C (Fig. 4), larger than the percentage in either the estimated extracellular or intracellular compartments at these temperatures.

Most of the shell lactate is postulated to exist in combined form rather than in solution as the simple lactate ion for two reasons. First, lactate combines with calcium when both ions are present at high concentration (Jackson & Heisler, 1982), a condition that surely exists in the shell of the anoxic turtle. The abundance of available calcium within the shell may, in fact, explain why this structure can act as a sink for

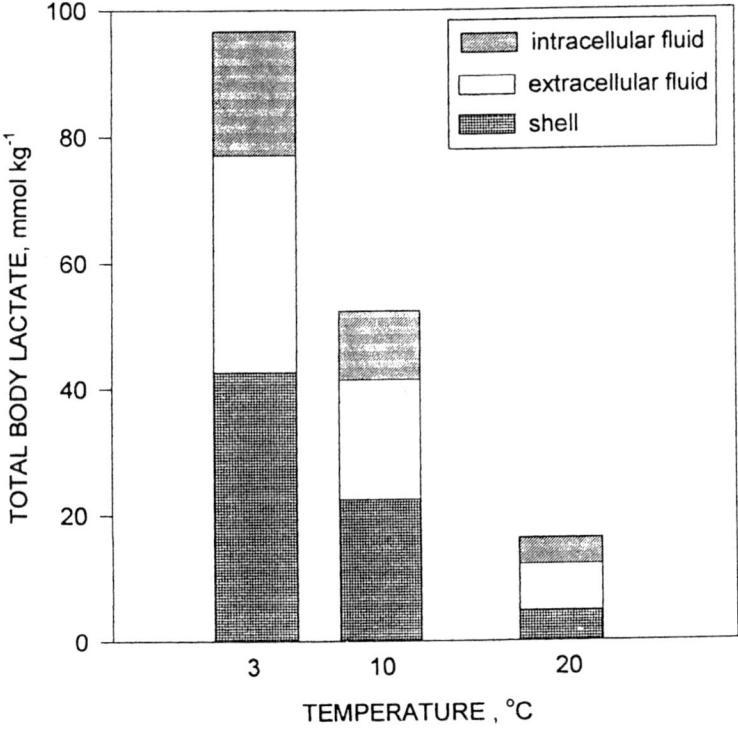

Fig. 4. Distribution of Lac⁻ in body compartments of turtles following submergence anoxia at 3 °C, 10 °C, and 20 °C. Note that at 3 °C and 10 °C, the shell is the chief repository for this ion.

lactate. Second, because of the small volume of shell water (only 32 percent of the shell mass), the total lactate would have to be at very high concentration (greater than 0.4 mol/l at 3 °C) if it were dissolved in this small volume. Such a high concentration strongly supports the notion that a sizable fraction of the lactate must be present in combined form, but whether in solution as the complex or as a precipitate is uncertain. Whatever its form, the lactate uptake is readily reversible, as shown by the recovery data of the 10 °C turtles.

Issues related to the results and methodologies

In vivo versus *in vitro* observations

Bone and shell are complex tissues with passive and active exchange processes. Measurements *in vivo*, while informative, are often difficult to interpret because of the integrative nature of the system. The shell (and bone generally) is a mixture of solids, cells, and distinct extracellular compartments, interacting in ways that are still not well understood, even in mammalian bone (Green, 1994). It is not a simple matter to secure samples of bone from a living animal, and samples may not be representative of the whole bone mass. A distinct advantage of the turtle shell is that it can be sampled readily *in vivo* and single samples are representative of the whole structure, at least in terms of lactate uptake (Jackson *et al.*, 1996).

In-vitro preparations can simplify the system and permit greater experimental control over important variables, but it is not always clear what has been lost in the process. For example, powdered turtle shell incubations *in vitro* can potentially permit an analysis of ionic exchanges between the shell substance and a defined bathing solution, but how meaningful is this physiologically? The normal structure of the shell, including possible barriers to exchange, as well as normal cellular function have been lost. The bathing solution may lack factors normally present *in vivo* that are important to the exchange processes. It is necessary to establish that the simple system reproduces, in important respects, the behavior of the tissue *in vivo*.

Contribution of the shell to total lactic acid buffering

The results presented indicate that the shell buffers lactic acid in two ways: first, by releasing buffers into the extracellular fluid; and second, by storing lactate in the substance of the shell itself. The technical challenges associated with quantifying these contributions involve determining both the total lactic acid production of the anoxic turtle and how much of it the shell buffers.

The total lactic acid load requires data on the volumes and lactate concentrations of the buffering compartments of the body. There are uncertainties in all these measurements and in the assumptions that underlie the calculations. A more direct approach that has been employed in some studies is to homogenize the entire animal and measure lactate on the homogenate (Gatten, 1974), but this is a terminal procedure and not readily performed on a relatively large animal with a calcified shell.

The contribution of mineral release from the shell to lactic acid buffering requires accurate measurements of the relevant strong cations in the fluid compartments they enter, and compartment volumes. The important compartment is the extracellular fluid, assuming little entry into the intracellular space. Published measurements of extracellular fluid volume in turtles, however, are quite variable (Stitt, Semple & Sigsworth, 1971; Smits & Kozubowski, 1985; Jackson *et al.*, 1996). Extracellular fluid volume could change during anoxia, and, as discussed earlier, this could account for the apparent discrepancy between the measured decrease in shell sodium without an accompanying increase in extracellular sodium.

For mechanism 2, the entry of lactate into the shell, the significance to acid buffering is uncertain because the observation that lactate concentration increases in the shell, *per se*, does not prove that the shell buffered the associated protons. Lactate may enter the shell accompanied by a strong cation (sodium, magnesium, calcium) or in exchange for a strong anion (chloride) and these are neutral exchanges from an acid–base point of view. However, strong circumstantial evidence diminishes the likelihood of these possibilities. Available evidence indicates that sodium, magnesium, and calcium all leave the shell during anoxia; for the anion, chloride, plasma levels fall substantially during anoxia (Herbert & Jackson, 1985a) so it is unlikely it is moving from the shell into the extracellular fluid. Based on these considerations and an application of Occam's razor, the simplest interpretation is that lactate entry into the shell is coupled to proton entry, and that the protons are buffered there by carbonate.

With these reservations for both mechanisms in mind, the contribution of the shell to lactic acid buffering in the anoxic turtle can be cautiously estimated. The calculation is based on the assumption that the increases in plasma calcium and magnesium occur throughout a known extracellular fluid volume and that the increase in shell lactate represents an equivalent amount of buffering by the shell. The distributions of buffering at 3 °C, 10 °C, and 20 °C are depicted in Figure 5. It is important to emphasize that these calculations may actually *underestimate* the true shell contribution because the limb bones and skull of the skeleton are not included and because the possible participation of shell sodium release is neglected.

Conclusions

The shell of the turtle is of crucial importance to this animal's remarkable anoxia tolerance. Without the great enhancement of total body buffering afforded by the shell, it is unlikely that such extended periods

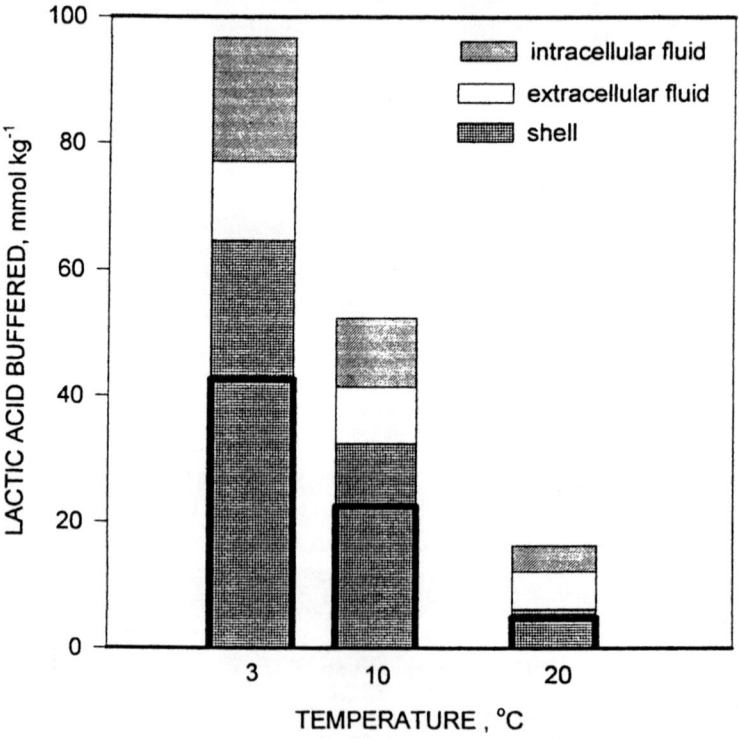

Fig. 5. Calculated distribution of lactic acid buffering by body compartments of turtles following prolonged submergence anoxia at 3 °C, 10 °C, and 20 °C. (See text for assumptions underlying these calculations.) The shell contribution (bottom panel at each temperature) is subdivided into the lactic acid uptake component (bottom portion with heavy border) and the buffer release component. The uptake component is the dominant one at each temperature.

of anoxia could be survived. The role of the shell is clearly an extreme case, but should nonetheless serve to remind us not to overlook the possible contribution of the skeleton in other organisms to buffering acid–base disturbances, particularly those of a chronic nature.

Acknowledgements

The author's work reported in this chapter was supported by the National Science Foundation (USA) grant IBN 94-20017. The contributions of the author's students and collaborators are gratefully acknowledged, as is the hospitality of Professor T.J. Dawson of the University of New South Wales, Sydney, where this chapter was written.

References

Bettice, J.A. (1984). Skeletal carbon dioxide stores during metabolic acidosis. *American Journal of Physiology* **247**, F326–30.

Burnell, J.M. (1971). Changes in bone sodium and carbonate in metabolic acidosis and alkalosis in the dog. *Journal of Clinical Investigation* **50**, 327–31.

Bushinsky, D.A. (1994). Acidosis and bone. *Mineral and Electrolyte Metabolism* **20**, 40–52.

Bushinsky, D.A., Goldring, J.M. & Coe, F.L. (1985). Cellular contribution to pH-mediated calcium flux in neonatal mouse calvariae. *American Journal of Physiology* **248**, F785–9.

Bushinsky, D.A., Krieger, N.S., Geisser, D.I., Grossman, E.B. & Coe, F.L. (1983). Effects of pH on bone calcium and proton fluxes *in vitro*. *American Journal of Physiology* **245**, F204–9.

Bushinsky, D.A. & Lechleider, R.J. (1987). Mechanism of proton-induced bone calcium release: calcium carbonate dissolution. *American Journal of Physiology* **253**, F998–1005.

Bushinsky, D.A., Levi-Setti, R. & Coe, F.L. (1986). Ion microprobe determination of bone surface elements: effect of reduced medium pH. *American Journal of Physiology* **250**, F1090–7.

Calder, W.A. III (1984). *Size, function, and life history.* Cambridge, MA: Harvard University Press.

Clark, N.B. (1970). The parathyroid. In *Biology of the reptilia*, Vol. 3, Morphology C, ed. C. Gans & T.S. Parsons., pp. 235–62. New York: Academic Press.

Daw, J.C., Wenger, D.P. & Berne, R.M. (1967). Relationship between cardiac glycogen and tolerance to anoxia in the western painted turtle, *Chrysemys picta bellii*. *Comparative Biochemistry and Physiology* **22**, 69–73.

Edgren, R.A. (1960). A seasonal change in bone density in female musk turtles, *Sternothaerus odoratus* (Latreille). *Comparative Biochemistry and Physiology* **1**, 213–17.

Eiamong, S. & Kurtzman, N.A. (1994). Metabolic acidosis and bone disease. *Mineral and Electrolyte Metabolism* **20**, 72–80.

Forbes, G.B. (1960). Studies on sodium in bone. *Journal of Pediatrics* **56**, 180–90.

Fuster, J.F., Pagés, T. & Palacios, L. (1997). Effect of temperature on oxygen stores during aerobic diving in the freshwater turtle *Mauremys caspica leprosa*. *Physiological Zoology* **70**, 7–18.

Gatten, R.E. Jr (1974). Effects of temperature and activity on aerobic and anaerobic metabolism and heart rate in the turtles *Pseudemys scripta* and *Terrapene ornata*. *Comparative Biochemistry and Physiology* **48**, 619–48.

Green, J. (1994). The physicochemical structure of bone: cellular and noncellular elements. *Mineral and Electrolyte Metabolism* **20**, 7–15.

Herbert, C.V. & Jackson, D.C. (1985a). Temperature effects on the responses to prolonged submergence in the turtle *Chrysemys picta bellii*. I. Blood acid–base and ionic changes during and following anoxic submergence. *Physiological Zoology* **58**, 655–69.

Herbert, C.V. & Jackson, D.C. (1985b). Temperature effects on the responses to prolonged submergence in the turtle *Chrysemys picta bellii*. II. Metabolic rate, blood acid–base and ionic changes, and cardiovascular function in aerated and anoxic water. *Physiological Zoology* **58**, 670–81.

Hochachka, P.W. (1982). Anaerobic metabolism. In *A companion to animal physiology*, ed. C.R. Taylor, K. Johansen & L. Bolis, pp. 138–47. Cambridge: Cambridge University Press.

Hochachka, P.W. & Mommsen, T.P. (1983). Protons and anaerobiosis. *Science* **219**, 1391–7.

Irving, L. & Chute, A.L. (1932). The participation of the carbonates of bone in the neutralization of ingested acid. *Journal of Cellular and Comparative Physiology* **2**, 157–76.

Jackson, D.C. (1968). Metabolic depresssion and oxygen depletion in the diving turtle. *Journal of Applied Physiology* **24**, 503–9.

Jackson, D.C. (1969). Buoyancy control in the freshwater turtle (*Pseudemys scripta elegans*). *Science* **166**, 1649–51.

Jackson, D.C. (1971). Mechanical basis for lung variability in the turtle. *American Journal of Physiology* **220**, 754–8.

Jackson, D.C. (1997). Lactate accumulation in the shell of the turtle *Chrysemys picta bellii*, during anoxia at 3 and 10 °C. *Journal of Experimental Biology* **200**, 2295–300.

Jackson, D.C. & Heisler, N. (1982). Plasma ion balance of submerged anoxic turtles at 3 °C: the role of calcium lactate formation. *Respiration Physiology* **49**, 159–74.

Jackson, D.C. & Heisler, N. (1983). Intracellular and extracellular acid–base and electrolyte status of submerged anoxic turtles at 3 °C. *Respiration Physiology* **53**, 187–201.

Jackson, D.C., Toney, V.I. & Okamoto, S. (1996). Lactate distribution and metabolism during and after anoxia in the turtle, *Chrysemys picta bellii*. *American Journal of Physiology* **271**, R409–16.

Poyart, C.F., Burseaux, E. & Fréminet, A. (1975). The bone CO_2 compartment: evidence for a bicarbonate pool. *Respiration Physiology* **25**, 85–99.

Schlesinger, P.H., Mattsson, J.P. & Blair, H.C. (1994). Osteoclastic acid transport: mechanism and implications for physiological and pharmacological regulation. *Mineral and Electrolyte Metabolism* **20**, 31–9.

Schwartz, W.B., Hall, P.W. III, Hays, R.M. & Relman, A.S. (1959). On the mechanism of acidosis in chronic renal disease. *Journal of Clinical Investigation* **38**, 39–52.

Smith, H.W. (1929). The inorganic composition of the body fluids of the Chelonia. *Journal of Biological Chemistry* **82**, 651–61.

Smits, A.W. & Kozubowski, M.M. (1985). Partitioning of body fluids and cardiovascular responses to circulatory hypovolaemia in the turtle, *Pseudemys scripta elegans*. *Journal of Experimental Biology* **116**, 237–50.

Stitt, J.T., Semple, R.E. & Sigsworth, D.W. (1971). Plasma sequestration produced by acute changes in body temperature in turtles. *American Journal of Physiology* **221**, 1185–8.

Ultsch, G.R. (1989). Ecology and physiology of hibernation and overwintering among freshwater fishes, turtles, and snakes. *Biological Reviews* **64**, 435–516.

Ultsch, G.R. & Jackson, D.C. (1982). Long-term submergence at 3 °C of the turtle, *Chrysemys picta bellii*, in normoxic and severely hypoxic water. I. Survival, gas exchange and acid–base status. *Journal of Experimental Biology* **96**, 11–28.

Warburton, S.J. & Jackson, D.C. (1995). Turtle (*Chrysemys picta bellii*) shell mineral content is altered by exposure to prolonged anoxia. *Physiological Zoology* **68**, 783–98.

Zangerl, R. (1969). The turtle shell. In *Biology of the reptilia*, Vol. 1, Morphology A, ed. C. Gans, A.d'A. Bellairs & T.S. Parsons, pp. 311–39. New York: Academic Press.

NIA M. WHITELEY

Acid–base regulation in crustaceans: the role of bicarbonate ions

Introduction

The maintenance of an appropriate acid–base status in crustaceans, as in all animals, takes place against a background of continuous production of acid metabolites, predominately CO_2 which hydrates to form H_2CO_3 and subsequently H^+ and HCO_3^-. The resultant acidosis, considered as $[H^+]$, is dependent on changes to strong anion *versus* cation concentrations, interactions with weak acids and ventilatory control of P_{CO_2} levels (see Tyler-Jones & Taylor, this volume). In aquatic crustaceans, compensation for acid–base perturbations is dominated by the branchial uptake of HCO_3^- from the surrounding water (Cameron, 1985; Wheatly & Henry, 1992). Some species, however, are able to survive bouts of aerial exposure by utilising an internal source of HCO_3^-, possibly from the calcified exoskeleton, to buffer an incipient acidosis (Truchot, 1975; Taylor & Wheatly, 1980, 1981; Taylor & Whiteley, 1989). This ability may dervive from the fact that the exoskeleton undergoes cyclical changes associated with the moult cycle, involving rapid and reversible changes in HCO_3^- fluxes. HCO_3^- transfer in aquatic crustaceans can therefore occur between the haemolymph and either the external medium or the calcified exoskeleton, and can rapidly switch direction, as shown during the moult cycle (Fig. 1).

The crustacean exoskeleton is chiefly composed of horizontal fibres of chitin and protein, interspersed with calcium carbonate held in a solid state by the presence of an alkaline pH (Cameron & Wood, 1985; Wheatly *et al.*, 1991). Consequently, the general body surface is relatively impermeable to ions, water, gases and solutes (Lignon, 1986). Over the gills, however, the exoskeleton is relatively thin (1–3 µm in the lamellae and filaments of aquatic species), uncalcified and, in several osmoregulating crustaceans, ion selective (Smith & Linton, 1971; Avenet & Lignon, 1985; Taylor & Taylor, 1992). As diffusion distances are also relatively small, the gills in aquatic crustaceans are the main route for gas and ion exchange between the animal and its environment. The branchial epithelial cells are composed of a number

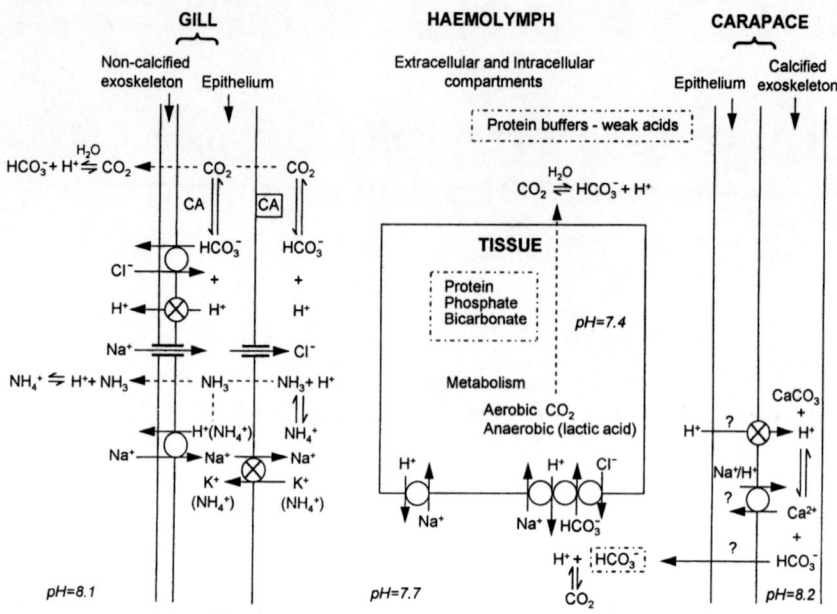

Fig. 1. Schematic summary of the possible mechanisms involved in the transport of acid–base equivalents between the water, haemolymph and intracellular compartments of a typical aquatic crustacean. The acidic end-products of tissue metabolism, CO_2 and lactic acid, are buffered intracellularly by protein, phosphate and HCO_3^-, with transfer of H^+ and HCO_3^- into the extracellular compartment by either electroneutral Na^+/H^+ ($2Na^+/H^+$ in hepatopancreas) or Na^+-dependent Cl^-/HCO_3^- exchange. Transfer of metabolic H^+ into the haemolymph is buffered by proteins and HCO_3^-, which both act as weak acids (see boxes outlined by the broken line). At the gill, CO_2 can either be excreted by passive diffusion or, in the presence of cytoplasmic carbonic anhydrase, be hydrated into HCO_3^- and H^+ for transfer by ion-exchange mechanisms. Protons can also be excreted across the gill in the form of NH_3/NH_4^+. Ion exchange is thought to take place in the mitochondria-rich cells or ionocytes, which are characterised by polarised ion transport with specific apical and basolateral ion exchangers. In hyperosmoregulating crustaceans, Cl^- uptake takes place *via* an apical electroneutral anion transporter in exchange for HCO_3^-. Compensation for an internal acidosis involves efflux of protons and/or uptake of HCO_3^- from the water (mechanisms unknown) or the calcified exoskeleton by mechanisms that remain to be identified but which involve acidification of the carapace fluid. Electrogenic proton pumps in the apical surface of gill epithelium cells are thought to drive Cl^-/HCO_3^- exchange across the gills (Onken &

of morphologically distinct cell types. Cells involved in active ion transport are characterised by abundant mitochondria, infoldings of the plasma membrane and the presence of carbonic anhydrase which catalyses the hydration/dehydration of $CO_2/H_2CO_3/HCO_3^-$. This enzyme is present in two functional forms: cytoplasmic carbonic anhydrase which catalyses CO_2 to HCO_3^-, providing HCO_3^- as a counter-ion for ion exchange, and membrane-bound carbonic anhydrase, which facilitates CO_2 excretion by catalysing HCO_3^- to CO_2 (Henry & Cameron, 1983; Henry, 1984; Burnett & McMahon, 1985). Although ion-transporting epithelia are also found in the antennal glands and the gut, their contribution to HCO_3^- exchange and acid–base regulation is relatively minor (Wheatly & Toop, 1989; Wheatly & Henry, 1992), emphasising the importance of the gills in HCO_3^- exchange.

Aquatic crustaceans provide interesting animal models in which to study the mechanisms of acid–base regulation. The restriction of base equivalent fluxes between tissues, haemolymph and gills, and the functional mobilisation of HCO_3^- to and from the exoskeleton, opens up many possibilities for an examination of the mechanisms responsible for transmembrane transfer of HCO_3^-. The aim of this chapter is to describe our current understanding of the role of HCO_3^- in acid–base regulation in aquatic decapod crustaceans. Attention is given to changes in circulating HCO_3^- levels which lead to adjustments in acid–base status in aquatic crustaceans during aerial exposure, when the branchial route of ion exchange is impaired; to the moult cycle when HCO_3^- fluxes are substantial and bidirectional; and in response to seasonal changes in temperature, as some crustaceans have been shown to maintain a constant pH with temperature in the winter by elevating $[HCO_3^-]$. Associated changes in intracellular pH (pH_i) are considered, including

---

Fig. 1 caption contd.

Putzenlechner, 1995). Proton pumps together with Ca^{2+}/H^+ exchangers may also be present in the epithelium lining the calcified exoskeleton. As a consequence of these regulatory mechanisms, whole-animal acid–base balance in settled, submerged aquatic crustaceans involves the net excretion of base equivalents (acid uptake), probably due to a herbivorous diet. However, some species can be in complete acid–base balance with zero excretion of base or acid (see Wheatly & Henry, 1992). Passive diffusion is represented by broken lines, electrogenic ion pumps by ─⊗►, and electroneutral ion exchange by ⟋O⟍. CA = carbonic anhydrase.

a brief description of the measurement techniques used to obtain these values. By describing three contrasting situations, the compensatory role of HCO_3^- during acid–base regulation is addressed at the level of the whole animal and then at the level of the cell. Our current understanding of the mechanisms involved in base equivalent transfer in crustaceans is limited and offers a fascinating area for further study.

Aerial exposure

The larger marine decapodan crustaceans inhabiting coastal waters are a valuable food resource and are highly sought after by fishermen. Live transport to market often exposes sublittoral species to long periods out of water, leading to disruptions of respiratory gas exchange and acid–base balance. On removal of these committed water breathers into air, their gills collapse, increasing diffusion distances and impairing respiratory gas exchange, resulting in immediate systemic hypoxia, an associated accumulation of lactic acid due to anaerobic metabolism, and progressive hypercapnia (Fig. 2). Despite these drastic changes in blood chemistry, species such as the European lobster, *Homarus gammarus*, are able to survive considerable periods out of water by virtue of an elevation in $[HCO_3^-]$ levels to compensate for the incipient respiratory and metabolic acidosis (Taylor & Whiteley, 1989). The accumulation of buffer base (HCO_3^-) in lobsters denied the branchial route for HCO_3^- exchange suggests that HCO_3^- is mobilised from internal stores, in particular from the $CaCO_3$ of the mineralised exoskeleton, as Ca^{2+} increases in stoichiometric proportion to the elevation in $[HCO_3^-]$ (Fig. 2). Compensation for the potential acidosis and progressive accumulation of lactate and Ca^{2+} which have specific allosteric effects on the haemocyanin molecule, increasing its affinity for O_2, contribute to a recovery in O_2 transport by the haemolymph. Consequently, O_2 content values rise between 3 and 14 hours in air (Taylor & Whiteley, 1989).

The ability to compensate for a progressive extracellular acidosis varies among species. The lobsters *H. gammarus* and *H. americanus* are able to survive prolonged periods out of water (72–96 hours) due to the elevation of haemolymph $[HCO_3^-]$ levels. Other marine crustaceans are less tolerant, as shown in the pH–HCO_3^- diagrams in Figure 3. The edible crab, *Cancer pagurus*, and the spider crab, *Maja squinado*, both develop an uncompensated acidosis after three hours in air, which is more pronounced in the latter due to the accumulation of lactic acid. The elevation in $[HCO_3^-]$ observed in *M. squinado* is insufficient to compensate for the metabolic acidosis, which proves fatal after 12 hours (Taylor & Innes, 1988). The velvet swimming crab, *Necora puber*, is

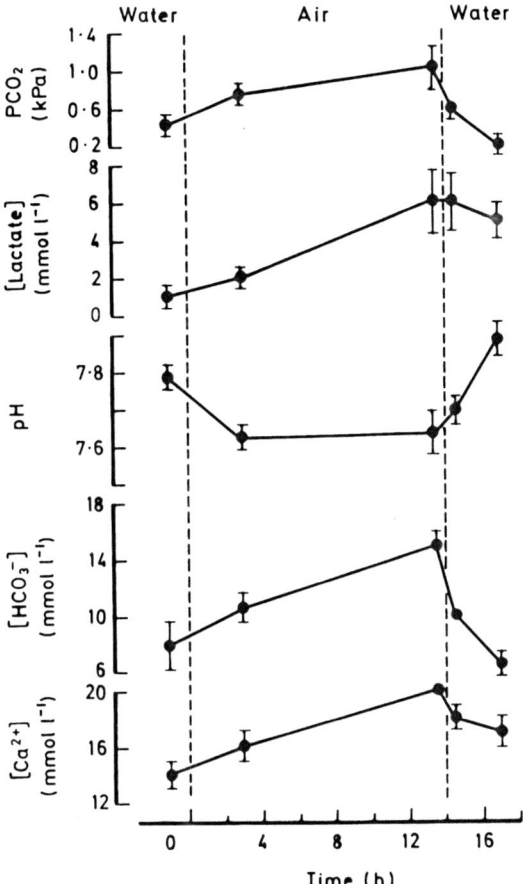

Fig. 2. Mean values (±SEM) for acid–base variables in the haemolymph of the lobster, submerged in water (time 0), after 3 hours' and 14 hours' aerial exposure, and on return to water for 0.5 hours and 3 hours. Acid–base variables include CO_2 partial pressure (P_{CO_2}), Lactate, pH and bicarbonate concentration [HCO_3^-]. Data taken from Taylor and Whiteley (1989).

particularly susceptible to aerial exposure and dies within four hours due to a rapid respiratory and metabolic acidosis which remains uncompensated as haemolymph [HCO_3^-] levels do not change from submerged levels (Fig. 3). These examples show that the ability to survive aerial exposure is closely linked to the ability to mobilise HCO_3^- from internal stores. Those crabs that are unable to mobilise HCO_3^- may not have the

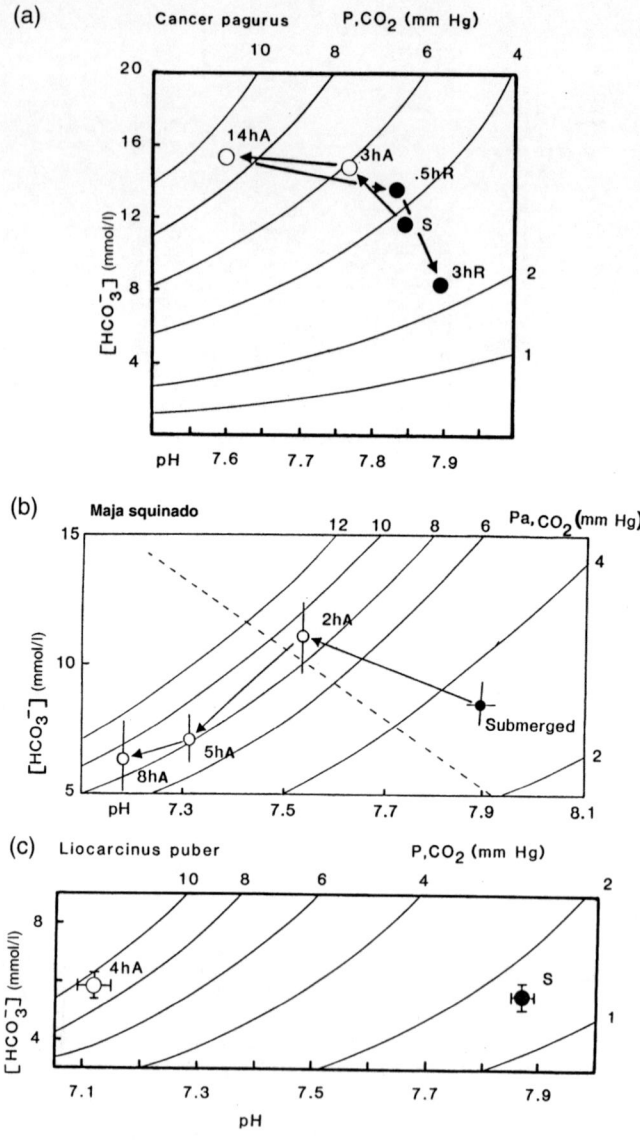

Fig. 3. Relationship between mean values (±SEM) of P_{CO_2}, HCO_3^- and pH in the haemolymph of three species of marine decapod crustaceans during aerial exposure, shown as pH-HCO_3^- diagrams. (a) *Cancer pagurus*, (b) *Maja squinado*, and (c) *Nercora puber* (formerly *Liocarcinus puber*). Closed circles represent values obtained from submerged animals (S) and animals returned to water (R); open sym-

mechanisms required to transfer HCO_3^- from the exoskeleton or there may be differences in the way in which these processes are controlled. Lobsters appear to be pre-adapted for aerial exposure. This ability may be related to their distribution in the intertidal zone before overfishing restricted numbers to the sublittoral zone. Therefore, the ability to survive aerial exposure may be related to the degree of aerial exposure that these crustaceans normally encounter in their natural environment, suggesting adaptive changes in the mechanisms involved in acid–base regulation.

Aquatic crustaceans exposed to highly variable environments, such as those found in the intertidal zone, can survive prolonged periods of aerial exposure by functioning as facultative air breathers. The shore crab, *Carcinus maenas*, manages to avoid low Po_2 levels encountered in small rock pools of warmed sea water by air breathing (Wheatly & Taylor, 1979). A similar response is shown in freshwater habitats where crayfish make excursions onto land in search of either food or new bodies of water (Huxley, 1880). Despite the ability of intertidal crabs and crayfish to respire equally well in water and in air (Wallace, 1972; Taylor & Wheatly, 1980, 1981; Houlihan & Innes, 1984), all primarily aquatic decapods become hypoxic and hypercapnic during aerial exposure due to the predominant diffusion limitation of the collapsed gill system. Even though some CO_2 elimination is possible over the gills into reservoirs of water retained in the branchial chambers (Burnett & McMahon, 1987), CO_2 accumulation usually exceeds CO_2 excretion during aerial exposure, and Pco_2 rises. Compensation for the ensuing acidosis results from an elevation in haemolymph $[HCO_3^-]$ as found in *H. gammarus*. This mechanism is effective but relatively slow, as complete compensation for an extracellular acidosis in *C. maenas* takes 100 hours (Truchot, 1975), after which time the crab can be regarded as being in a state of fully compensated hypercapnic acidosis (Dejours, 1981), with elevated $[HCO_3^-]$ at 14 mmol/l. In the blue crab, *C. sapidus*, Cameron and Iwama (1987) found that crabs subjected to a stepwise

---

Fig. 3 caption contd.

bols represent values measured during aerial exposure (x h A). The broken oblique line in (b) represents the non-bicarbonate buffering capacity of the haemolymph. Data from E. W. Taylor and P. J. Butler (unpublished observations), Taylor and Innes (1988), and N.M. Whiteley and E.W. Taylor (unpublished observations).

increase in environmental hypercapnia experienced further elevations in [HCO$_3^-$] to achieve a new steady-state set point for pH regulation. In these experiments, haemolymph [HCO$_3^-$] values of up to 50 mmol/l were recorded, challenging the idea of an upper limit for [HCO$_3^-$] levels in aquatic crustaceans.

During aerial exposure, restoration of haemolymph pH can also be mediated by the transfer of acid equivalents from extracellular to intracellular compartments. In the freshwater crayfish, *Austropotamobius pallipes*, haemolymph lactate levels increase on initial exposure to air but subsequently decline after three hours, reducing the metabolic acidosis (Taylor & Wheatly, 1981). On return to water, lactate ions are released back into the circulation, suggesting sequestration of lactate within internal compartments during aerial exposure. Measurements of pH$_i$ reveal that abdominal muscle pH$_i$ in the crayfish species, *A. pallipes* and *Pacifasticus leniusculus*, decreases during aerial exposure and remains low as extracellular pH (pH$_e$) levels recover (Tyler-Jones & Taylor, 1988; N.M. Whiteley, E.W. Taylor & A. Al-Wassia, unpublished observations). This could be due to sequestration of H$^+$ into the abdominal muscle or the slower kinetics of pH$_i$ regulation observed in this compartment (Rodeau, 1984). In contrast, heart pH$_i$ values in *P. leniusculus* are closely regulated during aerial exposure, as found in crayfish subjected to hyperoxia or hypercapnia (Gaillard & Malan, 1983; Wheatly *et al.*, 1991), whereas the hepatopancreas and the epidermis of the branchiostegite develop an alkalosis after 14 hours in air (N.M. Whiteley, E.W. Taylor & A. Al-Wassia, unpublished observations), suggesting an exchange of acid–base equivalents between these compartments and the haemolymph.

The mechanisms responsible for the mobilisation and transport of HCO$_3^-$ from the exoskeleton during aerial exposure are poorly understood. However, pH$_i$ is thought to be an important factor influencing the formation or dissolution of CaCO$_3$ to CO$_3^{2-}$ and HCO$_3^-$. Normally, the carapace fluid compartment is alkaline compared to the haemolymph and favours the formation and storage of CaCO$_3$ (Cameron & Wood, 1985; Wheatly *et al.*, 1991). During premoult, however, acidification of this compartment leads to the formation of Ca^{2+} and HCO$_3^-$ (Wheatly, 1997). A similar mechanism may be responsible for the mobilisation of exoskeletal HCO$_3^-$ during aerial exposure.

Moult cycle

Moulting in crustaceans involves the loss of one mineralised exoskeleton and the formation of another, to allow for continued growth and development. Ecdysis, the point at which the old exoskeleton is lost, is a physical event that involves morphological, physiological and biochemical changes. In the stages leading up to ecdysis, disruptions to acid–base balance are caused by the separation of the old and new exoskeletons (apolysis), resulting in an increase in diffusion distances and the isolation of ion-exchange mechanisms from the surrounding water. Studies on *C. maenas* (Truchot, 1976), *A. leptodactylus* (Dejours & Beckenkamp, 1978) and *C. sapidus* (Mangum *et al.*, 1985) have all shown that haemolymph pH values increase prior to ecdysis, accompanied by an elevation in [HCO_3^-] levels. In *C. sapidus* and *A. pallipes* (Fig. 4), haemolymph P_{CO_2} values also increase, and in *C. sapidus* this is followed by a virtual loss of the haemocyanin O_2 transport system and an elevation in lactate levels (Mangum *et al.*, 1985). Despite the potential acidosis, premoult *C. sapidus* develop an alkalosis due to a base excess in the haemolymph. The specific source of HCO_3^- remains obscure, but the impairment of branchial ion exchange suggests that HCO_3^- is mobilised from intracellular compartments, as found in some crustacean species during aerial exposure. The link between premoult increases in haemolymph [HCO_3^-] due to decalcification of the exoskeleton and the mobilisation of HCO_3^- in response to an acid–base disturbance around ecdysis is unclear. Mangum *et al.* (1985) suggested that in *C. sapidus*, the two processes are regulated in the exoskeleton independently of each other, but this remains to be tested in other crustaceans.

Immediately after ecdysis the general body surface is permeable and the rate of water uptake is greatly increased (Mykles, 1980), resulting in dilution of the haemolymph and exchange of respiratory gases across the general body surface. In *C. sapidus* there is a net uptake of CO_2, presumably in the form of HCO_3^-, in early postmoult which exceeds metabolic production for up to seven days after ecdysis (Cameron & Wood, 1985). At the same time there are massive reversals in the flux rates of several electrolytes, in particular Ca^{2+}, which is lost to the external environment during premoult but is taken up from the water for postmoult mineralisation of the exoskeleton (Wheatly, 1996). Flux experiments on freshwater crayfish have shown that the postmoult Ca^{2+} uptake is strongly coupled to HCO_3^- uptake (Greenaway, 1974; Wheatly & Ignaszewski, 1991; Wheatly & Gannon, 1993). Therefore, it is reasonable to assume that the large uptake of Ca^{2+} during postmoult calcification may affect acid–base balance due to an imbalance between rates of $CaCO_3$ uptake and deposition. However, changes in acid–base status in postmoult *C. sapidus* are minimal, consisting of a small transient decrease in pH immediately after ecdysis caused by a decrease in

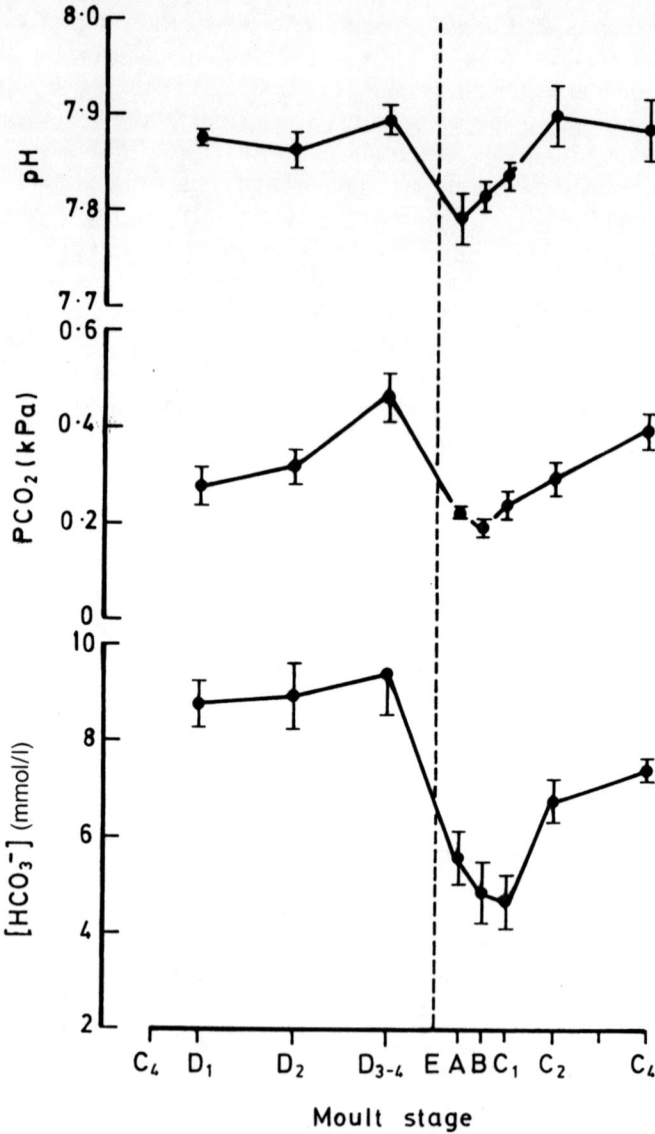

Fig. 4. Mean changes (±SEM) in haemolymph pH, P_{CO_2} and HCO_3^- in the crayfish, *A. pallipes*, over the moult cycle. The vertical broken line represents ecdysis (E), when the old exoskeleton is shed. Using the moult cycle stages originally described by Drach (1939), stages A, B, C_1–C_3 represent postmoult (metecdysis), C_4 represents intermoult, and D_1–D_4 represents premoult (proecdysis). Number of observations = 5–10.

buffer base at constant P_{CO_2}, and a transient accumulation of lactate ions (Mangum *et al.*, 1985). Both P_{CO_2} and lactate levels subsequently decrease and acid–base status returns towards intermoult levels by the early stages of postmoult, even though calcification is incomplete. A similar response occurs in *A. pallipes* as haemolymph P_{CO_2}, pH and [HCO$_3^-$] levels all decrease after ecdysis. Haemolymph pH and P_{CO_2} levels return to intermoult values by stage C_2 of postmoult, but [HCO$_3^-$] levels take longer to readjust (Fig. 4). Major disruptions to acid–base balance are restricted to individuals involved in abortive moults as these animals experience a severe acidosis and tissue hypoxia which prove fatal.

Temperature

Acid–base balance in aquatic crustaceans is affected by changes in environmental temperature in a variety of ways. Firstly, temperature has an important influence on the rate of various biological processes, including rates of oxidative metabolism reflected as changes in O_2 uptake and CO_2 excretion. Both rate processes vary directly with short-term changes in ambient temperature but show modulation of this response in the long term (Taylor, 1981). The increase in O_2 demand associated with rising temperature is met by an increase in ventilation and heart rate (Taylor, Burtler & Al-Wassia, 1977; McMahon *et al.*, 1978), resulting in increased O_2 uptake and delivery to the tissues. Aquatic crustaceans are also influenced by temperature-related changes in the gas-exchange conditions in the surrounding water. Acclimation to low temperatures in the crayfish, *A. leptodactylus*, results in a hyper-capnic acidosis in the haemolymph, due to hypoventilation in response to temperature-related increases in O_2 concentration in the surrounding water (Dejours & Armand, 1983).

Acid–base data collected from crustaceans held in the laboratory under controlled conditions have shown that there is an inverse relation-ship between haemolymph pH and temperature which approximates to the temperature-dependent change in the pH of pure water (pN) (Fig. 5). On the basis of this similarity, it is said that these animals maintain a constant difference bewteen pH and pN or, equivalently, a constant relative alkalinity (constant ratio of [OH$^-$] to [H$^+$]) as temperature varies (Rahn, 1967; Howell *et al.*, 1973). This is thought to confer the advan-tage of ensuring an unchanging degree of ionisation of proteins and of the α-imidazole group in particular, to maintain the integrity of protein function over a range of temperatures, as proposed by the alpha-stat hypothesis of Reeves (1972) and the 'Z'-stat model of Cameron

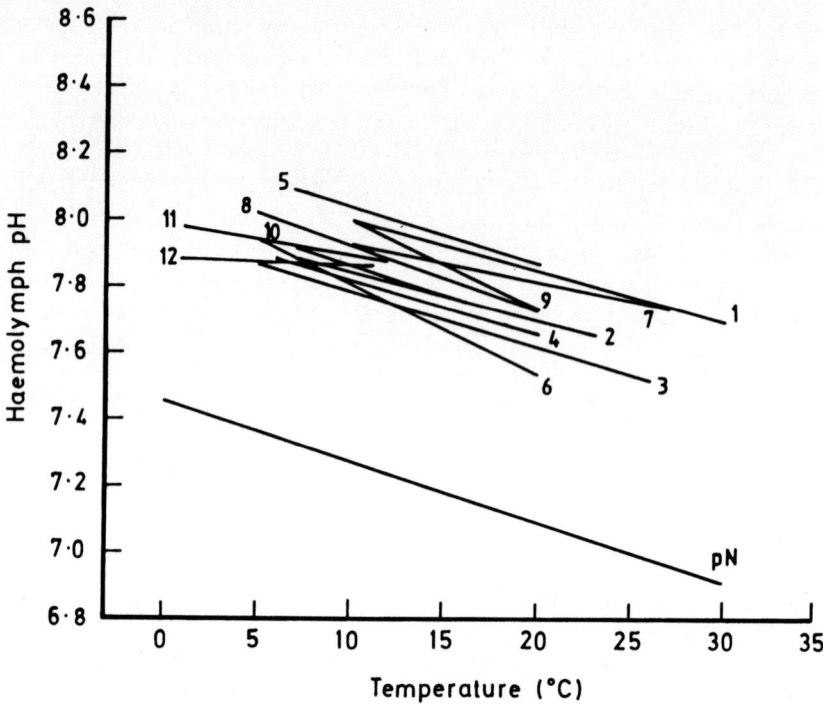

Fig. 5. Inverse relationship between haemolymph pH and temperature in different species of aquatic crustaceans from a range of different studies, compared to the change in the pH of water (pN) with temperature (broken line). Values include those obtained from: 1. *C. sapidus* (Wood & Cameron, 1985); 2. *C. maenas* (Truchot, 1978); 3. *C. maenas* (Truchot, 1973); 4. *A. leptodactylus* (Gaillard & Malan, 1985); 5. *A. leptodactylus* (Dejours & Armand, 1983); 6. *C. sapidus* (Mauro & Mangum, 1982); 7. *C. sapidus* (Cameron & Batterton, 1978); 8. *C. maenas* (Howell *et al.*, 1973); 9. *C. sapidus* (Howell *et al.*, 1973); 10. *C. magister* (McMahon *et al.*, 1978); 11. *A. pallipes* (Whiteley *et al.*, 1995); 12. *A. pallipes* (Whiteley & Taylor, 1993).

(1989a). As temperature increases, HCO_3^- levels in aquatic crustaceans typically decline, while adjustments in P_{CO_2} are relatively unimportant due to the physical limitations imposed by using water as a respiratory medium (Rahn, 1966). As shown in Figure 5, the majority of the pH measurements vary along the line for constant relative alkalinity, with temperature coefficients ranging from −0.015 to −0.023 pH units/°C. However, two exceptions to this general trend have been included (lines

11 and 12 on Fig. 5). These data were collected from the freshwater crayfish *A. pallipes*, seasonally acclimatised to changes in temperature and examined both in the field and in the laboratory, as described below.

Examination of a wild population of *A. pallipes* inhabiting a pool in the West Midlands, England, revealed a complex relationship between acid–base balance and temperature (Whiteley & Taylor, 1993). In crayfish sampled *in situ*, haemolymph pH varied with summer temperatures to follow a constant relative alkalinity. As temperatures decreased in the autumn/winter from 12 °C to 1 °C, haemolymph pH remained unchanged at approximately 7.9, despite an increase in [HCO_3^-] at constant P_{CO_2}. Consequently, haemolymph pH was maintained at a constant pH instead of a constant relative alkalinity at low environmental temperatures. This response can be explained in terms of the changing relationship between pH, P_{CO_2} and HCO_3^- against a background of the temperature-dependent changes in physico-chemical constants as explained by Whiteley and Taylor (1993), indicating active regulation of pH_e. Active regulation of haemolymph pH is highly seasonal as crayfish caught in the winter at a water temperature of 5 °C and acclimated to 2 °C, 8 °C and 15 °C in the laboratory show little change in pH_e with temperature. In contrast, animals caught in September at a temperature of 11 °C and warmed to 18 °C are characterised by pH_e values which decrease along a line of constant relative alkalinity at a coefficient of −0.013 pH units/°C (Fig. 6).

The physiological significance of constant pH_e in winter crayfish is unclear. It raises interesting questions about protein function and enzyme activity in particular. Maintenance of a relatively acidotic pH_e, compared to the value expected if the line of constant relative alkalinity is extrapolated back to 0 °C, indicates suppression of metabolic rate. This response may serve to preserve glycogen stores when resources are low, as suggested in other crustacean species (Cameron, 1989b). A similar response has been observed in several fish species seasonally acclimatised to low temperatures, such as the American eel in which temperature coefficients for pH_e are also very low (Walsh & Moon, 1982). Just like the American eel, the crayfish *A. pallipes* is sluggish and relatively inactive during the winter, suggesting reduced rates of metabolism at low temperatures.

Studies on the relationship between pH_i and temperature in aquatic crustaceans are limited, but in crabs acclimated to temperature change, pH_i varies between tissues and also with temperature to maintain a constant relative alkalinity (Rodeau, 1984; Gaillard & Malan, 1985; Wood & Cameron, 1985). These laboratory-based studies have shown similar relationships for $\Delta pH/\Delta t$ in both the extracellular and intracellu-

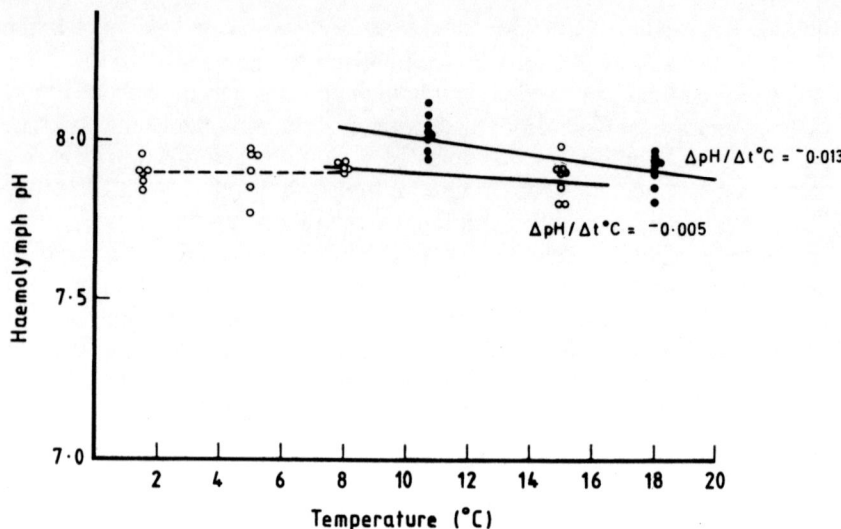

Fig. 6. Seasonal changes in the relationship between haemolymph pH and acclimation temperature in *A. pallipes*. Crayfish were taken from the wild, either in January at a water temperature of 5 °C (open symbols) or in September at a water temperature of 13 °C (closed symbols). In January, crayfish were acclimated to 1 °C, 8 °C or 15 °C for 4 weeks. In September, the animals were acclimated to 10 °C or 17 °C for the same amount of time. Individual values given.

lar compartments. In contrast, other studies have shown that pH_i can vary independently of pH_e (Tyler-Jones & Taylor, 1988; Wheatly & Henry, 1992). In *A. pallipes* caught in the winter and acclimated to 1 °C and 12 °C for four weeks, the relationship between pH_i and temperature varied between the tissues and was independent of changes in pH_e (Fig. 7). In claw muscle and hepatopancreas, pH_i remained unchanged with temperature, while in abdominal muscle, pH_i varied to follow a constant relative alkalinity (Whiteley, Naylor & Taylor, 1995). The maintenance of constant pH_i in some intracellular compartments and not in others is thought to correspond to the functional integrity of the respective tissues during the winter. For example, the claw muscles used for catching prey and defending territories in the summer are less in demand in the winter than the abdominal muscles, which may be needed in the tail-flick response to escape predators.

It appears that the maintenance of a constant relative alkalinity in the extracellular and intracellular compartments of aquatic crustaceans held

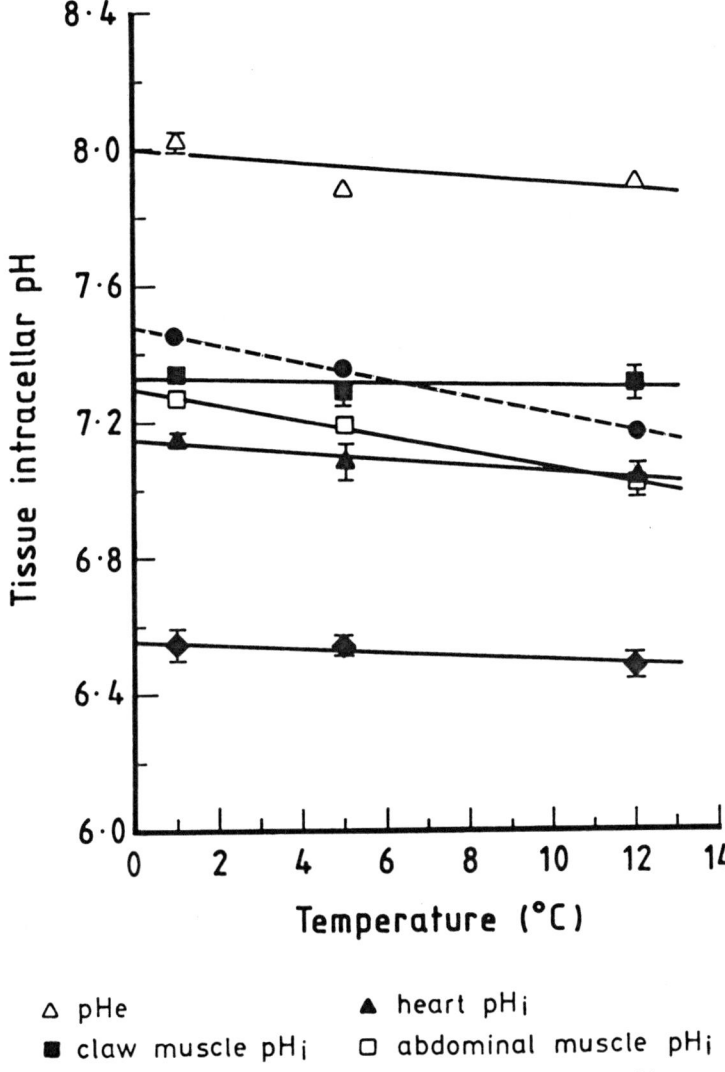

Fig. 7. Relationship between pH$_i$ and acclimation temperature in *A. pallipes* caught in the winter at 5 °C and returned to the laboratory for acclimation to 1 °C and 12 °C. Corresponding values for haemolymph pH (pH$_e$) are also included. In the claw muscle and hepatopancreas, there was little change in pH$_i$ with temperature (−0.003 and −0.006 pH units/°C, respectively), while heart pH$_i$ varied by −0.0087 pH units/°C to parallel changes in haemolymph pH. The pH$_i$ of the abdominal muscle varied by −0.0246 pH units/°C and showed the closest similarity with ΔpN/Δt (−0.019 pH units/°C). Mean values ±SEM (*n*=5–9).

in the laboratory may not apply to crustaceans sampled in their natural environment, particularly to those species that experience seasonal changes in temperature and low levels of activity in the winter. An interesting comparison for these data is the acid–base values measured in an Antarctic marine crustacean, the isopod *Glyptonotus antarcticus*, which lives continuously at a narrow range of temperatures from $-1.89\,°C$ to $+2.0\,°C$. Haemolymph pH values in *G. antarcticus* vary between 7.85 and 8.2 (Qvist *et al.*, 1977; Jokumsen *et al.*, 1981; Whiteley *et al.*, 1997). Towards the lower end of this range, the values resemble the pH_e levels measured in *A. pallipes* at $1\,°C$. This congruence is accompanied by similar HCO_3^- levels in both species, with $[HCO_3^-]$ ranging from 10 to 14 mmol/l. Consequently, the acid–base status of crustaceans living permanently at low temperatures may match those found in temperate crustaceans, subjected to seasonal bouts of low temperature. It is interesting to note that metabolic rates in *G. antarcticus* are relatively low, as this animal does not show metabolic cold adaptation (Whiteley, Taylor & El Haj, 1996). Therefore, the maintenance of a haemolymph pH value around 7.9 may be functionally related to the low rates of metabolism characteristic of aquatic crustaceans living close to the freezing point of water.

Methodology

A comparison of the pH_i values obtained in crustacean tissues by different measurement techniques reveals that pH_i values obtained by the homogenisation technique are lower than those obtained by other methods, such as the distribution of weak acids. For example, measurements of heart pH_i in *A. pallipes* at $12\,°C$ by the homogenisation of frozen tissue gives values of 7.03 (Whiteley *et al.*, 1995). These values are lower than the value of 7.41 obtained in *A. leptodactylus* at $13\,°C$ determined by the distribution of 5,5-dimethyl-2,4-oxazolidinedione (DMO) (Gaillard & Malan, 1983). However, this generalisation does not apply to all tissues, as shown by various measurements of pH_i in the claw muscle which are remarkably similar, regardless of measurement technique, being 7.3 in *A. pallipes* and 7.32 in *A. leptodactylus* under the conditions described above. In addition, measurement of pH_i by ^{31}P nuclear magnetic resonance spectroscopy of the abdominal muscle in the prawn *Palaemon serratus* gave a value of 7.13 ± 0.04 at $7\,°C$ (Thebault & Raffin, 1991), which was lower but similar to the value of 7.19 ± 0.04 obtained using the homogenisation method in *A. pallipes* held at $5\,°C$ (Whiteley *et al.*, 1995). From these comparisons, it can be seen that the pH_i values may be repeatable between studies for some tissues, possibly larger for anaerobic tissues, while for other tissues the measured values are best considered relative to one another rather than being taken as absolute values.

Mechanisms of HCO_3^- transfer between body compartments

Gas and ion exchange with the surrounding water occur across specialised epithelial cells in the permeable gills (Taylor & Taylor, 1992; Pequeux, 1995). The cells responsible for ion exchange (the ionocytes) are polarised. Cl^- uptake in hyper-osmoregulating crustaceans is associated with the efflux of HCO_3^- *via* an apical electroneutral Cl^-/HCO_3^- exchanger which is driven by an apical electrogenic proton pump (Onken & Putzenlechner, 1995; see Fig. 1). This branchial anion exchanger may also be involved in the regulation of acid–base balance as shown in the lobster, *H. americanus*, on return to water after a period of aerial exposure. In these crustaceans, there was an increase in haemolymph Cl^- and a decrease in haemolymph HCO_3^- levels (N.M. Whiteley, personal observations). Return to water in *H. gammarus* is also accompanied by an immediate haemolymph alkalosis due to the passive diffusion of CO_2 to the external water and longer term retention of HCO_3^-; further supporting the involvement of ion-translocating mechanisms in HCO_3^- efflux (Taylor & Whiteley, 1989). Alternatively, HCO_3^- may be re-incorporated into internal stores on recovery from aerial exposure. The specific involvement of branchial Cl^-/HCO_3^- exchangers in acid–base regulation, however, remains unresolved as measurements of H^+ fluxes in isolated gill preparations of *Carcinus maenas* demonstrated that Cl^-/HCO_3^- exchange, Na^+/H^+ exchange, Na^+/HCO_3^- co-transport and H^+-ATPases were not involved in the regulation of elevated pH levels (Siebers *et al.*, 1994). In contrast, pH regulation was dependent on the presence of carbonic anhydrase and HCO_3^- in the perfusion medium, indicating catalysed dehydration of HCO_3^- to diffusable CO_2. In addition, ion and acid–base adjustments in *Eriocheir sinensis* transferred from sea water to fresh water, and vice versa, occurred at different rates, indicating that ion-exchange mechanisms cannot be used to explain fully the acid–base changes that take place during variations in environmental salinity (Truchot, 1992).

The mechanisms responsible for the inward flux of HCO_3^- across the gills are even more obscure. One possibility is the involvement of a direct HCO_3^- uptake mechanism, as compensation for a hypercapnic acidosis in *C. sapidus* was accompanied by net apparent H^+ excretion due to the uptake of HCO_3^- and associated efflux of Cl^- (Cameron

& Iwama, 1987). Collectively, these experiments indicate that in hyper-osmoregulating crustaceans, net branchial HCO_3^- fluxes are a compromise between acid–base balance and ion regulation. The specific role of Cl^-/HCO_3^- exchangers in acid–base regulation in both osmoconforming and osmoregulating crustaceans is unclear and warrants further study.

Transmembrane electroneutral ion exchange is also involved in the regulation of pH_i in crustaceans (Taylor, Whiteley & Wheatly, 1991; Wheatly & Henry, 1992). Two mechanisms have been identified in crayfish neurons: a Na^+-dependent Cl^-/HCO_3^- exchanger which exchanges external Na^+ and HCO_3^- for internal Cl^- and H^+; and a Na^+/H^+ exchanger which is the main route for acid extrusion in this tissue (Moody, 1981; Thomas, 1989; see Fig. 1). Both exchangers have also been identified in crustacean muscle, but in this tissue the Cl^-/HCO_3^- exchanger is more important for acid extrusion than the Na^+/H^+ mechanism (Gaillard & Rodeau, 1987). An electrogenic $2Na^+/H^+$ exchanger has been identified in the hepatopancreas of several crustaceans which is used as a means of Na^+ uptake from dilute media across the lumen of the intestine (Ahearn, Franco & Clay, 1990). Both Na^+/H^+ and $2Na^+/H^+$ exchangers have also been located in the gills, either performing a function similar to that described above in the hepatopancreas or as a means of regulating pH_i in the gill epithelia (Shetlar & Towle, 1989; Onken & Putzenlechner, 1995).

Acknowledgements

The author would like to thank S. Egginton, E.W. Taylor and J.A. Raven for the opportunity to present this work. The author is also grateful to E.W. Taylor and A. Al-Wassia for the use of unpublished data and to the Natural Environment Research Council for financial support.

References

Ahearn, G.A., Franco, P. & Clay, L.P. (1990). Electrogenic $2Na^+/1H^+$ transport in crustaceans. *Journal of Membrane Biology* **116**, 215–26.

Avenet, P. & Lignon, J.M. (1985). Ionic permeabilities of the gill lamina cuticle of the crayfish, *Astacus leptodactylus* (E.). *Journal of Physiology* **363**, 377–401.

Burnett, L.E. & McMahon, B.R. (1985). Facilitation of CO_2 excretion by carbonic anhydrase located on the surface of the basal membrane of crab gill epithelium. *Respiration Physiology* **62**, 341–8.

Burnett, L.E. & McMahon, B.R. (1987). Gas exchange, haemolymph acid–base status, and the role of branchial water stores during air exposure in three littoral crab species. *Physiological Zoology* **60**, 27–36.

Cameron, J.N. (1985). Compensation of hypercapnic acidosis in the aquatic blue crab, *Callinectes sapidus*: the predominance of external sea water over carapace carbonate as a proton sink. *Journal of Experimental Biology* **114**, 197–206.

Cameron, J.N. (1989a). Acid–base homeostasis: Past and present perspectives. *Physiological Zoology* **62**, 845–65.

Cameron, J.N. (1989b). Acid–base equilibria in invertebrates. In *Acid–base regulation in animals*, ed. N. Heisler, pp. 357–94. New York: Elsevier.

Cameron, J.N. & Batterton, C.V. (1978). Temperature and acid–base status in the blue crab, *Callinectes sapidus*. *Respiration Physiology* **35**, 101–10.

Cameron, J.N. & Iwama, G. (1987). Compensation of progressive hypercapnia in channel catfish and blue crabs. *Journal of Experimental Biology* **133**, 183–97.

Cameron, J.N. & Wood, C.M. (1985). Apparent H^+ excretion and CO_2 dynamics accompanying carapace mineralization in the blue crab (*Callinectes sapidus*) following moulting. *Journal of Experimental Biology* **114**, 181–96.

Dejours, P. (1981). *Principles of comparative respiratory physiology*. New York: Elsevier.

Dejours, P. & Armand, J. (1983). Acid–base balance of crayfish haemolymph; effects of simultaneous changes of ambient temperature and water oxygenation. *Journal of Comparative Physiology* **149**, 463–8.

Dejours, P. & Beckenkamp, H. (1978). Crayfish respiration as a function of water oxygenation. *Respiratory Physiology* **30**, 241–51.

Drach, P. (1939). Mue et cycle d'intermue chez les Crustaces Decapodes. *Annals of the Institute of Oceanography, Paris* **19**, 103–391.

Gaillard, S. & Malan, A. (1983). Intracellular pH regulation in response to ambient hypoxia or hypercapnia in the crayfish. *Molecular Physiology* **4**, 231–43.

Gaillard, S. & Malan, A. (1985). Intracellular pH-temperature relationship in a water breather, the crayfish. *Molecular Physiology* **7**, 1–16.

Gaillard, S. & Rodeau, J.L. (1987). Na^+/H^+ exchange in crayfish neurones: dependence on extracellular sodium and pH. *Journal of Comparative Physiology* **157**, 435–44.

Greenaway, P. (1974). Calcium balance at the postmoult stage of the freshwater crayfish *Austropotamobius pallipes* (Lereboullet). *Journal of Experimental Biology* **64**, 149–57.

Henry, R.P. (1984). The role of carbonic anhydrase in blood ion and acid–base regulation. *American Zoologist* **24**, 241–51.

Henry, R.P. & Cameron, J.N. (1983). The role of carbonic anhydrase in respiration, ion regulation and acid–base balance in the aquatic crab, *Callinectes sapidus*, and the terrestrial crab, *Gecarcinus lateralis*. *Journal of Experimental Biology* **101**, 205–23.

Houlihan, D.F. & Innes, A.J. (1984). The cost of walking in crabs: aerial and aquatic oxygen consumption during activity of two species of intertidal crabs. *Comparative Biochemistry and Physiology* **77A**, 325–34.

Howell, B.J., Rahn, H., Goodfellow, D. & Herreid, C. (1973). Acid–base regulation and temperature in selected invertebrates as a function of temperature. *American Zoologist* **13**, 557–63.

Huxley, T.H. (1880). *The crayfish: an introduction to the study of zoology*. London: Kegan Paul.

Jokumsen, A., Wells, R.M.G., Ellerton, H.D. & Weber, R.E. (1981). Hemocyanin of the giant Antarctic isopod, *Glyptonotus antarcticus*: structure and effects of temperature and pH on its oxygen affinity. *Comparative Biochemistry and Physiology* **70A**, 91–5.

Lignon, J.M. (1986). Structure and permeability of decapod Crustacea cuticle. In *Comparative physiology of environmental adaptations to salinity and dehydration*, Vol. 1, ed. R. Kirsh & B. Lahlou, pp. 178–87. Basel: Karger.

Mangum, C.P., McMahon, B.R., deFur, P.L. & Wheatly, M.G. (1985). Gas exchange, acid–base balance and oxygen supply to the tissues during a molt of the blue crab, *Callinectes sapidus*. *Journal of Crustacean Biology* **5**, 188–206.

Mauro, N.A. & Mangum, C.P. (1982). The role of the blood in the temperature dependence of oxidation metabolism in decapod crustaceans. I. Intra-specific responses to seasonal differences in temperature. *Journal of Experimental Zoology* **219**, 179–88.

McMahon, B.R., Sinclair, F., Hassall, C.D., deFur, P.L. & Wilkes, P.R.H. (1978). Ventilation and control of acid–base status during temperature acclimation in the crab, *Cancer magister*. *Journal of Comparative Physiology* **128**, 109–16.

Moody, W.J. (1981). The ionic mechanism of intracellular pH regulation in crayfish neurones. *Journal of Physiology* **316**, 293–308.

Mykles, D.L. (1980). The mechanism of fluid absorption at ecdysis in the American lobster, *Homarus americanus*. *Journal of Experimental Biology* **84**, 89–101.

Onken, H. & Putzenlechner, M. (1995). A V-ATPase drives active, electrogenic and Na^+ independent Cl^- absorption across the gills of *Eriocheir sinensis*. *Journal of Experimental Biology* **198**, 767–74.

Pequeux, A. (1995). Osmotic regulation in crustaceans. *Journal of Crustacean Biology* **15(1)**, 1–60.

Qvist, J., Weber, R.E., De Vries, A.L. & Zapol, W.M. (1977). pH and haemoglobin oxygen affinity in blood from the Antarctic cod *Dissostichus mawsoni*. *Journal of Experimental Biology* **67**, 77–88.

Rahn, H. (1966). Aquatic gas exchange: theory. *Respiration Physiology* **1**, 1–12.

Rahn, H. (1967). Gas transport from the external environment to the cell. In *Development of the lung* (Ciba Foundation Symposium), ed. A.V.S. de Reuck & R. Porter, pp. 3–23. London: Churchill.

Reeves, R.B. (1972). An imidazole alpha stat hypothesis for vertebrate acid–base regulation: tissue carbon dioxide content and body temperature in bullfrogs. *Respiration Physiology* **14**, 219–36.

Rodeau, J.L. (1984). Effects of temperature on intracellular pH in crayfish neurons and muscle fibres. *American Journal of Physiology* **246**, C45–9.

Shetlar, R.E. & Towle, D.W. (1989). Electrogenic sodium–proton exchange in membrane vesicles from crab (*Carcinus maenas*) gill. *American Journal of Physiology* **257**, R924–33.

Siebers, D., Lucu, C., Böttcher, K. & Jürss, K. (1994). Regulation of pH in isolated perfused gills of the shore crab, *Carcinus maenas*. *Journal of Comparative Physiology* **164**, 16–22.

Smith, D.S. & Linton, J.R. (1971). Potentiometric evidence for active transport of sodium and chloride across excised gills of *Callinectes sapidus*. *Comparative Biochemistry and Physiology* **39A**, 367–78.

Taylor, E.W. (1981). Some effects of temperature on respiration in decapod crustaceans. *Journal of Thermal Biology* **6**, 239–48.

Taylor, E.W., Burtler, P.J. & Al-Wassia, A. (1977). Some responses of the shore crab, *Carcinus maenas* (L.), to progressive hypoxia at different acclimation temperatures and salinities. *Journal of Comparative Physiology* **122**, 391–402.

Taylor, E.W. & Innes, A. (1988). A functional analysis of the shift from gill- to lung-breathing during the evolution of land crabs (Crustacea, Decapoda). *Biological Journal of the Linnaean Society* **33**, 229–47.

Taylor, E.W. & Wheatly, M.G. (1980). Ventilation, heart rate and respiratory gas exchange in the crayfish *Austropotamobius pallipes* (Lereboullet) submerged in normoxic water and following 3h exposure in air at 15°C. *Journal of Comparative Physiology* **138**, 67–78.

Taylor, E.W. & Wheatly, M.G. (1981). The effect of long-term aerial exposure on heart rate, ventilation, respiratory gas exchange and acid–base status in the crayfish, *Austropotamobius pallipes*. *Journal of Experimental Biology* **92**, 109–24.

Taylor, E.W. & Whiteley, N. M. (1989). Oxygen transport and acid–base balance in the haemolymph of the lobster, *Homarus gammarus*, during aerial exposure and resubmersion. *Journal of Experimental Biology* **144**, 417–36.

Taylor, E.W., Whiteley, N.M. & Wheatly, M.G. (1991). Respiratory gas exchange and the regulation of acid–base status in decapod crustaceans. In *Physiological strategies for gas exchange and metabolism*, ed. A.J. Woakes, M.K. Grieshaber & C.R. Bridges, pp. 79–106. Society for Experimental Biology Seminar Series No. 41. Cambridge: Cambridge University Press.

Taylor, H.H. & Taylor, E.W. (1992). Gills and lungs: The exchange of gases and ions. In *Microscopic anatomy of invertebrates*, Vol. 10, *Decapod Crustacea*, ed. F.W. Harrison & A.G. Humes, pp. 203–93. New York: Wiley-Liss.

Thebault, M.T. & Raffin, J.P. (1991). Seasonal variations in *Palaemon serratus* abdominal muscle metabolism and performance during exercise, as studied by ^{31}PNMR. *Marine Ecology Progress Series* **74**, 175–83.

Thomas, R.C. (1989). Intracellular pH regulation and the effects of external acidification. In *Acid toxicity and aquatic animals*, ed. R. Morris, E.W. Taylor, D.J.A. Brown & J.A. Brown, pp. 113–24. Society for Experimental Biology Seminar Series No. 34. Cambridge: Cambridge University Press.

Truchot, J-P. (1973). Temperature and acid–base regulation in the shore crab *Carcinus maenas*. *Respiration Physiology* **17**, 11–20.

Truchot, J-P. (1975). Blood acid–base changes during experimental emersion and re-immersion of the intertidal crab, *Carcinus maenas* (L.). *Respiration Physiology* **23**, 351–60.

Truchot, J-P. (1976). CO_2 combining properties of the blood of the shore crab, *Carcinus maenas* (L.): CO_2 dissociation curves and Haldane effect. *Journal of Comparative Physiology* **112**, 283–93.

Truchot, J-P. (1978). Mechanisms of extracellular acid–base regulation as temperature changes in decapod crustaceans. *Respiration Physiology* **33**, 161–76.

Truchot, J-P. (1992). Acid–base changes on transfer between sea water and freshwater in the chinese crab, *Eriocheir sinensis*. *Respiration Physiology* **87(3)**, 419–27.

Tyler-Jones, R. & Taylor, E.W. (1988). Analysis of the haemolymph and muscle acid–base status during aerial exposure in the crayfish *Austropotamobius pallipes*. *Journal of Experimental Biology* **134**, 409–22.

Wallace, J.C. (1972). Activity and metabolic rate in the shore crab *Carcinus maeans* (L.). *Comparative Biochemistry and Physiology* **41(A)**, 523–33.

Walsh, P.J. & Moon, T.W. (1982). The influence of temperature on extracellular and intracellular pH in the American eel, *Anguilla rostrata* (Le Sueur). *Respiration Physiology* **50**, 129–42.

Wheatly, M.G. (1996). An overview of calcium balance in crustaceans. *Physiological Zoology* **69**, 351–82.

Wheatly, M.G. (1997). Crustacean models for studying calcium transport: the journey from whole organisms to molecular mechanisms. *Journal of the Marine Biology Association UK* **77**, 107–25.

Wheatly, M.G. & Gannon, A.T. (1993) The effect of external electrolytes on postmoult calcification and associated ion fluxes in the freshwater crayfish *Procambarus clarkii* (Girard). In *Freshwater crayfish*, Vol. 9, ed. D.M. Holdich & G.F. Warner, pp. 200–12. Lafayette: University of Southwestern Louisiana.

Wheatly, M.G. & Henry, R.P. (1992). Extracellular and intracellular acid–base regulation in crustaceans. *Journal of Experimental Zoology* **263**, 127–42.

Wheatly, M.G. & Ignaszewski, L.A. (1991). Electrolyte and gas exchange during the moulting cycle of a freshwater crayfish. *Journal of Experimental Biology* **151**, 469–83.

Wheatly, M.G. & Taylor, E.W. (1979). Oxygen levels, acid–base status and heart rate during emersion of the shore crab *Carcinus maenas* (L.) into air. *Journal of Comparative Physiology* **132**, 305–11.

Wheatly, M.G. & Toop, T. (1989). Physiological responses of the crayfish *Pacifasticus leniusculus* to environmental hyperoxia. II. The role of the antennal gland in acid–base and ion regulation. *Journal of Experimental Biology* **143**, 53–70.

Wheatly, M.G., Toop, T., Morrison, R.J. & Yow, L.C. (1991). Physiological responses of the crayfish *Pacifasticus leniusculus* to environmental hyperoxia. III. Intracellular acid–base balance. *Physiological Zoology* **64**, 323–43.

Whiteley, N.M., Naylor, J.K. & Taylor, E.W. (1995). Extracellular and intracellular acid–base status in the freshwater crayfish *Austropotamobius pallipes* between 1 and 12°C. *Journal of Experimental Biology* **198**, 567–76.

Whiteley, N.M. & Taylor, E.W. (1993). The effects of low temperature on extracellular and intracellular acid–base regulation in the freshwater crayfish, *Austropotamobius pallipes* (L.). *Journal of Experimental Biology* **181**, 295–311.

Whiteley, N.M., Taylor, E.W., Clarke, A. and El Haj, A.J. (1997). Haemolymph oxygen transport and acid–base status in *Glyptonotus antarcticus* Eights. *Polar Biology* **18**, 10–15.

Whiteley, N.M., Taylor, E.W. & El Haj, A.J. (1996). A comparison of the metabolic cost of protein synthesis in stenothermal and eurythermal isopod crustaceans. *American Journal of Physiology* **40(5)**, R1295–303.

Wood, C.M. & Cameron, J.N. (1985). Temperature and the physiology of intracellular and extracellular acid–base regulation in the blue crab *Callinectes sapidus*. *Journal of Experimental Biology* **114**, 151–79.

ROD W. WILSON

A novel role for the gut of seawater teleosts in acid–base balance

Introduction

Due to the physicochemical properties of the aquatic medium, the ability to regulate acid–base balance by manipulation of P_{CO_2} is severely limited in fish. Instead, the transfer of acid–base relevant ions between the extracellular fluid and external medium is utilised as the primary process for pH regulation (Heisler, 1993). There are four potential sites for the transfer of acid–base relevant ions to occur in fish: the gills, kidney, skin and gut. The relative and potential roles of each of these are outlined below, with emphasis on the areas least known about, before focusing on the newly discovered role of the gut in piscine acid–base balance.

Sites for the transfer of acid–base relevant ions in fish

The gills

Traditionally, the gills are considered to be the principal site of acid–base regulation. This has been repeatedly demonstrated in both freshwater-adapted and seawater-adapted fishes (Cameron, 1976; Claiborne & Heisler, 1984; Wood, Wheatly & Hobe, 1984; Tang, McDonald & Boutilier, 1989; Goss et al., 1992), and overall the gills are normally considered to contribute more than 90 per cent to the transfer of acid–base relevant ions to and from the external environment (Heisler, 1984, 1993; Wood, 1988). This is thought to be effected through a combination of (i) apical ion exchange mechanisms (Na^+/acidic equivalents and Cl^-/basic equivalents), and (ii) diffusion of acid–base relevant and 'strong' ions through the paracellular pathway (Wood, 1988; McDonald, Tang & Boutilier, 1989). Recent experiments in the freshwater rainbow trout suggest that although the perfusion rate within the central venous sinus of the gills is low (about 7 per cent of cardiac output; Ishimatsu, Iwama & Heisler, 1988), this may be the main site of ionic acid–base regulation in the gills (Iwama, Ishimatsu & Heisler,

1993). This is possibly related to the close proximity of the central venous sinus with the major fraction of branchial chloride cells (Laurent, 1984). However, this claim should be treated with some caution because (i) chloride cells can be found in large numbers on the lamellae as well (e.g. when fish are acclimated to low Ca^{2+} fresh water; Perry & Wood, 1985), and (ii) recent molecular evidence has shown proton pumps to be present in the lamellar pavement cells (e.g. Sullivan, Fryer & Perry, 1995).

The kidney

The involvement of the kidney in piscine acid–base regulation has been less well investigated, but the limited number of studies have shown that usually less than 8 per cent of the net acid–base relevant fluxes can be attributed to ions excreted *via* the urine in freshwater fish (Wood, 1988; Heisler, 1993). Furthermore, this proportion falls to less than 1 per cent in seawater-adapted teleosts, mainly due to their notoriously low urine flow rates (Evans, 1982; McDonald *et al.*, 1982), even though urine acid content tends to be somewhat higher in seawater-adapted teleosts (Hickman & Trump, 1969; McDonald *et al.*, 1982).

The skin

The skin of fish is normally considered relatively inactive with respect to acid–base balance. However, this assumption is based upon relatively little empirical evidence in teleost fish (Heisler, 1993), and despite the fact that the importance of the skin has been demonstrated in the transport of certain non-acid–base relevant ions, for example 50 per cent of total Ca^{2+} uptake in freshwater trout (Perry & Wood, 1985), and 65 per cent of total Cl^- efflux in the seawater shanny (*Blennius pholis*; Nonnette, Nonnette & Kirsch, 1979). Chloride cells are found in the skin of many fish species (Nonnette *et al.*, 1979; Hwang, 1989), and particularly in larvae and juveniles (Li *et al.*, 1995). However, with the exception of one study by Ishimatsu *et al.* (1992), the technical difficulty in sampling fluid from the low-pressure secondary circulatory system which supplies the skin (Vogel, 1985) has prevented a thorough investigation of its role in ion and acid–base regulation. Ishimatsu *et al.* (1992), by cannulating the lateral cutaneous vessel in freshwater rainbow trout, revealed a significant involvement of the skin in HCO_3^- accumulation (probably *via* Cl^-/HCO_3^- exchange) during hypercapnia. Although they judged the skin's overall contribution to acid–base regulation to be small (approximately 4 per cent) this was based on flow rates in the secondary system estimated under control conditions. They pointed out that this may underestimate the role of the skin during acid–base disturbances. In addition, it is worth noting that the skin becomes increasingly involved in nitrogen excretion at higher salinities, accounting for up to 53 per cent of the total amount of ammonia excreted in some marine teleosts (Sayer & Davenport, 1987). If a substantial portion of this cutaneous ammonia excretion occurs as ionised NH_4^+ (either by diffusion or Na^+/NH_4^+ exchange), then the skin of marine teleosts may have a greater role in determining acid–base balance than that of their freshwater counterparts. Further investigations are required to quantify the precise role of the skin in acid–base balance *in vivo*.

The gut

Two decades ago, Cameron (1978) highlighted that nothing is known about the possible role of the gastrointestinal tract of fish in acid–base regulation, even though it must be a good candidate given the known capacity for acid and base secretion from the stomach and pancreatic tissue, respectively. Based on this, Cameron suggested that the gut was worthy of investigation, especially in seawater teleosts in which the gut is constantly processing ingested seawater as part of their osmoregulatory strategy (see below). Despite his advice, this area has continued to be overlooked. Instead, the gills and kidney are typically treated as the only organs of major importance in acid–base regulation, with the gills performing the majority of this acid–base regulatory work (Wood, 1988; Heisler, 1993). However, it has recently been discovered that in the seawater-acclimated rainbow trout, the intestine does play a significant role in whole-animal acid–base balance by excreting substantial amounts of HCO_3^-/CO_3^{2-} base (Wilson *et al.*, 1996). The remainder of this chapter describes the evidence that has led to this conclusion, and discusses the possibility that it may be a feature common to all seawater-acclimated teleosts. In addition, our knowledge regarding the sites within the intestine and the mechanisms involved are highlighted, and the possibility that intestinal base excretion may function additionally, or even primarily, as part of the hypo-osmoregulatory strategy of marine teleosts is discussed.

Evidence for a role of the gut in acid–base regulation in seawater teleosts

Seawater-adapted teleosts are hyposmotic to the sea water in which they live and are constantly faced with an osmotic loss of water across their outer body surface. In order to compensate for this dehydration, they drink seawater at rates of 1–10 ml/kg/h per hour (Smith, 1930;

Shehadeh & Gordon, 1969; Evans, 1993). The intestine then plays an essential role in osmoregulation by actively absorbing NaCl, which is followed osmotically by a volume of water, thereby compensating for the aforementioned fluid losses. This process inevitably results in a salt load which in turn must be excreted by the gills. Thus, the intestine of seawater-adapted teleosts has a vital role in osmoregulation in addition to its more conventional role in digestion and nutrient absorption.

Of the imbibed seawater, approximately 70–85 per cent is absorbed during its passage through the gastrointestinal tract (Smith, 1930; Hickman, 1968; Shehedah & Gordon, 1969; Wilson et al., 1996), such that there is normally a steady excretion of rectal fluid (approximately 0.5 ml/kg/h in seawater trout). During their extensive investigation of the osmoregulatory function of the gut in seawater-adapted trout, Shehedah & Gordon (1969) reported carbonate deposits present in 'mucus tubes' found along the length of the intestine which were excreted with the rectal fluid. However, it was not until over 20 years later that Walsh et al. (1991) reported similar carbonate deposits in the intestine of gulf toadfish (*Opsanus beta*), and that measurements of rectal fluid in the same species revealed them to be alkaline (pH 8.6), with elevated total CO_2 (68 mmol/l; see also Wood et al., 1995). In a subsequent study on seawater-acclimated rainbow trout (Wilson et al., 1996), rectal fluid had similar values in fed fish, but significantly higher values in fish that had been starved for seven days (pH = 8.91 ± 0.04 and $[HCO_3^- + CO_3^{2-}]$ = 112 ± 9 mmol/l when measured by titration). Mucus tubes from the same trout were found to contain very high levels of $HCO_3^- + CO_3^{2-}$ (676 ± 124 mmol/kg), which is probably present as precipitated Ca and Mg salts (Walsh et al., 1991). This indicated that significant amounts of base as HCO_3^- and CO_3^{2-} may be excreted *via* the intestine in seawater teleosts.

Wilson et al. (1996) measured the rate of base excretion from the intestine in seawater-acclimated rainbow trout by collecting the excreted rectal fluid and mucus tubes *via* rectal catheters. The mean intestinal base excretion rate was 114 ± 15 μmol/kg/h. By far the majority of this excreted base arose endogenously. This was apparent from measurements of drinking rate and $[HCO_3^- + CO_3^{2-}]$ content of seawater which showed that only 6 per cent of that excreted from the intestine could be accounted for by that ingested in seawater (Fig. 1). Interestingly, even this small amount of base taken up with imbibed seawater is reduced to zero during passage through the stomach (Fig. 2), such that *all* base subsequently appearing in the intestine must arise from endogenous sources. It is also worth noting that although the base content of mucus tubes is up to six times higher than that in rectal

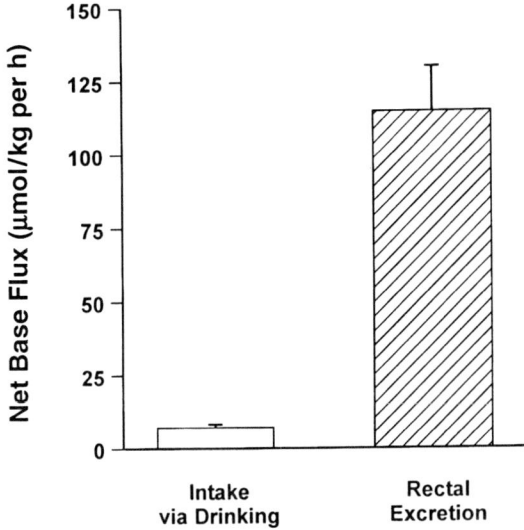

Fig. 1. The rates of base intake *via* drinking and base excretion *via* rectal fluid and mucus tube excretion in seawater-acclimated trout. Number of animals = 7.

fluids (on a per gram basis), the excretion of precipitated bicarbonates/carbonates only contributes a minor fraction (less than 18 per cent) of the total base excreted by the intestine.

To put the scale of this intestinal base excretion into perspective, whole-animal net acid–base fluxes are normally assessed by measuring two variables, the titratable acid flux (J^{TA}) and the net ammonia flux (J^{Amm}) between the fish and the external water (see Wood, 1988). In freshwater rainbow trout, these have values typically in the range of +100 to +300 µmol/kg/h for J^{TA}, and −100 to −300 µmol/kg/h for J^{Amm}. Negative values indicate excretion and positive indicate uptake, i.e. positive J^{TA} values imply acid uptake or base excretion. The sum of these two variables (signs considered) gives the net acid flux, which is often slightly negative (between 0 and −100 µmol/kg/h), at least in starved fish used in laboratory experiments, probably reflecting the metabolic production of fixed acidic equivalents in the oxidation of proteins (e.g. H_2SO_4 from cysteine). This indicates a small net excretion of acid under resting conditions.

The mean rate of base excretion from the intestine in seawater trout (114 µmol/kg/h) was approximately one-third of the whole-animal J^{Amm} flux in the same fish, and hence a significant component of the whole-

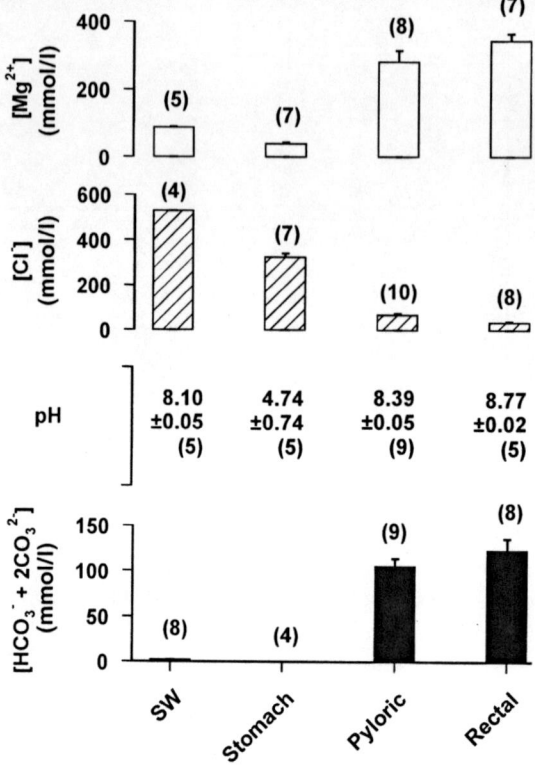

Fig. 2. Measured concentrations of [HCO₃⁻ + 2CO₃²⁻] base, Cl⁻ and Mg²⁺, as well as pH of fluid sampled from different positions along the gastrointestinal tract of seawater-acclimated rainbow trout. Numbers of samples are indicated in parentheses.

animal acid–base balance equation. When rectal catheters were allowed to drain outside of the chamber holding the seawater trout, it was found that this rate of intestinal base excretion was balanced by a similar net rate of acid excretion by the rest of the body, which helps to maintain the animals in approximate acid–base equilibrium (Wilson *et al.*, 1996). Thus, it would appear that the intestine of seawater-acclimated trout does have a considerable and hitherto unrecognised role in whole-animal acid–base balance. In addition, there appears to be a degree of anatomical separation for net acid and base fluxes in the seawater trout.

The source(s) and site(s) of this 'balancing' acid excretion are currently under investigation. The most likely site is the gills, and an intriguing possibility is that net acid excretion could be accounted for simply by an increased proportion of NH_4^+ excreted as opposed to NH_3. Freshwater trout probably excrete all their ammonia as NH_3 (Avella & Bornancin, 1989; Wilson *et al.*, 1994), which is acid–base neutral (Wood, 1993). In contrast, branchial diffusion of NH_4^+ will be enhanced in seawater teleosts by the considerably higher ionic permeabilities (Payan, Girard & Mayer-Gostan, 1984) and the tendency for strongly positive (blood side) transepithelial potentials (Potts, 1984). In addition, there is an increased availability of the external counter-ion for Na^+/NH_4^+ exchange in seawater (Tang *et al.*, 1989; Wilson & Taylor, 1992; Wood, 1993). Marine teleosts are therefore likely to excrete a larger proportion of the ammonia transferred across the gills as NH_4^+. Extrusion of NH_4^+ by diffusion and/or electroneutral Na^+/NH_4^+ exchange would contribute to net acid excretion, and may partly assist in countering the net base excretion from the intestine in seawater trout.

Although it is clear that intestinal base excretion forms a substantial part of acid–base balance in seawater trout, it is still unclear whether it is modified in response to internal acid–base status, and therefore whether or not it actually complements the gills as a site of true acid–base *regulation*. Experiments are currently underway to investigate this possibility using standard methods of inducing acid–base disturbances in fish (e.g. infusion of acid or base, exposure to hypercapnia, hyperoxia) coupled with measurements of acid–base fluxes *via* the intestine, kidney, gills and skin.

Gut base excretion as a general feature of seawater teleosts

Walsh *et al.* (1991) considered the formation of intestinal carbonates to be an osmoregulatory feature of marine teleosts in general because they found precipitates similar to those from toadfish in the intestine of four marine species which had been starved for several days (cod, rockfish, quillback rockfish, and midshipman), whereas the precipitates disappear in rainbow trout or toadfish transferred to lower salinities (Shehedah & Gordon, 1969; Walsh *et al.*, 1991). In support of this claim, Figures 3 and 4 show the acid–base status of rectal fluid and mucus tubes sampled from four euryhaline species and three truly marine species following acclimation to either fresh water or seawater (Table 1). These data confirm that carbonate deposits are present in seawater-acclimated teleosts and practically disappear when fish are held (or caught in the case of mullet) in fresh water. However, they also demonstrate for the first

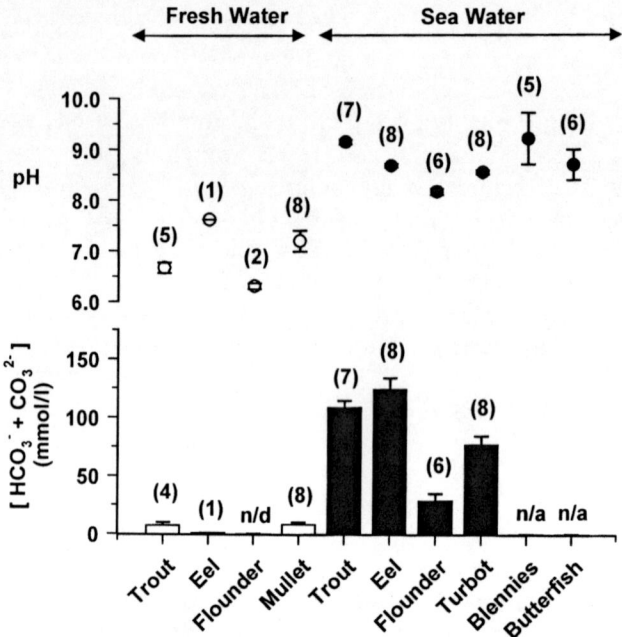

Fig. 3. Measured pH and $[HCO_3^- + CO_3^{2-}]$ base concentration of rectal fluid in four species of euryhaline teleost and three species of marine teleost (see Table 1 for a description of acclimation salinities). Number of animals sampled is indicated in parentheses. n/d = not detectable, n/a = not applicable due to insufficient volume for measurement.

time that the intestinal fluids (when present in sufficient amounts to be measured) are either neutral or slightly acidic in freshwater fish and have dissolved $[HCO_3^- + CO_3^{2-}]$ similar to or much lower than the expected plasma concentrations.

The low rectal fluid $[HCO_3^- + CO_3^{2-}]$ from all the freshwater species (less than 10 mmol/l) coupled with their much lower drinking rates (average for five different freshwater species was less than 0.1 ml/kg/h; Perrot et al., 1992) implies that intestinal base excretion will be negligible in freshwater teleosts (less than 1 μmol/kg/h). Thus, it seems reasonable to conclude that a role for the gut in acid–base balance is confined to teleosts acclimated to hyperosmotic salinities.

Site of intestinal base secretion

The intestine of trout can be divided into three approximately equal portions: pyloric, mid and posterior. The pyloric (anterior) intestine is

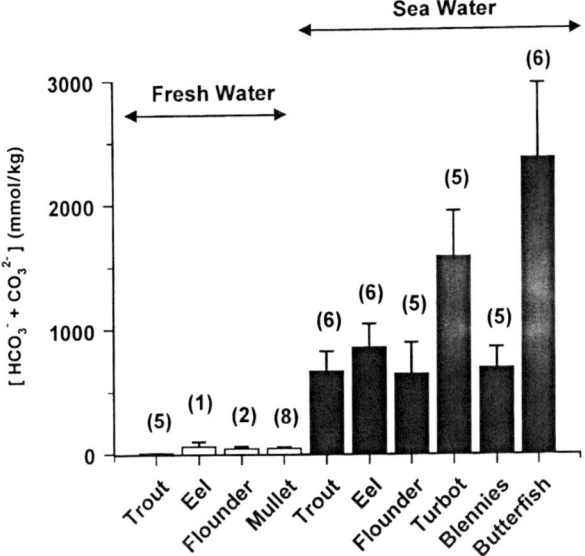

Fig. 4. Measured [HCO_3^- + CO_3^{2-}] base content of intestinal mucus tubes from four species of euryhaline teleost and three species of marine teleost (see Table 1 for a description of acclimation salinities). Number of animals sampled is indicated in parentheses.

Table 1. *Teleost species acclimated to either fresh water (FW) or sea water (SW) for sampling intestinal fluid and mucus tubes*

Rainbow trout	(*Oncorhynchus mykiss*)	FW and SW
Eel	(*Anguilla anguilla*)	FW and SW
Flounder	(*Platichthys flesus*)	FW and SW
Grey mullet	(*Chelon labrosus*)	FW*
Turbot	(*Rhombus maximus*)	SW
Blenny	(*Blennius pholis*)	SW
Butterfish	(*Centronotus gunnellus*)	SW

The first four species are all euryhaline and therefore capable of acclimation to either fresh water or sea water. Rainbow trout were obtained from a local fish farm; all other species were caught from the wild, either by electrofishing or with small nets. All fish (except mullet) were held in the laboratory at the appropriate salinity and kept unfed for at least a week prior to sampling.

*Mullet were electrofished from the River Clyst estuary in South Devon and sampled at the collection site. The salinity they were found in was equivalent to fresh water, but the recent salinity history of these animals was not known.

characterised by numerous (60–70) blind-ended caeca in rainbow trout, which open into the intestinal lumen. The mid intestine is relatively thin and separated from the wider, darker and annular-ringed posterior intestine by a muscular sphincter (Loretz, 1995). Thus, the fluids present in the latter two sections are effectively isolated into separate compartments. Fluid sampled from different portions of the gastrointestinal tract in seawater trout shows a very dramatic increase in HCO_3^-/CO_3^{2-} base when moving from the stomach into the pyloric intestine (see Fig. 2). This suggests that this most anterior third of the intestine is the primary site of endogenous base secretion in seawater trout. The pyloric intestine also appears to be where the majority of fluid is absorbed in salmonids (Bogé, Lopez & Pérès, 1988; Kerstetter & White, 1994), which explains the simultaneous sevenfold increase in $[Mg^{2+}]$ (relative to the stomach; see Fig. 2). Magnesium ions are very poorly absorbed during passage through the gastrointestinal tract of fish, which makes them a useful endogenous marker for net water movements (Hickman, 1968; Evans, 1993; Loretz, 1995). In addition, the concentration of $[Cl^-]$ ions was reciprocally related to the concentration of $[HCO_3^- + CO_3^{2-}]$ (see Fig. 2), which is relevant to the discussion of possible ion-transport mechanisms involved (see below).

The concentration of $[HCO_3^- + CO_3^{2-}]$ in gut fluid increases slightly (by 20 per cent) as it passes from the pyloric to the posterior intestine. However, this may simply be accounted for by subsequent water reabsorption (rather than further base excretion) because the concentration of Mg^{2+} rises by a similar degree between the two ends of the intestine (see Fig. 2). Thus, the pyloric intestine may be the sole site for the net secretion of base, and appears to coincide with the primary site of water absorption in seawater-acclimated salmonids.

One small caveat should be added, however, because the luminal fluid does become progressively more alkaline along the intestine (see Fig. 2). As a result, there is a change in the relative amounts of HCO_3^- and CO_3^{2-} present, with the calculated $[CO_3^{2-}]$ doubling between the pyloric intestine and the terminal rectal fluid (Wilson et al., 1996). This suggests that although the pyloric intestine does perform the majority of base secretion, the more distal regions may have an alkalinising effect, but in the absence of further increases in lumen $[HCO_3^-]$. Possible explanations include: changes in P_{CO_2} (not measured), secretion of OH^- or CO_3^{2-}, or a selective permeability to HCO_3^- in the posterior intestine (i.e. increased permeabilty to HCO_3^- relative to CO_3^{2-} allowing greater retention of secreted CO_3^{2-} in the lumen compared to secreted HCO_3^-). Further experiments are required to determine which of these operates in the more distal segments of the intestine.

There are at least four tissues that could be responsible for the secretion of basic equivalents into the pyloric region of the intestine: (i) the epithelium lining the lumen of the pyloric intestine itself, (ii) the numerous blind-ended pyloric caeca, (iii) the diffuse exocrine pancreatic tissue in the fat surrounding the pyloric caeca and the hepatic portal vein (Kurokawa & Susuki, 1995), and (iv) the gall bladder. The gall bladder has been excluded as a potential source of base due the neutral pH (7.2) and low $[HCO_3^-]$ (less than 8 mmol/l) of bile (Wilson *et al.*, 1996). At present it is unclear which of the first three might be responsible for net base secretion into the pyloric intestine lumen.

Measurements of the base content of the fluid/mucus sampled from inside the pyloric caeca show a concentration of 52.5 ± 7.8 mmol/l, i.e. considerably higher than plasma concentrations but less than half the concentration found in the lumen of the pyloric intestine itself. This could be accounted for by secretion within the caeca and then absorption of water following movement into the pyloric lumen. Conversely, it may simply be indicative of secretion of basic equivalents elsewhere followed by limited diffusion of HCO_3^- into the pyloric caeca. However, the fact that high concentrations of base are found in the intestines of fish which lack pyloric caeca (e.g. seawater-acclimated eels) suggests that they are certainly not essential for intestinal base secretion.

Given the obvious anatomical comparison with the duodenum of mammals, it is tempting to draw an analogy between HCO_3^- secretion from the mammalian pancreas, which drains into the duodenum, and base secretion in the teleost pyloric intestine. However, one key difference is that fluid and HCO_3^- are both secreted in mammalian pancreatic ducts, but move in opposite directions in the teleost pyloric intestine. In addition, the alkaline pancreatic secretions of mammals serve only to neutralise stomach acid, and duodenal contents rarely exceed pH 6–7 (Ganong, 1989).

Ion transport mechanisms involved in intestinal base secretion

In theory, net HCO_3^- secretion could be explained by primary movement of any one of the components of the CO_2/HCO_3^- buffer system (i.e. HCO_3^-, CO_3^{2-}, OH^-, H^+ or CO_2). However, studies on bicarbonate-secreting epithelia in mammals suggest that either proton uptake or HCO_3^- secretion is the primary transport process (Case & Argent, 1993). Although the precise location of base secretion within the pyloric intestine has not yet been identified, perfusion of the pyloric intestine *in vivo* (i.e. perfusion of the lumen of the pyloric intestine with its blood supply intact; Wilson *et al.*, 1996) has given some insight into the possible ion-transport mechanisms involved, whichever tissue might be the ultimate source.

As stated earlier, the increasing $[HCO_3^-/CO_3^{2-}]$ within the gastro-intestinal tract *in situ* coincides with a reciprocal and marked reduction in intestinal $[Cl^-]$ (see Fig. 2). Subsequent experiments using the *in vivo* perfusion technique revealed high rates of base secretion (550 μmol/kg/h under optimal conditions due to the very low $[HCO_3^-]$ (1 mmol/l) in the inflowing luminal perfusate, which were correlated with net Cl^- absorption rates (Fig. 5). In addition, net base secretion was inhibited by 62 per cent when Cl^- was removed from the luminal perfusate (Wilson *et al.*, 1996). This all suggests the involvement of a mucosal (lumen side) Cl^-/HCO_3^- exchange mechanism. However, further investigations demonstrated that base secretion was insensitive to mucosally applied 10^{-4} M 4-acetamido-4'-isothiocyanostilbene-2,2'-disulphonic acid (SITS) or 2×10^{-5} M 4-4'-diisothiocyanostilbene-2,2'-disulphonic acid (DIDS) (inhibitors of erythrocyte anion exchangers). Mucosal application of amiloride (10^{-4} M; an inhibitor of Na^+/H^+ exchange) caused a 76 per cent stimulation of base secretion.

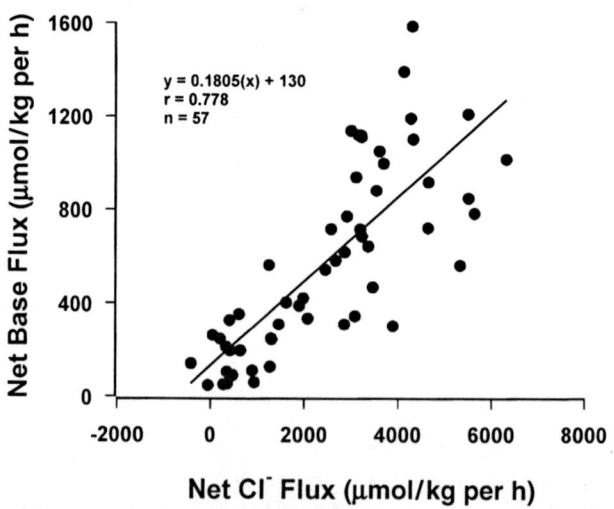

Fig. 5. Relationship between net base ($HCO_3^- + 2CO_3^{2-}$) fluxes and net Cl^- fluxes in the *in situ* perfused pyloric intestine of seawater-acclimated rainbow trout. Positive numbers on the x and y axes represent net Cl^- absorption and net base secretion, respectively. Data from Wilson *et al.* (1996).

The enormous lumen-to-blood [HCO_3^-] gradient would require an equilibrium potential of -63 mV (blood side negative). Reports of the transepithelial potential for the intestine of marine teleosts vary from -3 to -8 mV (Ando, 1990; Kirschner, 1991). Thus, it appears that lumen [HCO_3^-] is well above the transepithelial equilibrium potential. This, together with the above results, has led to the tentative conclusion that intestinal base secretion may be driven by an active Cl^-/HCO_3^- ATPase and is actually working against a background of significant H^+ secretion *via* a mucosal Na^+/H^+ exchanger. This agrees with the *in vitro* results of Dixon and Loretz (1986), who concluded that there must be an active Cl^-/HCO_3^- exchange mechanism on the basolateral membrane and a conductive exit pathway for HCO_3^- or OH^- (or entry of H^+) on the apical membrane of posterior intestinal cells of the goby. Without direct measurements of membrane potential and intracellular ion concentrations, it is impossible to say with certainty whether an active Cl^-/HCO_3^- ATPase is required to explain base secretion in the seawater trout. In general, little is known about anion-translocating ATPases, with the exception of recent studies by Gerencser and colleagues on a Cl^-/HCO_3^- stimulated ATPase in the intestinal mucosa of the mollusc *Aplysia* (e.g. Gerencser & Zelezna, 1993).

The presence of luminal acetazolamide (10^{-4} M) only reduced net base secretion by 20 per cent using the *in vivo* perfusion of the seawater trout intestine (Wilson *et al.*, 1996). This, together with the relatively low levels of carbonic anhydrase activity within all portions of the intestine of the same fish, suggests that the majority of secreted HCO_3^- is transported across the entire epithelium in this form, and that intracellular production from CO_2 hydration is minor (Wilson *et al.*, 1996). Although this is similar to the conclusions of Dixon and Loretz (1986) for cells from the posterior intestine of the goby, a full characterisation of this transport system in seawater trout will have to await further experiments involving the application of inhibitors to both the basolateral and apical aspects of the epithelium.

Possible functions and repercussions of intestinal base secretion

The reason(s) why the intestine of seawater-acclimated teleosts displays such high rates of base secretion is currently unknown. However, the absence of this phenomenon in their freshwater counterparts suggests it may be related to the osmoregulatory strategy of teleosts living in a hyperosmotic environment. Given the high content of Ca^{2+} and Mg^{2+} in the mucus tube 'solids' found in the intestine of seawater trout (approximately 600 and 300 mmol/kg, respectively), it is interesting to speculate that base secretion may facilitate the osmoregulatory function of the intestine in seawater-adapted teleosts by precipitating these divalent ions as carbonates. This would reduce the effective osmolality of fluid along the intestine and in turn promote further water absorption. However, a precise analysis of all components of the imbibed and excreted fluid osmolality needs to be conducted to test this hypothesis.

The highly alkalkine intestinal contents of seawater teleosts will undoubtedly have a number of significant and as yet unrecognised repercussions for other physiological functions. For example, the pH requirements/sensitivity of digestive enzymes of the intestine will presumably be different in freshwater-acclimated and seawater-acclimated fishes, and different enzymes may be induced following migration of euryhaline species between these two environments. The role of intestinal base excretion in acid–base balance, osmoregulation and digestive functions of seawater teleosts, clearly warrants further investigation.

Conclusions

In summary, we can now confirm that base excreted *via* the intestine of seawater-acclimated trout arises endogenously and contributes significantly to the overall acid–base balance. This should therefore be taken into account in studies of acid–base balance in seawater teleosts, but is probably quantitatively unimportant in freshwater species. The net excretion of HCO_3^-/CO_3^{2-} base by the intestine is countered by a similar net loss of acid from other parts of the body, most likely the gills. The majority of this HCO_3^-/CO_3^{2-} base appears to be secreted from somewhere within the pyloric (most anterior) region of the intestine, though the precise location remains unknown. The mechanism of base secretion within the pyloric intestine appears to involve Cl^-/HCO_3^- exchange, which is insensitive to inhibitors of passive Cl^-/HCO_3^- exchangers, suggesting the presence of a primary active HCO_3^--transporting ATPase. Future experiments should elucidate whether intestinal base excretion varies in response to changes in internal acid–base status and therefore can be considered a site of true acid–base regulation.

Acknowledgements

The data reported here are primarily from a study performed at Bamfield Marine Station, Vancouver Island, BC, Canada, in 1995. The author would therefore like to acknowledge his co-workers in that study – Chris Wood, Katie Gilmour and Ray Henry – and the NSERC support to Chris Wood for that work. In addition, the author would like to thank Debbie Tunnicliffe for technical assistance in the comparison of freshwater and seawater species carried out at Exeter University.

References

Ando. M. (1990). Effects of bicarbonate on salt and water transport across the intestine of the seawater eel. *Journal of Experimental Biology* **150**, 367–79.

Avella, M. & Bornancin, M. (1989). A new analysis of ammonia and sodium transport through the gills of the freshwater rainbow trout (*Salmo gairdneri*). *Journal of Experimental Biology* **142**, 155–75.

Bogé, G., Lopez, L. & Pérès, G. (1988). An *in vivo* study of the role of pyloric caeca in water absorption in rainbow trout (*Salmo gairdneri*). *Comparative Biochemistry and Physiology* **91A**, 9–13.

Cameron, J.N. (1976). Branchial ion uptake in arctic grayling: resting values and effects of acid–base disturbances. *Journal of Experimental Biology* **64**, 711–25.

Cameron, J.N. (1978). Regulation of blood pH in teleost fish. *Respiration Physiology* **33**, 129–44.

Case, R.M. & Argent, B.E. (1993). Pancreatic duct cell secretion: Control and mechanisms of transport. In *The pancreas: pathobiology and disease*, 2nd edition, ed. W. Vay Liang, pp. 301–50. New York: Raven Press.

Claiborne, J.B. & Heisler, N. (1984). Acid–base regulation and ion transfers in the carp (*Cyprinus carpio*) during and after exposure to environmental hypercapnia. *Journal of Experimental Biology* **108**, 25–43.

Dixon, J.M. & Loretz, C.A. (1986). Luminal alkalinization in the intestine of the goby. *Journal of Comparative Physiology* **156**, 803–11.

Evans, D.H. (1982). Mechanisms of acid extrusion by two marine fishes: the teleost, *Opsanus beta*, and the elasmobranch, *Squalus acanthis. Journal of Experimental Biology* **97**, 289–99.

Evans, D.H. (1993). Osmotic and ionic regulation. In *The physiology of fishes*, ed. D.H. Evans, pp. 315–41. New York: CRC Press.

Ganong, W.F. (1989). Regulation of gastrointestinal function. In *Review of medical physiology*, 14th edition, pp. 408–35. London: Prentice-Hall.

Gerencser, G.A. & Zelezna, B. (1993). Reaction sequence and molecular mass of a Cl^--translocating P-type ATPase. *Proceedings of the National Academy of Sciences USA* **90**, 7970–4.

Goss, G.G., Perry, S.F., Wood, C.M. & Laurent, P. (1992). Mechanisms of ion and acid–base regulation at the gills of fresh-water fish. *Journal of Experimental Zoology* **263**, 143–59.

Heisler, N. (1984). Acid–base regulation in fishes. In *Fish physiology*, Vol. XA, ed. W.S. Hoar & D.J. Randall, pp. 315–401. New York: Academic Press.

Heisler, N. (1993). Acid–base regulation. In *The physiology of fishes*, ed. D.H. Evans, pp. 343–78. New York: CRC Press.

Hickman, C.P. Jr (1968). Ingestion, intestinal absorption, and elimination of seawater and salts in the southern flounder, *Paralicthys lethostigma*. *Canadian Journal of Zoology* **46**, 457–66.

Hickman, C.P. Jr & Trump, B.R. (1969). The kidney. In *Fish physiology*, Vol. I, ed. W.S. Hoar & D.J. Randall, pp. 91–239. New York: Academic Press.

Hwang, P.P. (1989). Distribution of chloride cells in teleost larvae. *Journal of Morphology* **200**, 1–8.

Ishimatsu, A., Iwama, G.K., Bentley, T.B. & Heisler, N. (1992). Contribution of the secondary circulatory system to acid–base regulation during hypercapnia in rainbow trout. *Journal of Experimental Biology* **170**, 43–56.

Ishimatsu, A., Iwama, G.K. & Heisler, N. (1988). *In vivo* analysis of partitioning of cardiac output between systemic and central venous sinus circuits in rainbow trout: a new approach using chronic cannulation of the branchial vein. *Journal of Experimental Biology* **137**, 75–88.

Iwama, G.K., Ishimatsu, A. & Heisler, N. (1993). Site of acid–base relevant ion transfer in the gills of rainbow trout exposed to environmental hypercapnia. *Fish Physiology and Biochemistry* **12**, 269–80.

Kerstetter, T.H. & White, R.J. (1994). Changes in intestinal water absorption in coho salmon during short-term seawater adaptation: a developmental study. *Aquaculture* **121**, 171–80.

Kirschner, L.B. (1991). Water and ions. In *Environmental and metabolic animal physiology: comparative environmental physiology*, 4th edition, ed. C.L. Prosser, pp. 67–9. New York: Wiley-Liss.

Kurokawa, T. & Susuki, T. (1995). Structure of the exocrine pancreas of flounder (*Paralichthys olivaceus*): immunological localization of zymogen granules in the digestive tract using anti-trypsinogen antibody. *Journal of Fish Biology* **46**, 292–301.

Laurent, P. (1984). Gill internal morphology. In *Fish physiology*, Vol. XA, ed. W.S. Hoar & D.J. Randall, pp. 79–183. New York: Academic Press.

Li, J., Eygensteyn, J., Lock, R.A.C. *et al.* (1995). Branchial chloride cells in larvae and juveniles of fresh-water tilapia *Oreochromis mossambicus*. *Journal of Experimental Biology* **198**, 2177–84.

Loretz, C.A. (1995). Electrophysiology of ion transport in teleost intestinal cells. In *Fish physiology*, Vol. 14: *Cellular and molecular approaches to fish ionic regulation*, ed. C.M. Wood & T.J. Shuttleworth, pp. 25–99. New York: Academic Press.

McDonald, D.G., Tang, Y. & Boutilier, R.G. (1989). Acid and ion transfer across the gills of fish: mechanisms and regulation. *Canadian Journal of Zoology* **67**, 3046–54.

McDonald, D.G., Walker, R.L., Wilkes, P.R.H. & Wood, C.M. (1982). Acid excretion in the marine teleost, *Parophrys vetulus*. *Journal of Experimental Biology* **98**, 403–14.

Nonnette, G., Nonnette, L. & Kirsch, R. (1979). Chloride cells and chloride exchange in the skin of a seawater teleost, the shanny (*Blennius pholis* L.). *Cell and Tissue Research* **199**, 387–96.

Payan, P., Girard, J.P. & Mayer-Gostan, N. (1984). Branchial ion movements in teleosts: the roles of respiratory and chloride cells. In *Fish physiology*, Vol. XA, ed. W.S. Hoar & D.J. Randall, pp. 39–63. New York: Academic Press.

Perrot, M.N., Grierson, C.E., Hazon, N. & Balment, R.J. (1992). Drinking behaviour in sea water and fresh water teleosts, the role of the renin-angiotensin system. *Fish Physiology and Biochemistry* **10**, 161–8.

Perry, S.F. & Wood, C.M. (1985). Kinetics of branchial calcium uptake in the rainbow trout: effects of acclimation to various external calcium levels. *Journal of Experimental Biology* **116**, 411–33.

Potts, W.T.W. (1984). Transepithelial potentials in fish gills. In *Fish physiology*, Vol. XA, ed. W.S. Hoar & D.J. Randall, pp. 105–28. New York: Academic Press.

Sayer, M.D.J. & Davenport, J. (1987). The relative importance of the gills to ammonia and urea excretion in five seawater and one freshwater teleost species. *Journal of Fish Biology* **31**, 561–70.

Shehadeh, Z.H. & Gordon, M.S. (1969). The role of the intestine in salinity adaptation of the rainbow trout, *Salmo gairdneri*. *Comparative Biochemistry and Physiology* **30**, 397–418.

Smith, H.W. (1930). The absorption and excretion of water and salts by marine teleosts. *American Journal of Physiology* **93**, 480–505.

Sullivan, G.V., Fryer, J.N. & Perry, S.F. (1995). Immunolocalisation of proton pumps (H$^+$-ATPase) in pavement cells of rainbow trout gill. *Journal of Experimental Biology* **198**, 2619–29.

Tang, Y., McDonald, D.G. & Boutilier, R.G. (1989). Acid–base regulation following exhaustive exercise: a comparison between freshwater and seawater-adapted rainbow trout. *Journal of Experimental Biology* **141**, 407–18.

Vogel, W.O.P. (1985). Systemic vascular anastomoses, primary and secondary vessels in fish, and the phylogeny of lymphatics. In *Cardiovascular shunts: phylogenetic, ontogenetic and clinical aspects*, ed. K. Johansen & W. Burrgren, Copenhagen: Munksgaard. pp. 143–59.

Walsh, P.J., Blackwelder, P., Gill, K.A., Danulat, E. & Mommsen, T.P. (1991). Carbonate deposits in marine fish intestines: a new source of biomineralization. *Limnology and Oceanography* **36**, 1227–32.

Wilson, R.W. Gilmour, K.M., Henry, R.P. & Wood, C.M. (1996). Intestinal base excretion in the seawater-adapted rainbow trout: a role in acid–base balance? *Journal of Experimental Biology* **199**, 2331–43.

Wilson, R.W. & Taylor, E.W. (1992). Transbranchial ammonia gradients and acid–base responses to high external ammonia in rainbow trout (*Oncorhynchus mykiss*) acclimated to different salinities. *Journal of Experimental Biology* **166**, 95–112.

Wilson, R.W., Wright, P.M., Munger, S. & Wood, C.M. (1994). Ammonia excretion in freshwater rainbow trout (*Oncorhynchus mykiss*) and the importance of gill boundary layer acidification: lack of evidence for Na^+/NH_4^+ exchange. *Journal of Experimental Biology* **191**, 37–58.

Wood, C.M. (1988). Acid–base and ionic exchanges at the gills and kidney after exhaustive exercise in the rainbow trout. *Journal of Experimental Biology* **136**, 461–81.

Wood, C.M. (1993). Ammonia and urea excretion and metabolism. In *The physiology of fishes*, ed. D.H. Evans, pp. 379–425. New York: CRC Press.

Wood, C.M., Hopkins, T.E., Hogstrand, C. & Walsh, P.J. (1995). Pulsatile urea excretion in the ureagenic toadfish *Opsanus beta*: an analysis of rates and routes. *Journal of Experimental Biology* **198**, 1729–41.

Wood, C.M., Wheatly, M.G. & Hobe, H. (1984). The mechanisms and acid–base and ionoregulation in the freshwater rainbow trout during environmental hyperoxia and subsequent normoxia. III. Branchial exchanges. *Respiration Physiology* **55**, 175–92.

SUSAN WRAY, ANTHONY J. BULLOCK,
NAOMI BUTTELL and ROBERT A. DUQUETTE

pH and smooth muscle: regulation and functional effects

Introduction

Smooth muscle is found within the walls of hollow organs and tubes of the body such as the uterus, stomach, ureter and blood vessels. Its main function is to move liquids or objects through these tubes, which it does by altering its contractile state. Its correct functioning is therefore vital to many activities, for example breathing, reproductive ability, voiding urine and controlling blood pressure. The control of the contraction of smooth muscle is accomplished by neuronal, hormonal and local mechanisms. This chapter focuses on one such local mediator – pH – and considers how pH may be altered by normal and pathological processes. Given its potent ability to modify smooth muscle contractile activity, it is clearly necessary that pH is regulated. Therefore, what is known about this process in smooth muscle is discussed, and how changes in pH affect contractility is described.

Techniques used to measure pH in smooth muscles

Measurement of extracellular pH (pH_e) in smooth muscles is generally straightforward. In single cell studies or perfused organs pH_e will be that of the perfusate. This is usually 7.4, to match that of plasma. For measurement of surface pH, microelectrodes can be placed against the cell or tissue (Thomas, 1984).

The measurement of intracellular pH (pH_i) in smooth muscle has now become routine, mainly because of the development of pH-sensitive fluorescent indicators. The small size of smooth muscle cells coupled with their contractile nature had largely prevented the successful use of pH-sensitive microelectrodes (Aickin, 1984). The technique of nuclear magnetic resonance (NMR, discussed elsewhere in this book) provided some of the earliest measurements of pH_i in smooth muscle (Vogel, Lilja & Hellstrand, 1983; Vermuë & Nicolay, 1983; Dawson & Wray, 1985). These were *in vitro* measurements on excised tissues. Recently,

^{31}P-NMR spectroscopy has provided the first *in vivo* measurements of pH$_i$ for smooth muscles (Harrison *et al.*, 1994). Limitations to using NMR are its time resolution, technical difficulty and expense. The poor time resolution (about 5 minutes) is a reflection of the low signal obtained from smooth muscles, which is due to the small amounts of tissue usually available, and the relatively low-concentration phosphorus metabolites present in smooth muscle (Dawson & Wray, 1987; Wray, 1991). The calibration of the spectrum requires titrating the shift in the inorganic phosphate (Pi) resonance position, for known changes of pH, in solutions of the correct ionic strength, [Mg^{2+}] and temperature. The first two of these may be tissue specific, leading to possible errors unless the calibration was performed for the same tissue. The strength of the NMR technique lies in it providing a simultaneous measurement of several metabolites important to cellular energetics, including ATP, Pi, and phosphocreatine (PCr), along with pH$_i$.

The advent of pH-sensitive fluorophores has greatly increased our knowledge of smooth muscle pH$_i$, pH regulation and pH buffering power. The majority of these studies have used either BCECF, a dual excitation, single emission fluorophore, or SNARF, a single excitation, dual emission fluorophore (Rink, Tsien & Pozzan, 1982; Buckler & Vaughan-Jones, 1990; Owen, 1992; Mason, 1993). By incubating tissues or single smooth muscle cells with the membrane-permeant acetoxymethylester forms of these fluorophore, it is possible to introduce them into the cytoplasm without damaging the cell. Endogenous intracellular esterases cleave the ester bond to leave a charged, and relatively non-lipid-soluble, fluorophore in the cytoplasm. This cleavage process may form potentially damaging by-products, but their concentration is considered generally to be too low to produce any significant damage (Mason, 1993). Certainly, when we have compared contractile function in smooth muscles before and after loading with indicators, we have seen no decrement in contractile ability, supporting the assertion that no cellular impairment occurs. It is also assumed that the indicator is cleaved before it can cross the membranes of intracellular organelles; this can be verified by sequentially removing these membranes with different solvents and examining the effect on the fluorescent signal. It is also possible, but difficult, to micro-inject the cell with non-esterified indicator if organellar accumulation is problematic.

In our experience, fluorophore-loading conditions vary to a small extent with different smooth muscles (Austin & Wray, 1993b; Taggart & Wray, 1993; Taggart, Austin & Wray, 1994; Burdyga, Taggart & Wray, 1996). Thus, thin (100–300 μm) strips of muscle can be loaded with SNARF at a concentration of 5–25 μM for two to four hours at 20 °C or 12–18 hours at 4 °C. Useful signal, i.e. sufficiently above background, can be obtained for approximately two hours working at 37 °C. Single cells can be loaded in 20–30 minutes (Eiesland *et al.*, 1991). Calibration of the pH_i can be achieved *in situ* (in some conditions) or *in vitro* in all cases. The *in situ* methods use the K^+-H^+ ionophore nigericin to set pH_i to known values (Buckler & Vaughan-Jones, 1990). This technique does not readily work with intact tissues, presumably because of the inability of nigericin to perforate all the cells in the preparation. However, even in single-cell experiments it is generally too time consuming for routine use. Another complication is that nigericin has been reported to persist in the tubing and baths associated with many experimental systems (Richmond & Vaughan-Jones, 1993). Therefore, many calibrations are performed on *in vitro* solutions designed to simulate intracellular conditions (Taggart & Wray, 1993). When compared to *in situ* calibrations, good agreements have often been reported (Taggart & Wray, 1993; Sun, Leem & Vaughan-Jones, 1996), leading to further reliance on *in vitro* methods despite some technical reservations (Opitz, Merten & Acker, 1994). It can also be noted that in smooth muscles where more than one technique has been used to measure pH_i, good agreements have occurred.

Table 1 collates some values for resting pH_i made with fluorescent indicators, under comparable experimental conditions. It can be seen that pH_i is around 7.10 in HEPES buffer at 37 °C.

Intracellular buffering power

Buffering power (β) is defined as the amount of H^+ (moles) required to produce a one-unit change in pH in 1 litre of solution. Thus, the larger the value of β, the smaller will be the pH change produced by a certain acid or alkaline load. The total intracellular β is conveniently divided into that component due to HCO_3^-/CO_2 and the remainder which is termed intrinsic buffering power (β_i). In the main, β_i arises from intracellular proteins which are able to buffer because they have large numbers of acidic or basic groups which can accept or donate H^+, and their overall pK is close to physiological pH, although β has been found to be greatest at acidic pH (see below). In smooth muscle and many other tissues, β_i contributes more to total β than HCO_3^-/CO_2 (Aickin, 1984; Vaughan-Jones & Wu, 1988; Szatkowski & Thomas, 1989).

Measurements of β_i are best made under conditions in which pH regulation is prevented, in order that the change in pH_i produced by an acid or alkaline load is not underestimated. Table 1 gives some examples of β_i values found in smooth muscles. It can be seen that the

Table 1. *pH_i and β_i values in smooth muscles*

Tissue	pH_i	β (mM/ pH unit)	Notes	References
Uterus	7.18 ± 0.04	13	Non-pregnant rat	Eiesland *et al.* (1991) Taggart & Wray (1993)
Uterus	7.06 ± 0.03	n.a.	Non-pregnant human	Parratt *et al.* (1994)
Ureter	7.22 ± 0.04	14	Guinea-pig	Aickin (1994a) Burdyga *et al.* (1996)
Vas deferens	7.06	9	Guinea-pig, HCO_3^-	Aickin (1994a)
Portal vein	7.06 ± 0.03	n.a.	Rat	Taggart *et al.* (1994)
Coronary, vascular*	6.97 ± 0.03	n.a.	Rat, single cells	Ramsey *et al.* (1994)
Mesenteric, vascular	7.10 ± 0.04	30–40	Rat	Aalkjaer & Cragoe (1988) Austin & Wray (1993b, 1995)
Bladder*	7.11 ± 0.10	22	Ferret, HCO_3^-	Liston *et al.* (1991)

Resting values of pH_i obtained using the fluorescent pH-sensitive indicator SNARF or BCECF in perfused tissues or cells at external pH 7.4, with HEPES buffer unless indicated otherwise. Temperature 37 °C except for * where it was room temperature. n.a. = not available, β_i = buffering power.

range is between 10 and 40 mmol/pH unit. It is of interest to note that in some studies of β in smooth muscle, little or no increase was found by the presence of HCO_3^-/CO_2 (Baro *et al.*, 1989; Aickin, 1994a). This is in contrast to the increase of two to three-fold produced by HCO_3^-/CO_2 in other tissues (Roos & Boron, 1981). One explanation for this may lie in the reported absence from smooth muscle of carbonic anhydrase (Muhleisen & Kreye, 1985). Differences in β have recently been reported between adult and neonatal smooth muscles (Bullock *et al.*, 1998). β_i increases as pH_i becomes acidic, as shown in vascular smooth muscle (Austin & Wray, 1995) in which the behaviour of β_i with pH_i could be described by assuming the presence of a single buffer at a concentration of 310 mmol with a pK to 6.0. The sensitivity of β_i to pH_i means that the smooth muscle cells possess an extra degree of protection to acidic loads. Given that many of the processes involved in contraction are pH sensitive (see below), the short-term effect of buffering will help to maintain normal activity.

pH regulation in smooth muscle

As outlined above, in the short term, acid or alkaline loads will be buffered. pH_i-regulating mechanisms are then activated to restore pH_i to resting values. There has not been an extensive study of just what these mechanisms are for many smooth muscles. The best-studied ones are the ureter and vascular smooth muscle, and these will therefore be focused on here.

Ureteric smooth muscle

Figure 1 summarises the pH-regulating mechanisms considered to be operating in the ureter (and vascular smooth muscle). The experimental evidence for the proposed mechanisms is given below. For convenience, the mechanisms are divided into those mainly (or solely) concerned with regulation from acid or alkaline loads.

Recovery from an acid load
Both HCO_3^--dependent and HCO_3^--independent mechanisms for pH_i regulation have been shown to be present in guinea-pig ureter (Aickin,

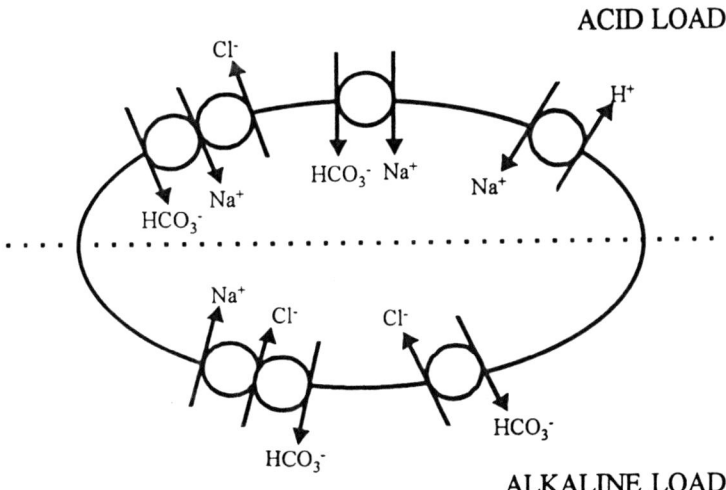

Fig. 1. Summary of pH-regulating mechanisms identified in smooth muscle. The top half of the diagram shows those mechanisms considered to bring about pH recovery from an acid load, and the bottom half shows mechanisms to bring about pH recovery from an alkaline load.

1994a, 1994b). In the absence of HCO_3^-, recovery from an acid load was shown to be dependent on the presence of Na^+ and inhibited by amiloride. This indicates the involvement of Na^+–H^+ exchange. This is supported by the authors' recent work on both rat and guinea-pig ureters using the fluorescent indicator SNARF to monitor pH_i. It was found, in the absence of HCO_3^-, that recovery from acidification was inhibited by the potent and selective blocker of the Na^+–H^+ exchanger, 5-*N*-ethyl-*N*-isopropylamiloride (EIPA). EIPA also produced a lowering of resting pH_i, supporting a role for Na^+–H^+ exchange in the maintenance of resting pH_i. Recovery from acid load was faster in the presence of HCO_3^- than in its absence, indicating that HCO_3^--dependent mechanisms also play a role in this pH_i regulation. Furthermore, in the presence of HCO_3^-, amiloride was shown not to affect the recovery from acid loads, indicating that Na^+–H^+ exchange may not be important under these conditions. However, in the presence of HCO_3^-, removal of the Na+ was seen to cause an inhibition in the recovery from acidification. This therefore led to the suggestion that the HCO_3^--dependent mechanism is Na^+–HCO_3^- co-transport (see Fig. 1). This transporter was unaffected by DIDS and considered to be electroneutral (Aickin, 1994a). Steady-state pH_i has also been shown to fall in the absence of HCO_3^-, supporting the suggestion that Na^+–HCO_3^- co-transport is important in maintaining pH_i.

Recovery from alkaline load

A transient alkalinisation was observed on removal of Cl^- in the presence of HCO_3^-, indicating Cl^-–HCO_3^- exchange (see Fig. 1). Recovery from alkalinisation was inhibited by DIDS, showing that a rise in pH_i stimulates the Cl^-–HCO_3^- exchange to restore pH_i. The Cl^--free conditions also inhibited recovery from an alkaline load. No evidence was found for reversal of this exchanger to aid recovery from acidic load, i.e. HCO_3^- entry.

Vascular smooth muscle

Recovery from an acid load

The major mechanisms for acid removal from vascular smooth muscle are the Na^+–H^+ exchanger (Lucchesi, Deroux & Berk, 1994; Lucchesi & Berk, 1995) and Na^+–HCO_3^- co-transporter. As in ureter, in the absence of HCO_3^-, the Na^+–H^+ exchanger has been shown to be unable to maintain pH_i in an active tissue (Aalkjaer & Mulvany, 1991). Na^+–H^+ exchange is triggered by acid loads as well as by a variety of vasoconstricting agonists (see section on 'Agonists and pH_i'). Amiloride and its derivatives significantly impede pH_i regulation from acid load in a variety of vascular tissues, showing the importance of Na^+–H^+ exchange. A Na^+–HCO_3^- co-transporter (Aalkjaer & Hughes, 1991; Aickin, 1994c; Carr, McKinnon & Poston, 1995) has been shown to be present in virtually all smooth muscles including vascular tissues; the exceptions being some smooth muscle cell lines (Vigne *et al.*, 1988; Korbmacher *et al.*, 1988). As with the ureter, it seems to play a significant role in the control of pH_i in vascular tissues and recovery from an acid load. However, the characteristics of the transporter can vary between vessels: a Cl^- dependence of the co-transporters in canine femoral artery has been reported (Kahn *et al.*, 1990), while others have provided evidence for a Cl^--independent Na^+–HCO_3^- co-transport in subcutaneous (Carr *et al.*, 1995) and mesenteric arteries (Aalkjaer & Hughes, 1991). The relative importance of Na^+–H^+ exchange and Na^+–HCO_3^- co-transport probably varies between different blood vessels. In femoral artery it was concluded that Na^+–H^+ exchange was largely responsible for the effective extrusion of an acid load, as HCO_3^- made no difference to recovery. In mesenteric resistance vessels, HCO_3^- accelerated the recovery from an acid load and amiloride did not prevent the recovery if HCO_3^- was present (Aalkjaer & Cragoe, 1988). It has been suggested that there may be a functional separation between large vessels like aorta and femoral and smaller vessels such as mesenteric and coronary, with the former being more reliant on Na^+–H^+ exchange and the latter on Na^+–HCO_3^- co-transport (Aickin, 1994c).

Recovery from an alkaline load

The Na^+-independent Cl^-–HCO_3^- exchanger is thought to be ubiquitous in vascular smooth muscle and has been demonstrated in many vascular smooth muscle types (Gerstheimer *et al.*, 1987; Aalkjaer & Hughes, 1991; Aickin, 1994c; Carr *et al.*, 1995). Its role, as in the ureter, is to protect against alkaline load. There is also evidence for a Na^+-dependent Cl^-–HCO_3^- exchanger (Little *et al.*, 1995), shown to be both SITS and EIPA sensitive. This form may be reversible, that is, bring HCO_3^- into the cell and aid recovery from acidification.

The importance of the Na^+–H^+ exchanger in smooth muscle

As detailed in the preceding sections, a major mechanism underlying the recovery of pH_i from an acid load is the Na^+–H^+ exchanger of the surface membrane. The Na^+–H^+ exchanger regulates pH_i by extruding one H^+ in exchange for one Na^+. When decreases in pH_i occur, the exchanger activity increases. At its maximum velocity, the Na^+–H^+ exchanger has probably the highest transport capacity of all the plasma membrane transport systems (Fliegel & Dyck, 1995). There are four known isoforms of the mammalian Na^+–H^+ exchanger, referred to as NHE-1 to NHE-4. NHE-1 was isolated by Sardet, Franchi and Pouysse-gur (1989). The protein encoded is the housekeeping isoform and is the most sensitive isoform to the blocker amiloride. NHE-1 mRNA has been shown to be present in most tissues (although in varying quantities) as a single major gene transcript of approximately 4.8 kbp length with a molecular mass of approximately 90 000 Da (Orlowski, Kandasamy & Shiull, 1992). NHE-2 was cloned from rabbit ileal villus cells and shows approximately 50 per cent homology with rabbit NHE-1. The NHE-2 mRNA has a similar membrane topology to NHE-1, with a molecular mass also of approximately 90 000 Da. NHE-2 mRNA has also been found to be expressed in the kidney. It is much more (25-fold) resistant to the blocker EIPA than NHE-1 (Tse *et al.*, 1993). NHE-3 and NHE-4 were cloned by Orlowski *et al.* (1992). They have 38 per cent and 43 per cent primary structure homology with rat NHE-1 and 39 per cent homology to each other. They also have molecular masses of approximately 90 000 Da, NHE-3 mRNA being approximately 5.6 kbp and NHE-4 being approximately 4.2 kbp in length. NHE-3 mRNA is expressed at high levels in the colon and small intestine, with significant levels also present in the kidney and stomach. NHE-4 mRNA is most abundant in the stomach, with smaller amounts in the small intestine, colon, kidney, brain, uterus and skeletal muscle. These characteristics are summarised in Table 2.

Isoforms have varying numbers of hydrophobic putative trans-

Table 2. *Comparison of the isoforms of the Na^+-H^+ exchangers*

Isoform	Molecular mass	Amino acid length	mRNA tissue specificity
NHE-1	91 506	820	All tissues
NHE-2	90 787	809	Kidney > ileum
NHE-3	92 997	831	Large > small intestine Kidney > stomach
NHE-4	81 427	717	Stomach > large and small intestine Uterus, brain, kidney, skeletal muscle

membrane spanning regions, ten being the most usual number, these being in the *N*-terminal region of the molecule and having the highest homology between isoforms (55–95 per cent). The *C*-terminal region of the molecule has the lowest homology between isoforms, it is highly hydrophilic and is therefore probably situated on the cytoplasmic side of the membrane. All the exchanger isoforms have one or more potential serine/threonine kinase phosphorylation sites and may be phosphorylated by protein kinase C, cAMP-dependent protein kinase, or both. (See also section on 'Agonists and pH_i'.)

Differences in mRNA isoform expression in varying tissues point to a physiological significance to a family of $Na^+–H^+$ exchanger isoforms. They may exhibit differences in cellular localisation, membrane targeting, ion specificities, kinetic parameters and response to various molecular stimuli, all of which will potentially lead to functional diversity (Orlowski *et al.*, 1992). However, mRNA expression may not be an indicator of final functional protein, and therefore protein expression data would give a better indication of the link between isoform expression and function of the various $Na^+–H^+$ exchangers. It will be of interest to discover how different smooth muscles vary in their expression of the $Na^+–H^+$ exchanger, and how they may be modulated.

A comparison of the role of $Na^+–H^+$ exchange in three smooth muscles

A limitation to the understanding of pH_i regulation in smooth muscle arises from the different experimental conditions between studies. Thus, it is often difficult to draw direct conclusions from studies in one particular smooth muscle compared to another. We have therefore studied three different tissues, under comparable conditions, to ask two questions. (1) What contribution does $Na^+–H^+$ exchange make to the maintenance of resting pH_i? (2) How much is regulation from an acid load dependent upon $Na^+–H^+$ exchange? The three tissues studied were uterine, gastric and ureteric smooth muscles. They were all loaded with SNARF for pH_i measurement, superfused with oxygenated, HEPES-buffered physiological saline at pH 7.4 and 35–37 °C. The $Na^+–H^+$ exchanger was blocked with EIPA (50 µmol). Figure 2A shows the effect on resting pH_i of EIPA. It can be seen in all three muscles that blocking $Na^+–H^+$ exchange causes a significant lowering of resting pH_i. The extent of this, however, was more marked in uterus and gastric smooth muscle than in the ureter (Fig. 2A). In addition, pH_i reaches a new steady state much more rapidly in the ureter. In the absence of HCO_3^- and the presence of EIPA, it is unclear what mechanism is responsible for setting this new lower pH_i value.

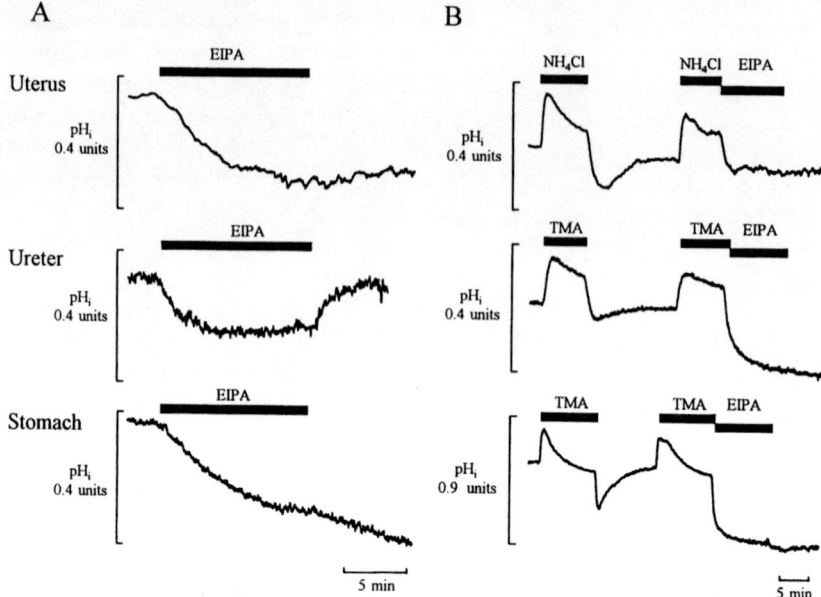

Fig. 2. The effect of blocking Na⁺–H⁺ exchange with EIPA (50 mM) on (**A**) resting pH$_i$, and on (**B**) regulation from an acid load in uterine (top panel), ureteric (middle panel) and gastric (bottom panel) smooth muscle. In (**B**), weak base NH$_4$Cl or trimethylamine (TMA) was removed to produce the acidic rebound. All experiments were buffered with HEPES, 37 °C, pH 7.4. EIPA is not readily reversible, hence the lack of recovery after its use. pH$_i$ was measured from the ratio of the signals at 590 nm and 560 nm from the fluorescent pH-sensitive indicator SNARF.

Figure 2B shows the effect on pH$_i$ regulation from an acid load produced by withdrawal of a weak base. In all three tissues, recovery and regulation can be seen during both the alkalinisation produced by weak base (NH$_4$⁺ or trimethylamine, TMA) and the acid rebound on washout of the base (Fig. 2B). As can also be seen, the rates of regulation varied between the three tissues (gastric >uterus > ureter). In all three tissues, however, EIPA completely blocked regulation from the acid load, showing the general importance of the Na⁺–H⁺ exchange.

The effects of species, development stage and gestational state on pH$_i$ regulation

The above experiments have shown that different smooth muscles differ in the contribution Na⁺–H⁺ exchange makes to resting pH$_i$ and regulation.

The differences in pH_i-regulatory mechanisms and buffering power between smooth muscles have also been described. In this section, the significance of three factors – species, development and gestational state – will be addressed; little is so far known about any of them, but there are preliminary data indicating a role for them.

Species

Little is known concerning pH_i regulation in gastric smooth muscle and therefore the authors have begun to investigate this, and whether species differences exist in the regulatory mechanisms. Expression of different isoforms of the Na^+–H^+ exchanger has been found in varying levels of the rat gastrointestinal tract (Orlowski *et al.*, 1992; see previous section), but this has not been investigated in other species. In guinea-pig gastric tissues the preliminary data show EIPA prevented the normal pH_i regulation following an acid load (Fig. 2B) and also blocked the alkaline rebound upon removal of butyrate found in control tissues. In contrast, in rat gastric smooth muscle, EIPA does not appear to prevent pH_i regulation. These preliminary experiments therefore suggest that in gastric smooth muscle major differences exist in the mechanisms underlying regulation from acid loads, or a non-EIPA-sensitive isoform of the Na^+–H^+ exchanger predominates in rat gastric smooth muscle.

Development

As yet, there are very few data on the development of pH-regulation mechanisms in smooth muscle tissues. Nakanishi *et al.* (1990) have suggested that the ability of the neonatal rabbit myocardium to maintain function during hypoxia may in part be due to an increased ability to regulate the acidification associated with hypoxia. Similarly, neonatal guinea-pig ureter shows a remarkable resistance to metabolic inhibition (A.J. Bullock & S. Wray, unpublished observations), which may be due to a similar involvement of pH regulation. It is anticipated that developmental differences may occur in ureter from the neonate to adult. Studies have therefore been initiated to elucidate any changes in the types and characteristics of the transporters present with age and species. Initial data show Na^+–H^+ exchange and Na^+/HCO_3^- co-transport are present in neonatal guinea-pig ureters (A.J. Bullock & S. Wray, unpublished findings), similar to those found in adult guinea-pig ureter (Aickin, 1994b). However, regulation from acid load may be more dependent upon Na^+–H^+ exchange in neonatal guinea-pig ureter compared to adult.

Gestation

There are currently no available data on the effect of gestational state on uterine pH regulation. It would be logical to propose that pregnant tissue would have a more active pH regulation than non-pregnant tissue due to the importance of maintaining resting pH_i levels of the uterus during pregnancy (see later). Parratt, Taggart and Wray (1995) have shown that as term approaches, the human uterus becomes more alkaline. This new 'set point' and regulation to it may be due to changes in either expression or activity of pH-regulating proteins.

Preliminary data show that pregnant myometrium is less susceptible to blockade of $Na^+–H^+$ exchange with EIPA after acid loading than non-pregnant rat myometrium. This suggests that $Na^+–H^+$ exchange is more important in the recovery from an acid load in the pregnant rat than in the non-pregnant rat. This may be due to an increased activity of the exchangers normally present, an increase in the number of exchangers present, or an increase in expression of an isoform either not normally present or only present in non-pregnant myometrium at a very low level. The two different mRNA isoforms present in the uterus from studies by Orlowski *et al.* (1992) were NHE-1, the ubiquitous housekeeping form of the exchanger, and the more tissue-specific NHE-4 isoform. Whether changes in relative expression of these isoforms is responsible for the functional changes observed is not yet known. Even the particular functional characteristics of the isoforms are not yet established However, it is interesting to theorise about the different isoforms having subtly different roles in pH_i regulation in the uterus, especially during pregnancy.

Agonists and pH_i

Agonists play an important role in modulating the contraction of smooth muscles. Much is now known about the second messenger systems which are activated by agonists. What is also apparent is that the action of several agonists, in particular those which act as growth factors, includes alteration in the activity of the $Na^+–H^+$ exchanger (Danthuluri *et al.*, 1987; Berk *et al.*, 1987; Boyarsky *et al.*, 1990; Kahn *et al.*, 1992). This activation of the $Na^+–H^+$ exchanger can lead to an intracellular alkalinisation. Most of this work has been performed on cultured vascular cells in HEPES buffer, but Aalkjaer and Mulvany (1991) demonstrated activation of the exchanger in intact vessels with vasopressin stimulation. As $[Ca^{2+}]$ rises when agonists stimulate smooth muscle cells, it has been speculated that it is $[Ca^{2+}]$ which stimulates the exchanger (Mitsuhashi & Ives, 1988). Other studies have implicated protein kinase C and cGMP (Carmelo, Lopezfarre & Riesco, 1994; Lucchesi & Berk, 1995). Stimulation of the exchanger may be an important response, as acidification occurs with contraction in many smooth muscles (see below). Thus, by agonists co-stimulating the contractile machinery and the exchanger, the pH_i change may be minimised. Modulation of the pH_i-regulating mechanisms by extracellular agents has also been reported in cardiac muscle (Guo, Wasserstrom & Rosenthal, 1992; Wu & Vaughan-Jones, 1994).

Contraction and pH_i

With the application of fluorescent indicators to smooth muscle strips, it is possible to measure pH_i and tension simultaneously. These measurements have shown that contraction itself may alter pH_i. This was first identified in the rat myometrium (Taggart & Wray, 1993). During spontaneous contractions, it was observed that small (0.04 pH unit) transient acidifications were associated with the contractile activity. Their peak lagged the peak of the contraction by several seconds. If contraction was increased, for example by high K^+ depolarisation, then the size of the acidification increased (Taggart & Wray, 1993). Recent work has shown that these acid transients can also be seen in other smooth muscles including portal vein and ureter. The acidification was shown not to be caused by lactic acid production consequent to contraction (Taggart & Wray, 1993), but may be due to the surface membrane Ca^{2+}-ATPase counter-transporting H^+ (Niggli, Sigel & Carafoli, 1982; Schwiening, Kennedy & Thomas, 1993). Thus, in many cells, activation of cellular processes by $[Ca^{2+}]$ will stimulate the Ca^{2+}-ATPase and produce transient changes of pH_i, which in turn will have both energetic and functional consequences.

Extracellular pH

In the body, changes of the pH of the extracellular fluid occur as a consequence of normal physiological responses, for example increased activity, change of respiratory rate, as well as for pathophysiological reasons such as occlusions, ischaemia and kidney disease. The consequences of altered pH_e may include: (i) an induction of a change in pH_i, (ii) effects on pH_i-regulatory mechanisms, and (iii) a direct effect on contractile ability.

Until recently, little was known of how altered pH_e affected pH_i and it was therefore generally assumed, from studies on other cell types, that any effects would be small and slow. The work of Austin and Wray (1993b), however, showed this to be far from correct. In a study on rat mesenteric vessels, they showed that when pH_e was altered it induced a rapid change in pH_i, i.e. within one to five minutes. Furthermore, about 70 per cent of the pH_e change was transmitted to the cytoplasm, compared to around 30 per cent in cardiac cells (Ellis & Thomas, 1976). Since this study, other smooth muscles have been examined and it is apparent that there is a spectrum of responses: in ureter >10 per cent (Burdyga et al., 1996), in vas deferens 20 per cent (Aickin, 1984), in portal vein 35 per cent and in coronary blood vessel cells 70–80 per cent (Ramsey et al., 1994) of the pH_e change is induced in pH_i (Taggart et al., 1994). Consideration of why these differences should arise involves differing proton permeabilities, mechanisms of regulating pH or background acid-loading mechanisms. In mesenteric vessels, a particularly high permeability of the surface membrane to H^+ (0.4 cm/s) was suggested as the underlying mechanism (Austin & Wray, 1993a).

Changes of pH_e will also influence pH_i if they significantly affect pH_i regulating mechanisms. When pH_e is reduced, the $Na^+–H^+$ exchanger will be inhibited and this will presumably contribute to some of the acid load during extracellular acidosis. This has not, however, been much studied in smooth muscle. When it comes to considering what a change of pH_e will do to contraction, it is clear from the above discussion that in order to understand the mechanism of its effects, it will be necessary to distinguish between the effects of pH_e alone on contraction and those arising from the induced pH_i change. It is clear that the effects of pH_e on smooth muscle contraction are tissue specific. It is not surprising, therefore, that when the mechanisms underlying how pH_e alters contraction have been studied, several different ones have been found. In mesenteric vessels, it appears that when pH_e is altered it changes pH_i, which in turn changes $[Ca^{2+}]_i$ and hence tension (Austin et al., 1996). In cerebral vessels, the changes in tone are associated with alteration of membrane potential as a result of pH_e affecting outward K^+ current or inward Ca^{2+} current (Harder & Madden, 1985; West, Leppla & Simard, 1992). More work is needed to further our understanding of how much alteration of pH_i regulation by pH_e is responsible for changes in pH_e affecting pH_i in smooth muscles, as well as the mechanisms responsible for the alteration of contraction.

Contractile responses to pH_i change

That pH_i is an important physiological modulator of force in smooth muscle is now well appreciated. It is not the intention here to review the whole of this literature, but rather to focus on the same three smooth muscles compared for pH_i regulation, i.e. uterus, stomach and ureter (see Fig. 3). This will highlight the range of responses seen. Where known, the mechanisms underlying the effects seen will be briefly discussed. Given the powerful effects produced by pH_i alteration on contractile activity, it is not surprising that smooth muscles possess a variety of pH_i-regulating mechanisms and substantial pH-buffering power capacity. The resting pH_i in smooth muscle, as shown in table 1, is around 7.1. This value will change *in vivo* as activity is altered, CO_2 changes, pH_e changes, there is activation by agonists, and during hypoxia or ischaemia. The effect on contraction may be beneficial to the organism; for example, if hypoxia causes a fall in pH_i that can cause decreased contraction of (some) blood vessels, the dilatation will increase blood flow to the hypoxic area. In other cases, the effects of pH_i on contraction may not be helpful, for example if it increases contractile activity when the metabolic demands cannot be matched and tissue damage occurs.

A comparison of the effects of pH_i alteration in three smooth muscles

Uterus

Intracellular pH is now recognised as being a potent modulator of force production within the smooth muscle of the uterus (myometrium). Recent work has been devoted to elucidating the mechanisms underlying its effects on uterine contraction. The uterus contracts in a phasic manner, as pacemaker activity in some myometrial cells leads to depolarisation of the surface membrane. This results in Ca^{2+} entry *via* L-type Ca^{2+} channels and contraction. Agonists can also modify or initiate this Ca^{2+} entry. Calcium may also be elevated in myometrium as Ca^{2+} is released from the internal store, the sarcoplasmic reticulum.

Intracellular acidification can abolish spontaneous uterine contractions in rat (see Fig. 3) and human tissues (Taggart & Wray, 1992; Parratt *et al.*, 1994). Intracellular alkalinisation converts phasic contractions to a maintained (tonic) contraction. Measurements of $[Ca^{2+}]_i$ suggest that this effect is due to pH_i altering Ca^{2+} entry *via* the L-type Ca^{2+} current, i.e. acidification depresses the Ca^{2+} transient (Taggart *et al.*, 1996). Studies on patched, clamped single uterine cells provide direct evidence consistent with this: the Ca^{2+} current produced by depolarisation was decreased by acidic pH (Shmigol *et al.*, 1995). Using permeabilised preparations to study directly the relationship between Ca^{2+} and force, it was found that pH_i produced no significant effect on this relationship in the uterus, in other words, pH_i does not affect the sensitivity of the Ca^{2+}–force relationship (Crichton *et al.*, 1993).

Ureter

The ureter is a fibromuscular tube composed of smooth muscle. Its unique function is to transport urine from the renal pelvis of the kidney to the bladder. Continuous peristaltic contractions are required

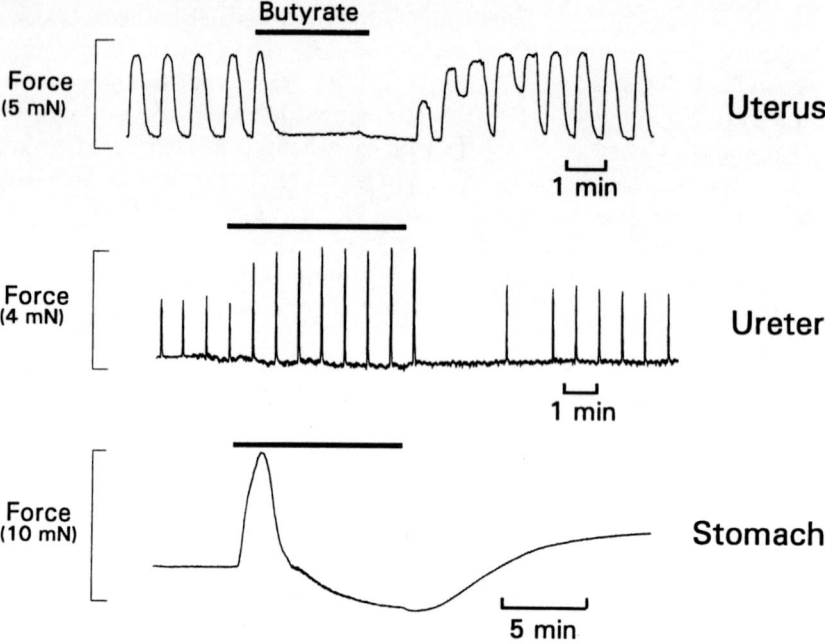

Fig. 3. The effect of acidification on contraction in uterus (top), ureter (middle) and stomach (bottom) smooth muscles. The acidification was produced by addition of 20–40 mM butyrate, for the period indicated by the bar.

to maintain the flow of urine and hence renal function. Force production in the ureter can be affected by changes in pH_i (Burdyga et al., 1996); in guinea-pig ureter, the effects of acidification and alkalinisation on electrically stimulated phasic contractions were potentiation and inhibition, respectively, i.e. these are functionally the opposite of those found in the uterus (described above and in Fig. 3). Acidification prolonged the plateau phase of the action potential, increasing the magnitude of both the Ca^{2+} transient and force of contraction. It is thought that this effect of acidification is likely to be due to a blockade of K^+ currents. Again, this can be contrasted with the uterus, where little or no effect on K^+ currents could be seen but the Ca^{2+} current was pH sensitive (Shmigol et al., 1995). Alkalinisation shortens the action potential and reduces the Ca^{2+} and force transients. Recent findings in the rat ureter show dramatically different effects from those in the guinea-pig: the magnitude and frequency of phasic contractions were increased by alkalinisation, while acidification caused a reduction in contraction. The mechanism of this is unclear, although it may again be due to effects on electrical activity.

Stomach

The smooth muscle of the stomach plays a key role in upper gastrointestinal physiology. First it stores boluses of food passed through the oesophagus and then slow, sustained contractions gradually force the contents towards the distal stomach and duodenum. Peristaltic waves aid in this propulsion and mix the contents with gastric juices. The mucosa protects the smooth muscle from the acidic environment of the lumen. However, if it is damaged or the circulation to the stomach is impaired, then alterations of pH may be experienced by these cells. To date, few studies have investigated the role pH_i plays in the contraction of gastric smooth muscle. In canine antral and corpal smooth muscle, the acidification induced by butyrate occurred rapidly and preceded a decrease in tone. The resting tone of the tissue is dependent on Ca^{2+} entry, thus acidification may be inhibiting voltage-dependent Ca^{2+} entry mechanisms. Gastric rat and guinea-pig tissues elicited similar responses, but also frequently produced a transient initial increase in force followed by relaxation (see Fig. 3). This also occurs in portal vein (Taggart *et al.*, 1994). It could be due to protons displacing Ca^{2+} from internal binding sites, releasing them into the cytosol, and thus resulting in the transient increase.

Summary and future research

Compared to ten years ago, we know an enormous amount more about pH_i regulation in smooth muscles and the effects of both pH_i and pH_e on contraction. What is also clear is that it is hard to discuss smooth muscle as if it were a single entity; the responses seen in different smooth muscles can be enormously different. Thus, in the case of Na^+–H^+ exchanger, all smooth muscle (and, in fact, all mammalian cells) possesses the exchanger, but its isoform distribution and contribution to maintaining resting pH_i and recovery from an acid load vary between tissues. In the case of contraction, the basic mechanism of excitation–contraction coupling is the same in all smooth muscles, but the relative importance of the various membrane ionic channels and their susceptibility to pH_i and the type of intracellular Ca^{2+} store present, all summate to produce differing effects on the end-process – contraction. Areas that require further investigation have been mentioned throughout this chapter and include: a better understanding of how pH_e affects pH_i regulation, the characterisation of pH_i-regulatory mechanisms present in smooth muscles such as gut and respiratory tract, and the effects of pH_i on biochemical pathways of contraction, especially those not involving the traditional Ca^{2+}-calmodulin path (Horowitz *et al.*, 1996).

Acknowledgements

The authors are grateful to the MRC, Wellcome Trust and BSAVA for supporting this work, to Colette Duvall for secretarial help, and to Bill Franks for technical help.

References

Aalkjaer, C. & Cragoe, E.J. (1988). Intracellular pH regulation in resting and contracting segments of rat mesenteric resistance vessels. *Journal of Physiology* **402**, 391–410.

Aalkjaer, C. & Hughes, A. (1991). Chloride and bicarbonate transport in rat resistance arteries. *Journal of Physiology* **436**, 57–73.

Aalkjaer, C. & Mulvany, M.J. (1991). Steady-state effects of arginine vasopressin on force and pH_i of isolated mesenteric resistance arteries from rats. *American Journal of Physiology* **261**, C1010–17.

Aickin, C.C. (1984). Direct measurement of intracellular pH and buffering power in smooth muscle cells of guinea-pig vas deferens. *Journal of Physiology* **349**, 571–85.

Aickin, C.C. (1994a). Regulation of intracellular pH in the smooth muscle of guinea-pig ureter: HCO_3^- dependence. *Journal of Physiology* **479**, 317–30.

Aickin, C.C. (1994b). Regulation of intracellular pH in the smooth muscle of guinea-pig ureter: Na^+ dependence. *Journal of Physiology* **479**, 301–16.

Aickin, C.C. (1994c). Regulation of intracellular pH in smooth muscle cells of the guinea-pig femoral artery. *Journal of Physiology* **479**, 331–40.

Austin, C., Dilly, K., Eisner, D. & Wray, S. (1996). Simultaneous measurement of intracellular pH, calcium, and tension in rat mesenteric vessels: effects of extracellular pH. *Biochemical and Biophysical Research Communications* **222**, 537–40.

Austin, C. & Wray, S. (1993a). Changes of intracellular pH in rat mesenteric vascular smooth muscle with high-H^+ depolarization. *Journal of Physiology* **469**, 1–10.

Austin, C. & Wray, S. (1993b). Extracellular pH signals affect rat vascular tone by rapid transduction into intracellular pH changes. *Journal of Physiology* **466**, 1–8.

Austin, C. & Wray, S. (1995). An investigation of intrinsic buffering power in rat vascular smooth muscle cells. *Pflugers Archiv* **429**, 325–31.

Baro, I., Eisner, D.A., Raimbach, S.J. & Wray, S. (1989). Intracellular pH regulation and buffering power in single, isolated vascular and intestinal smooth muscle cells. *Journal of Physiology* **417** (Abstract).

Berk, B.C., Aronow, M.S., Brock, T.A., Cragoe, E., Gimbrone, M.A. & Alexander, R.W. (1987). Angiotensin II-stimulated Na^+/H^+ exchange in cultured vascular smooth muscle cells. *Journal of Biological Chemistry* **262**, 5057–64.

Boyarsky, G., Ganz, M.B., Cragoe, E.J. & Boron, W.F. (1990). Intracellular-pH dependence of Na–H exchange and acid loading in quiescent and arginine vasopressin-activated mesangial cells. *Proceedings of the National Academy of Sciences USA* **87**, 5921–4.

Buckler, K.J. & Vaughan-Jones, R.D. (1990). Application of a new pH-sensitive fluoroprobe (carboxy-SNARF-1) for intracellular pH measurement in small, isolated cells. *Pflugers Archiv* **417**, 234–9.

Bullock, A.J., Duquette, R.A., Buttell, N. & Wray, S. (1998). Developmental changes in intracellular pH buffering power in smooth muscle. *Pflugers Archiv* **435**, 578–80.

Burdyga, T.V., Taggart, M.J. & Wray, S. (1996). An investigation into the mechanism whereby pH affects tension in guinea-pig ureteric smooth muscle. *Journal of Physiology* **493**, 865–76.

Carmelo, C., Lopezfarre, A. & Riesco, A. (1994). Atrial natriuretic peptide and cGMP inhibit Na^+/H^+ antiporter in vascular smooth muscles in culture. *Kidney International* **45**, 66–75.

Carr, P., McKinnon, W. & Poston, L. (1995). Mechanisms of pH_i control and relationships between tension and pH_i in human subcutaneous small arteries. *American Journal of Physiology* **268**, C580–9.

Crichton, C.A., Taggart, M.J., Wray, S. & Smith, G.L. (1993). The effects of pH and inorganic phosphate on force production in α-toxin permeabilized isolated rat uterine smooth muscle. *Journal of Physiology* **465**, 629–45.

Danthuluri, N.R., Berk, B.C., Brock, T.A., Cragoe, E.J. & Deth, R.C. (1987). Protein kinase C-mediated intracellular alkalinization in rat and rabbit aortic smooth muscle cells. *European Journal of Pharmacology* **141**, 503–6.

Dawson, M.J. & Wray, S. (1985). The effects of pregnancy and parturition on phosphorus metabolites in rat uterus studied by 31-P nuclear magnetic resonance. *Journal of Physiology* **368**, 19–31.

Dawson, M.J. & Wray, S. (1987). Studies of uterine bioenergetics using ^{31}P nuclear magnetic resonance spectroscopy. In *Magnetic resonance of the reproductive system*, ed. S. McCarthy & F. Haseltine, pp. 97–115. New Jersey: Slack Inc.

Eiesland, J., Baro, I., Raimbach, S., Eisner, D.A. & Wray, S. (1991). Intracellular pH and buffering power measured in isolated single cells from pregnant rat uterus. *Experimental Physiology* **76**, 815–18.

Ellis, D. & Thomas, R.C. (1976). Direct measurement of the intracellular pH of mammalian cardiac muscle. *Journal of Physiology* **262**, 755–71.

Fliegel, L. & Dyck, J.R.B. (1995). Molecular biology of the cardiac sodium/hydrogen exchanger. *Cardiovascular Research* **29**, 155–9.

Gerstheimer, F.P., Muhleisen, M., Nehring, D. & Kreye, V.A.W. (1987). A chloride–bicarbonate exchanging anion carrier in vascular smooth-muscle of the rabbit. *Pflugers Archiv – European Journal of Physiology* **409**, 60–6.

Guo, H., Wasserstrom, J.A. & Rosenthal, J.E. (1992). Effect of catecholamines on intracellular pH in sheep cardiac purkinje fibres. *Journal of Physiology* **458**, 289–306.

Harder, D.R. & Madden, J.A. (1985). Cellular mechanism of force development in cat middle cerebral artery by reduced pCO_2. *Pflugers Archiv* **403**, 402–4.

Harrison, N., Larcombe-McDouall, J. B., Earley, L. & Wray, S. (1994). An *in vivo* study of the effects of ischaemia on uterine contraction, intracellular pH and metabolites in the rat. *Journal of Physiology* **476**, 349–54.

Horowitz, A., Menice, C.B., Laporte, R. & Morgan, K.G. (1996). Mechanisms of smooth muscle contraction. *Physiological Reviews* **76**, 967–1003.

Kahn, A.M., Bishara, M., Cragoe, E.J. *et al.* (1992). Effects of serotonin on intracellular pH and contraction in vascular smooth muscle. *Circulation Research* **71**, 1294–304.

Kahn, A.M., Cragoe, E.J., Allen, J.C., Halligan, R.D. & Shelat, H. (1990). Na^+-H^+ and Na^+-dependent $Cl^--HCO_3^-$ exchange control pH_i in vascular smooth muscle. *American Journal of Physiology* **259**, C134–43.

Korbmacher, C., Helbig, H., Stahl, F. & Wiederholt, M. (1988). Evidence for Na/H exchange and Cl/HCO_3 exchange in 10 vascular smooth-muscle cells. *Pflugers Archiv – European Journal of Physiology* **412**, 29–36.

Little, P.J., Neylon, C.B., Farrelly, C.A., Weissberg, P.L., Cragoe, E.J. & Bobik, A. (1995). Intracellular pH in vascular smooth muscle: regulation by sodium–hydrogen exchange and multiple sodium dependent HCO_3^- mechanisms. *Cardiovascular Research* **29**, 239–46.

Lucchesi, P.A. & Berk, B.C. (1995). Regulation of sodium–hydrogen exchange in vascular smooth muscle. *Cardiovascular Research* **29**, 172–7.

Lucchesi, P.A., Deroux, N. & Berk, B.C. (1994). Na^+-H^+ exchanger expression in vascular smooth-muscle of spontaneously hypertensive and wistar-kyoto rats. *Hypertension* **24**, 734–8.

Mason, W.T. (1993). *Fluorescent and luminescent probes for biological activity.* London: Academic Press.

Mitsuhashi, T. & Ives, H.E. (1988). Intracellular Ca^{2+} requirement for activation of the Na^+/H^+ exchanger in vascular smooth muscle. *Journal of Biological Chemistry* **263**, 8790–5.

Muhleisen, M. & Kreye, V.A.W. (1985). Lack of soluble carbonic anhydrase in aortic smooth muscle of the rabbit. *Pflugers Archiv* **405**, 234–6.

Nakanishi, T., Seguchi, M., Tsuchiya, T., Yasukouchi, S. & Takao, A. (1990). Effect of acidosis on intracellular pH and calcium concentration in the newborn and adult rabbit myocardium. *Circulation Research* **67**, 111–23.

Niggli, V., Sigel, E. & Carafoli, E. (1982). The purified Ca^{2+} pump of human erythrocyte membranes catalyzes an electroneutral Ca^{2+}–H+ exchange in reconstituted liposomal systems. *Journal of Biological Chemistry* **257**, 2350–6.

Opitz, N., Merten, E. & Acker, H. (1994). Evidence for redistribution-associated intracellular pK shifts of the pH-sensitive fluorophore carboxy-SNARF-1. *Pflugers Archiv* **427**, 332–42.

Orlowski, J., Kandasamy, R.A. & Shiull, G.E. (1992). Molecular cloning of putative members of the Na/H exchanger gene family – cDNA cloning, deduced amino acid sequence, and messenger-RNA tissue expression of the rat Na/H exchanger NHE-1 and 2 structurally related proteins. *Journal of Biological Chemistry* **267**, 9331–9.

Owen, C.S. (1992). Comparison of spectrum-shifting intracellular pH probes 5′(and 6′)-carboxy-10-dimethylamino-3-hydroxyspiro [7H-benzo[c]xanthese-7,1′(3′H)-isobenzofuran-3′-one and 2′,7′-biscarboxyethyl-5(and 6)-carboxyfluorescein. *Analytical Biochemistry* **204**, 65–71.

Parratt, J., Taggart, M.J. & Wray, S. (1994). Abolition of contractions in the myometrium by acidification *in vitro*. *Lancet* **344**, 717–18.

Parratt, J.R., Taggart, M.J. & Wray, S. (1995). Changes in intracellular pH close to term and their possible significance to labour. *Pflugers Archiv* **430**, 1012–14.

Ramsey, J., Austin, C. & Wray, S. (1994). Differential effects of external pH alteration on intracellular pH in rat coronary and cardiac myocytes. *Pflugers Archiv* **428**, 674–6.

Richmond, P. & Vaughan-Jones, R.D. (1993). K^+–H^+ exchange in isolated carotid body type-1 cells of the neonatal rat is caused by nigericin contamination. *Journal of Physiology* **467** (Abstract).

Rink, T.J., Tsien, R.Y. & Pozzan, T. (1982). Cytoplasmic pH and free Mg in lymphocytes. *Journal of Cell Biology* **95**, 189–92.

Roos, A. & Boron, W.F. (1981). Intracellular pH. *Physiological Reviews* **61**, 296–434.

Sardet, C., Franchi, A. & Pouyssegur, J. (1989). Molecular cloning, primary structure and expression of the human growth factor – activatable Na$^+$/H$^+$antiporter. *Cell* **56**, 271–80.

Schwiening, C.J., Kennedy, H.J. & Thomas, R.C. (1993). Calcium–hydrogen exchange by the plasma membrane Ca-ATPase of voltage-clamped snail neurons. *Proceedings of the Royal Society London* **253**, 285–9.

Shmigol, A.V., Smith, R.D., Taggart, M.J., Wray, S. & Eisner, D.A. (1995). Changes of pH affect calcium currents but not outward potassium currents in rat myometrial cells. *Pflugers Archiv* **431**, 135–7.

Sun, B., Leem, C.H. & Vaughan-Jones, R.D. (1996). Novel chloride-dependent acid loader in the guinea-pig ventricular myocyte: part of a dual acid-loading mechanism. *Journal of Physiology* **495**, 65–82.

Szatkowski, M.S. & Thomas, R.C. (1989). The intrinsic intracellular H$^+$ buffering power of snail neurones. *Journal of Physiology* **409**, 89–101.

Taggart, M., Austin, C. & Wray, S. (1994). A comparison of the effects of intracellular and extracellular pH on contraction in isolated rat portal vein. *Journal of Physiology* **475**, 285–92.

Taggart, M.J., Burdyga, T.H., Heaton, R.C. & Wray, S. (1996). Stimulus-dependent modulation of smooth muscle intracellular calcium and force by altered intracellular pH. *Pflugers Archiv* **432**, 803–11.

Taggart, M.J. & Wray, S. (1992). The relation between intracellular pH (pH$_i$) and force in isolated rat uterus: simultaneous measurements of pH$_i$ and force. *Journal of Physiology* **452**, 232P.

Taggart, M. & Wray, S. (1993). Simultaneous measurement of intracellular pH and contraction in uterine smooth muscle. *Pflugers Archiv* **423**, 527–9.

Thomas, R.C. (1984). Review lecture. Experimental displacement of intracellular pH and the mechanism of its subsequent recovery. *Journal of Physiology* **354**, 3P–22P.

Tse, C-M., Levine, S.A., Yin, C.H.C. *et al.* (1993). Cloning and expression of a rabbit cDNA encoding a serum-activated ethyliso-propylamiloride-resistant epithelial Na$^+$/H$^+$ exchanger isoform (NHE-2). *Journal of Biological Chemistry* **268**, 11917–24.

Vaughan-Jones, R.D. & Wu, M. (1988). pH dependence of intrinsic H$^+$ buffering power in the sheep cardiac purkinje fibre. *Journal of Physiology* **425**, 429–48.

Vermuë, N.A. & Nicolay, K. (1983). Energetics of smooth muscle taenia caecum of guinea-pig: a ^{31}P NMR study. *FEBS Letters* **156**, 293–7.

Vigne, P., Breittmayer, J.P., Frelin, C. & Lazdunski, M. (1988). Dual control of the intracellular pH in aortic smooth-muscle cells by a Camp-sensitive HCO_3^-/Cl^- antiporter and a protein kinase C-sensitive Na^+/H^+ antiporter. *Journal of Biological Chemistry* **263**, 18023–9.

Vogel, H.J., Lilja, H. & Hellstrand, P. (1983). Phosphorus-31 NMR studies of smooth muscle from guinea-pig taenia coli. *Bioscience Reports* **3**, 863–70.

West, G.A., Leppla, D.C. & Simard, J.M. (1992). Effects of external pH on ionic currents in smooth muscle cells from the basilar artery of the guinea pig. *Circulation Research* **71**, 201–9.

Wray, S. (1991). Biomedical NMR spectroscopy and transport. In *Cell membrane transport*, ed. D.L. Yudilevich, R. Deves, S. Peran, & Z.L. Cabantchik, pp. 273–96. New York: Plenum Press.

Wu, M. & Vaughan-Jones, R.D. (1994). Effect of metabolic inhibitors and second messengers upon Na^+–H^+ exchange in the sheep cardiac purkinje fibre. *Journal of Physiology* **478**, 301–13.

BRUCE L. TUFTS

Regulation of pH in vertebrate red blood cells

Introduction

Studies examining the regulation of intracellular pH (pH_i) in vertebrate red blood cells can be generally divided into two research areas. Because blood is relatively easy to obtain and handle in a laboratory setting, it is an ideal tissue for studies of ion-transport processes in vertebrate membranes. Many investigators have therefore used red blood cells as a model tissue for *in vitro* studies to elucidate the ion-transport mechanisms involved in pH_i regulation in vertebrates. The fact that different phylogenetic groups of vertebrates possess red blood cells with quite different membrane characteristics has led to a proliferation of this research and to the detailed description of a variety of ion-transport mechanisms that are involved in the transport of acid–base equivalents across vertebrate membranes. In contrast, other studies in this area have been oriented towards understanding the functional significance of the factors affecting red blood cell pH in vertebrates. Vertebrate red blood cells have a central role in the transport of O_2 and CO_2 between the tissues and respiratory surfaces. Since their pH and the carriage of these respiratory gases in blood are intimately related *via* the Bohr/Haldane and Root effects, the factors determining red blood cell pH in vertebrates have also been intensely studied by respiratory physiologists, whose main goal is to understand the mechanisms involved in the respiratory physiology of different vertebrate groups at both the cellular and organismal levels. Clearly, important insights into the subject of pH regulation in vertebrate red blood cells may be derived from research in each of these areas. This chapter attempts to integrate the work in these different research areas and therefore discusses the different mechanisms involved in the regulation of pH in vertebrate red blood cells as well as their possible functional significance in the different vertebrate groups.

Methods used to measure red blood cell pH

The two most commonly used approaches to measure red blood cell pH are the freeze–thaw method and the 5,5-dimethyl-2,4-oxazolidinedione

(DMO) method. Each of these methods has advantages and disadvantages, but both are relatively simple and do not require much in the way of specialized equipment. The freeze–thaw method is described in detail by Zeidler and Kim (1977). Briefly, this method involves the collection and centrifugation of whole blood (volume approximately 500 µl) at the time of sampling to obtain a red blood cell pellet at the base of a plastic centrifuge tube. After the plasma supernatant is removed by aspiration, the pellet is immediately frozen in liquid nitrogen. Next, the pellet is thawed at room temperature, refrozen, and thawed again. This process lyses the cells and produces a hemolysate which can be drawn into a micro-pH unit associated with a pH meter. The pH measured using this approach is the mean of all cellular contents, including the lysed organelles and nuclei in nucleated red blood cells, and it is therefore not a true indication of cytoplasmic pH. Due to the high protein content of the hemolysate, measurements using this method may also be affected by a 'suspension effect' at the liquid junction potential which causes a minor artifact in the measurement of pH in some red blood cells (Albers & Goetz, 1985; Tufts & Randall, 1988). Nonetheless, the freeze–thaw method has been used extensively and produces highly repeatable measurements of red blood cell pH. One of the most significant advantages of this approach is the fact that it can be used to measure the instantaneous red blood cell pH of blood samples collected during either *in vitro* or *in vivo* experiments without the addition of chemicals or isotopes to the blood.

A recent study by Jorgensen (1995) has used an interesting variation of the freeze–thaw method to determine red blood cell pH following intracellular acidification in frog red blood cells. The freeze–thaw method normally requires relatively large sample volumes of blood (approximately 500 µl), which may not always be available or convenient in some experiments. To avoid this problem, Jorgensen (1995) initially determined the buffering capacity of frog red blood cells using the freeze–thaw method. In subsequent experiments, red blood cell pH at specific times during recovery from intracellular acidification was simply calculated from measurements of H^+ efflux obtained using a conventional pH electrode to measure extracellular pH (pH_e).

Use of the DMO method for the determination of pH_i in tissues was first introduced and described by Waddel and Butler (1959). Briefly, this approach is based on the principle that the distribution of the weak acid DMO will be determined by the pH gradient between the extracellular and intracellular fluid. Thus, measurements of pH_e and DMO levels in known volumes of extracellular fluid and red blood cells (which can be determined by liquid scintillation counting of ^{14}C-DMO in red blood cell pellets) can be used to calculate red blood cell pH. The DMO method has been used extensively to measure red blood cell pH in a variety of species. It should also be noted that while DMO is commonly used in most studies, other radiolabeled weak acids such as benzoic acid are equally as effective (Borgese *et al.*, 1994). It may be an important consideration that this method requires the use of radio-isotopes and a scintillation counter to determine DMO levels in plasma and red blood cells. The time required for the DMO distribution to come to equilibrium between the intracellular and extracellular fluid has also been viewed as a potential problem in studies attempting to measure transient changes in pH_i, although Milligan and Wood (1985) have shown that DMO distribution responds very quickly (less than 5 minutes at 15 °C) to changes in pH in the nucleated red blood cells of fish. Preliminary experiments to confirm the time required for the distribution of the weak acid to reach equilibrium across the red blood cell membrane under the chosen experimental conditions are highly recommended. The fact that this method requires smaller volumes of red blood cells for sample analysis may be a significant advantage for some studies.

Several studies have compared measured red blood cell pH values using both the DMO method and the freeze–thaw method. In a number of these investigations, the values obtained for red blood cell pH were not significantly different using these two approaches (Waddel & Bates, 1969; Roos & Boron, 1981; Milligan & Wood, 1985). In some types of red blood cells, however, the values obtained for red blood cell pH using the freeze–thaw method may be slightly lower than those obtained using the DMO method. This situation has been been observed in red blood cells from the carp (Albers & Goetz, 1985) and in amphibians (Tufts & Randall, 1988). Figure 1 shows how the distribution ratios for protons across the red blood cell membrane of the amphibian, *Bufo marinus*, can be affected by the method used to measure red blood cell pH. Interestingly, the differences in pH that are observed between these two methods in *Bufo* red blood cells can be entirely removed when a saline bridge is incorporated to eliminate the 'suspension effect' at the liquid junction potential (Table 1; Tufts & Randall, 1988). Thus, it is possible that the observed differences between these two methods may be caused by an extremely high protein content in the red blood cell hemolysates of certain species. Another potential explanation, however, may be that some pH electrodes are simply more sensitive to suspension effects.

Instantaneous measurements of red blood cell pH using fluorescence methods may also be possible. At present, however, the use of these

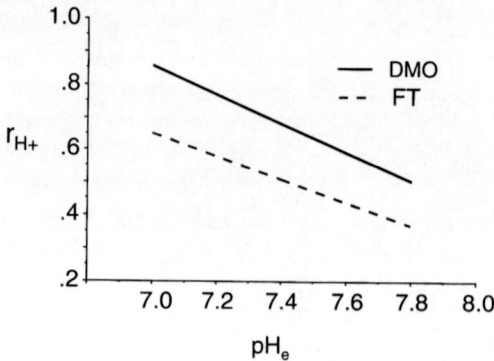

Fig. 1. Regression lines for proton (r_{H+}) distribution ratios across the red blood cell membrane *versus* pH_e in the amphibian, *Bufo marinus*. Proton distribution ratios were determined using both the freeze–thaw method ($r_{H+(FT)}$) and the DMO method ($r_{H+(DMO)}$) for the measurement of pH_i. Based on data from Tufts and Randall (1988).

Table 1. *Red blood cell pH (pH_i) determined using freeze-thaw method (FT), freeze-thaw method with saline bridge (FT + sal) and DMO distribution method (DMO)*

pH_e	pH_i	$pH_{i(DMO)}$
7.732 ± 0.037	7.314 ± 0.054(FT)	7.449 ± 0.037*
7.738 ± 0.031	7.469 ± 0.068(FT+sal)	7.477 ± 0.032

*Significant ($p < 0.05$) difference from the freeze–thaw value.
Values are means ± SD.
Data from Tufts and Randall (1988).

techniques for this purpose is not widespread. According to M. Nikinmaa (personal communication), a major obstacle that must be overcome before using this approach with red blood cells is the fact that the absorption maximum for hemoglobin is very close to the excitation/emission wavelengths for dyes such as 2′,7′-bis-(-2-carboxyethyl)-5,6-carboxyfluorescein acetoxy-methyl-ester (BCECF) that are commonly used to determine red blood cell pH with this method. P. Cala (personal communication) has also encountered problems with this approach, resulting from dye leakage into the extracellular fluid which produces a high background signal due to the higher pH_e. Thus, while it may be theoretically possible to measure red blood cell pH using fluorometric methods, the simplicity and accuracy of the DMO and freeze–thaw measurements currently make these the methods of choice for most studies of red blood cell pH regulation.

Theoretical considerations

Steady-state ion distributions and red blood cell pH

The factors influencing the steady-state distribution of ions across the red blood cell membrane have been thoroughly discussed in previous reviews (Hladky & Rink, 1977; Nikinmaa, 1990) and the reader should consult these articles for a detailed discussion of this subject.

Under steady-state conditions and in the presence of rapid anion exchange, the distribution ratios for H^+, HCO_3^- and Cl^- are linked and, theoretically, should be equal. Indeed, numerous investigations of vertebrate red blood cells have found the distribution ratios for these ions to be very similar (Albers & Goetz, 1985; Heming *et al.*, 1986, Tufts & Randall, 1988). As a further consequence of these relationships, the red blood cell pH (pH_i) of most vertebrates can also be calculated under steady-state conditions from the equation:

$$pH_i = pH_e + \log[Cl^-]_i - \log[Cl^-]_e \qquad (1)$$

Since all permeable ions such as H^+, HCO_3^-, OH^-, Cl^- will be distributed across the red blood cell membrane under steady-state conditions according to a Donnan equilibrium, factors that affect the net charge on the impermeable anions inside the cell will also have a significant influence on the steady-state red blood cell pH at any given pH_e. The most relevant factors in this regard will be the concentrations of organic phosphates within the cell and the oxygenation state of hemoglobin, both of which play an important role in determining the net charge of impermeable anions inside the red blood cell. Decreases in either red blood cell organic phosphate levels or hemoglobin oxygenation, for example, will cause a relative increase in pH.

Addition of CO_2 or metabolic acid to the blood

The most common physiological factors that alter red blood cell pH are increased levels of CO_2 or addition of metabolic acid to the blood. Carbon dioxide will diffuse easily into the red blood cell membrane, where it will be hydrated in the presence of carbonic anhydrase to form HCO_3^- and protons. Some of the resulting protons will be taken up by intracellular buffers, but in the absence of secondarily active proton extrusion (see next section), protons that are not buffered will cause a decrease in pH_i.

Increased levels of metabolic protons in the blood will also affect the red blood cell pH of most vertebrates due to the Jacobs–Stewart cycle (Fig. 2). In the plasma, these protons will combine with HCO_3^- to form molecular CO_2, which diffuses easily into the red blood cell. As explained above, CO_2 will then be hydrated within the cell to form HCO_3^- and a proton. The resulting HCO_3^- will normally be transferred to the plasma, but the protons have the potential to decrease the cell's pH if they exceed the capacity of intracellular buffers. Since most vertebrates lack carbonic anhydrase in the plasma, the rate that these metabolic acid loads can be transferred to the red blood cell *via* the Jacobs–Stewart cycle will be determined by the rate of the uncatalysed reaction forming CO_2 in the plasma, which is highy dependent upon temperature.

Potential for red blood cell pH regulation *via* secondarily active ion transport

The potential significance of secondarily active red blood cell ion-transport mechanisms such as Na^+/H^+ exchange towards the regulation

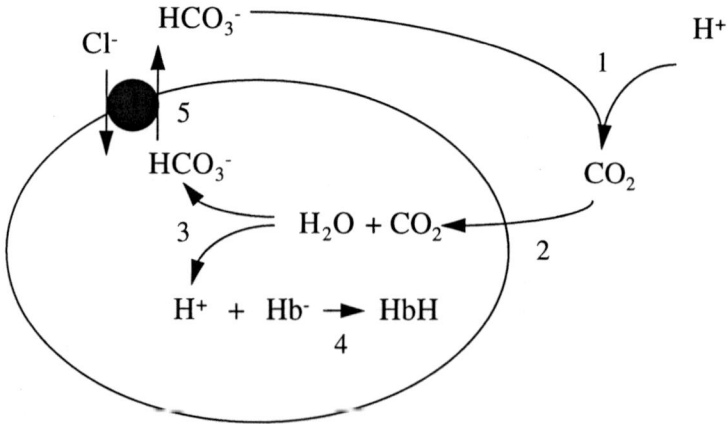

Fig. 2. The Jacobs–Stewart cycle. (1) Extracellular protons combine with HCO_3^- to form carbonic acid and then CO_2 at the uncatalysed rate. (2) CO_2 diffuses easily into the cell, where (3) it combines with water to form HCO_3^- and a proton. This intracellular CO_2 hydration reaction is catalysed by carbonic anhydrase. (4) Protons combine with intracellular buffers (eg. hemoglobin), and (5) HCO_3^- is exchanged for extracellular Cl^-. Redrawn, based on Nikinmaa and Tufts (1989).

of red blood cell pH is determined by several factors. In cells that possess rapid anion exchange, acid–base equivalents such as protons, which are extruded from the red blood cell to the plasma, will enter the Jacobs–Stewart cycle and re-equilibrate across the cell membrane. Normally, the rate-determining step in this process is the rate of the uncatalysed conversion of plasma HCO_3^- and protons to CO_2. In order to have a significant impact on red blood cell pH, the activity of any proton-extruding mechanism within the membrane must therefore exceed the rate of the uncatalysed formation of CO_2 from HCO_3^- and protons. In addition, it follows that another important prerequisite for the regulation of red blood cell pH by ion-transport mechanisms such as Na^+/H^+ exchange must be the absence of significant extracellular carbonic anhydrase activity in the plasma because catalysis of the CO_2 reactions in this compartment will short-circuit the system. Since the uncatalysed CO_2 reactions in the plasma will also be greatly influenced by temperature, the rates of proton extrusion necessary to have a significant impact on red blood cell pH will also increase as temperature increases. Finally, the potential impact of proton extrusion on pH will also be largely determined by the red blood cell buffering properties. If the buffer capacity of the red blood cell is extremely high, the potential impact of proton extrusion on the pH will probably be minimal.

The remaining sections review the current literature describing the mechanisms involved in red blood cell pH regulation in the different vertebrate groups.

Agnathans

In contrast to the situation in most vertebrates examined to date, the measured pH of lamprey red blood cells is very different from the pH_i calculated from the Cl^- distribution across the red blood cell membrane (Nikinmaa, 1986). The distribution ratios across the red blood cell membrane for protons (H^+_e/H^+_i) and chloride ions (Cl^-_i/Cl^-_e) are therefore very different in these animals (Nikinmaa, 1986; Tufts & Boutilier, 1989, 1990a). Moreover, the red blood cell pH of lampreys is relatively higher than that in most other vertebrates and may even be higher than the pH_e under certain conditions (Nikinmaa, 1986; Nikinmaa, Kunnamo-Ojala & Railo, 1986; Tufts & Boutilier, 1989, 1990a; Nikinmaa & Matsoff, 1992; Tufts, Bagatto & Cameron, 1992; Ferguson *et al.*, 1992). This uncoupling of the distribution ratios for protons and Cl^- and the peculiar relationship between pH_e and pH_i can be attributed to several unique characteristics of the lamprey red blood cell.

Unlike the red blood cells of other vertebrates examined to date,

Cl^-/HCO_3^- exchange across the membrane of some species of agnathans appears to be extremely limited, if not entirely absent (Ohnishi & Asai, 1985; Ellory, Wolowyk & Young, 1987; Nikinmaa & Railo, 1987). Thus, in contrast to the situation in other vertebrates, equilibration of acid–base equivalents across the red blood cell membrane probably cannot proceed rapidly *via* the Jacobs–Stewart cycle in lamprey because anion movements across the membrane are extremely slow. According to Heisler (1986), red blood cell pH should be significantly higher at any given blood P_{CO_2} in the absence of significant anion transfer between red blood cells and plasma. The retention of HCO_3^- within the lamprey red blood cell at any given P_{CO_2}, which results from the absence of functional anion exchange, is therefore an important factor contributing to the relatively high pH observed in these animals. Evidence to support this is provided by experiments in which the ionophore for anions, tri-n-propyl tin chloride, was added to the blood. Tufts and Boutilier (1990a) showed that addition of this artificial ionophore to sea lamprey blood *in vitro* significantly reduced the pH_i and caused the distribution ratios for protons and Cl^- to become very similar. A relatively large Haldane effect (Ferguson *et al.*, 1992; Nikinmaa & Matsoff, 1992) is another feature contributing to the relatively high red blood cell pH in lampreys. In addition to these factors, secondarily active ion transport across the red blood cell membrane may have a significant impact on red blood cell pH in these animals.

In lamprey, red blood cell pH is maintained by a Na^+-dependent, amiloride-sensitive mechanism which may even be active under steady-state conditions (Nikinmaa, 1986; Nikinmaa *et al.*, 1986; Tufts, 1992). This mechanism is probably an early vertebrate Na^+/H^+ exchanger. At present, however, there has been no immunological or amino acid sequence information to confirm the presence of a Na^+/H^+ exchanger (NHE-like protein) in these cells. Because rapid equilibration of acid–base equivalents cannot occur across the red blood cell membrane of agnathans due to the virtual lack of the anion-exchange pathway, the potential impact of proton movements *via* Na^+/H^+ exchange on red blood cell pH will be increased in lamprey, as compared to the situation in most vertebrates which possess rapid anion exchange. The activity of the lamprey Na^+/H^+ exchanger is markedly elevated following intracellular acidification (Nikinmaa, Tufts & Boutilier, 1993; Virrki & Nikinmaa, 1994). Blood oxygenation status may also modulate the activity of this mechanism in the sea lamprey, *Petromyzon marinus* (Ferguson *et al.*, 1992). In contrast to the Na^+/H^+ exchanger present in the red blood cells of teleost fish, the activity of the Na^+/H^+ exchanger in those of the lamprey is not markedly affected by adrenergic stimulation (Tufts, 1991; Virrki & Nikinmaa, 1994).

Regulation of pH in lamprey red blood cells has significant consequences for blood gas transport processes. Lamprey blood has recently been shown to have a Root effect (Nikinmaa, 1993) and a relatively large Bohr/Haldane effect when properly determined according to red blood cell pH (Ferguson *et al.*, 1992; Nikinmaa & Matsoff, 1992). Even under normal resting conditions, the relative increase in pH resulting from Na^+/H^+ exchange may therefore improve the uptake of O_2 by lamprey hemoglobin at the gill (Nikinmaa, 1993).

Because agnathans lack significant red blood cell anion exchange, the transport of CO_2 from the tissues to gills is largely dependent upon red blood cell CO_2 carriage (Tufts & Boutilier, 1989, 1990a; Tufts *et al.*, 1992; Ferguson *et al.*, 1992; Nikinmaa & Matsoff, 1992). One might therefore speculate that features increasing the amount of CO_2 that can potentially be carried within the red blood cell of agnathans could also be subject to considerable selective pressure. Red blood cell Na^+/H^+ exchange, for example, appears to have an important impact on the distribution of CO_2 between the plasma and the red blood cells in lampreys (Tufts, 1992; Cameron & Tufts, 1994). Extrusion of protons from the red blood cell *via* this mechanism effectively increases the apparent nonbicarbonate buffer value of the cell during CO_2 loading into the blood and thereby increases red blood cell CO_2 carriage (Cameron & Tufts, 1994). Thus, regulation of red blood cell pH by Na^+/H^+ exchange may have an important role in facilitating both O_2 and CO_2 transport in the blood of lampreys.

Hagfish red blood cells have also been shown to possess extremely low levels of anion-exchange activity (Ellory *et al.*, 1987). In contrast to the lamprey, however, the distribution of protons seems to approach equilibrium across the red blood cell membrane of the hagfish (Tufts & Boutilier, 1990b) and the pH is typically much lower than the pH_e in these animals (Tufts & Boutilier, 1989). Moreover, there is minimal recovery of red blood cell pH in hagfish following intracellular acidification as compared to that in lamprey (Fig. 3). In lamprey, red blood cell Na^+/H^+ exchange appears to be responsible for the rapid recovery of pH after acidification (Fig. 3; Nikinmaa *et al.*, 1993). It would therefore appear that some features of the red blood cell membrane in hagfish are quite different from those in the lamprey.

Gnathostome fishes

Teleosts

Under steady-state conditions, the distribution of protons probably approaches electrochemical equilibrium in teleost red blood cells

A. Hagfish

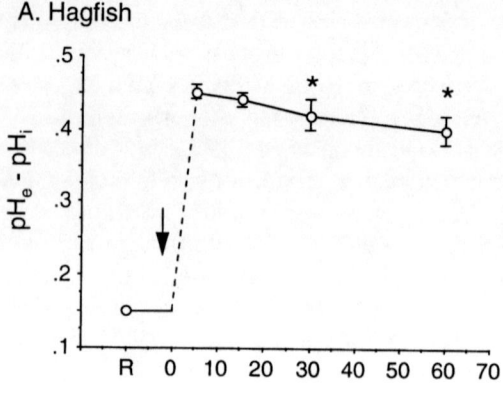

B. Lamprey

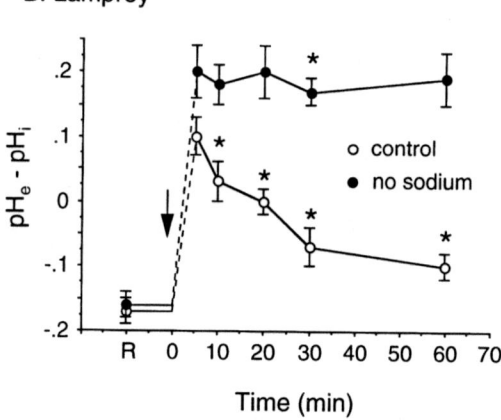

Time (min)

Fig. 3. The pH gradient across the red blood cell membrane (pH_e–pH_i) before (R), and at specific times after, washing away NH_4Cl from red blood cell suspensions of (**A**) hagfish, *Myxine glutinosa*, and (**B**) lamprey, *Lampetra fluviatilis*. An asterisk indicates that a recovery value is significantly different from the 5-minute value. Drawn from data presented in Nikinmaa et al. (1993).

because the pH_i is not affected by Na^+ removal (Nikinmaa et al., 1987), or treatment with either amiloride (Nikinmaa et al., 1987) or the protonophore 2,4-dinitrophenol (2,4-DNP) (Nikinmaa, Tiihonen & Paajaste, 1990). In the presence of elevated catecholamines, however, a secondarily active Na^+/H^+ exchange mechanism becomes activated within the red blood cell membrane of many teleost species (Nikinmaa & Huestis, 1984; Baroin et al., 1984; Cossins & Richardson,

1985). This exchanger causes an increase in red blood cell pH (pH_i) and a decrease in pH_e. Tetens, Lykkeboe and Christensen (1988) have found that noradrenaline is more effective than adrenaline in activating the Na^+/H^+ exchanger in trout and therefore concluded that the adrenergic receptors involved in the response in this species are probably the β1 type. Similar results have also been obtained for carp (Salama & Nikinmaa, 1990), but a recent study indicates that adrenaline may be a more potent activator of the response in cod red blood cells (Berenbrink & Bridges, 1994).

The mechanisms involved in the adrenergic regulation of pH in teleost red blood cells have now been well described and are illustrated in Figure 4 (see also Nikinmaa & Tufts, 1989; Nikinmaa, 1990, 1992). Briefly, binding of catecholamines to adrenergic receptors on the red blood cell membrane activates adenylate cyclase and causes an increase in intracellular cAMP (Mahe, Garcia-Romeu & Motais, 1985). Mahe *et*

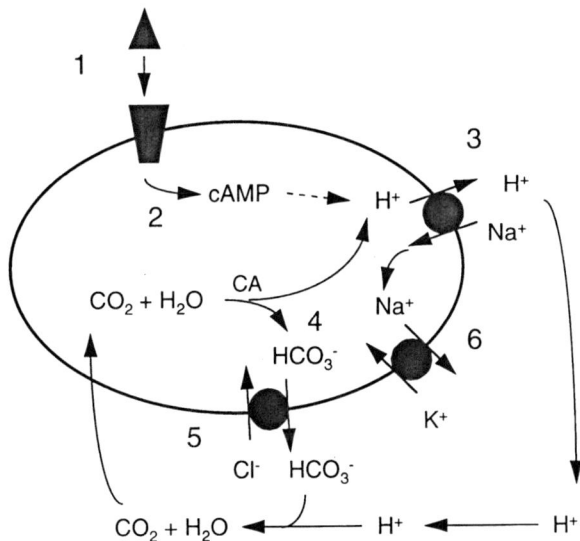

Fig. 4. Adrenergic response in teleost red blood cells. (1) β-agonists bind to a receptor which activates adenylate cyclase. (2) Intracellular cAMP levels increase. (3) The Na^+/H^+ exchanger is activated and increases red blood cell pH. (4) The CO_2 hydration reaction is shifted towards the formation of additional HCO_3^- within the red blood cell. (5) Extracellular Cl^- enters the cell in exchange for HCO_3^-. (6) The elevated intracellular Na^+ levels increase the activity of the Na^+/K^+ pump. Redrawn, based on Nikinmaa and Tufts (1989).

al. (1985) have shown that this increase in cAMP then activates the Na^+/H^+ exchanger. Activation of the Na^+/H^+ exchanger causes an increase in red blood cell pH and a decrease in pH_e. Thus, following adrenergic stimulation, protons are no longer distributed in an electrochemical equilibrium across the cell membrane. Protons can be displaced from electrochemical equilibrium in adrenergically stimulated teleost red blood cells because the rate of the Na^+/H^+ exchanger is faster than the Jacobs–Stewart cycle which equilibrates protons across the cell membrane (see above). If the rate of the extracellular hydration/dehydration reactions for CO_2 in the extracellular compartment are increased by the addition of carbonic anhydrase, the impact of the Na^+/H^+ exchanger on the pH gradient across the red blood cell membrane after adrenergic stimulation can be reduced (Motais, Garcia-Romeu & Thomas, 1989; Nikinmaa *et al.*, 1990). Addition of the protonophore 2,4-DNP (to increase the proton permeability of the red blood cell membrane) also abolishes the effects of adrenergic stimulation on the pH gradient across the membrane, but does not affect the sodium accumulation in adrenergically stimulated trout cells (Nikinmaa *et al.*, 1990). Activation of red blood cell Na^+/H^+ exchange by catecholamines shifts the intracellular CO_2 hydration/dehydration reactions towards the formation of increased levels of intracellular HCO_3^- (Tufts, Vincent & Currie, 1998). Efflux of some of this HCO_3^- down its concentration gradient in exchange for extracellular Cl^- therefore causes a rise in intracellular Cl^- levels during the adrenergic response in teleost red blood cells. Net proton efflux from adrenergically stimulated red blood cells stops when proton influx *via* passive diffusion and the Jacobs–Stewart cycle becomes equal to the proton efflux *via* Na^+/H^+ exchange.

Adrenergic stimulation is also associated with a significant decrease in cellular ATP levels and a significant increase in the O_2 consumption of teleost red blood cells (Ferguson & Boutilier, 1988; Tufts & Boutilier, 1991). Since this increase in aerobic metabolism is entirely abolished *in vitro* by the presence of ouabain, it appears to be due to a stimulation of the Na^+/K^+-ATPase as a result of the large influx of sodium *via* the activated Na^+/H^+ exchanger (Fig. 5; Tufts & Boutilier, 1991). Another important prerequisite for the presence of a secondarily active Na^+/H^+ exchanger with relatively high activity, such as that found in teleost red blood cells, may therefore be a relatively high rate of ATP production to fuel the active restoration of appropriate ion gradients following adrenergic stimulation. The nucleated red blood cells of lower vertebrates are capable of significant aerobic metabolism, but this is not the case for the non-nucleated red blood cells of mammals (Boutilier & Ferguson, 1989). Thus, the energetic support necessary for a second-

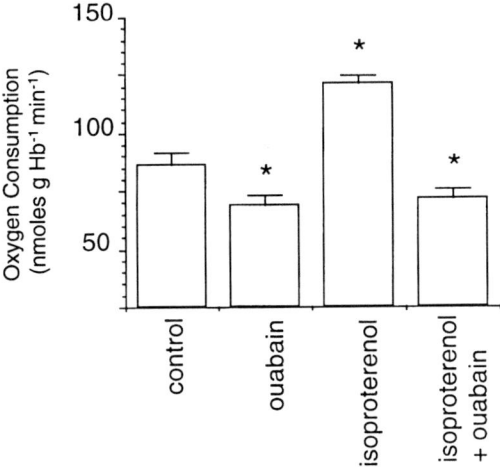

Fig. 5. Effects of isoproterenol (10^{-5} mol/l) and ouabain (10^{-4} mol/l) on the O_2 consumption of rainbow trout red blood cells at 15 °C. All values are means + SE (n=6). Asterisks denote a significant difference from control. Redrawn from Tufts and Boutilier (1991).

arily active Na^+/H^+ exchanger that is capable of influencing red blood cell pH may only be available in the nucleated cells of lower vertebrates.

The rate of Na^+/H^+ exchange in adrenergically stimulated teleost red blood cells is influenced by a number of factors. A reduction in atmospheric O_2 tension, for example, increases the activity of Na^+/H^+ exchange in the red blood cells of trout (Motais, Garcia-Romeu & Borgese, 1987) and carp (Salama & Nikinmaa, 1988). The activity of the red blood cell Na^+/H^+ exchanger in teleosts is also dependent upon the pH_e (Borgese, Garcia-Romeu & Motais, 1987) and temperature (Cossins & Kilbey, 1990). Because the potential impact of the teleost Na^+/H^+ exchanger on red blood cell pH is determined by the rate of the uncatalysed CO_2 hydration/dehydration reactions, which are highly temperature dependent, an increased rate of activity of the Na^+/H^+ exchanger observed at higher temperatures allows the activated Na^+/H^+ exchanger to keep pace with these extracellular CO_2 reactions and thereby still have a significant impact on red blood cell pH over a broad range of temperatures in poikilotherms such as the rainbow trout (Cossins & Kilbey, 1990). The magnitude of the adrenergic response in teleost fish red blood cells also varies markedly between species (Salama & Nikinmaa, 1989). Investigators have recently begun to characterize the molecular properties of the Na^+/H^+ exchanger (βNHE) in teleost red blood cells (Borgese *et al.*, 1992, 1994). It is likely that studies at this level may provide a more complete explanation for some of the observed interspecific and intraspecific differences in the magnitude of the adrenergic response in teleosts.

The functional significance of the adrenergic response in teleost fish red blood cells has been linked to the preservation of O_2 transport during stressful conditions. A lowering of the pH within the hemoglobin environment will reduce the O_2-carrying capacity of the blood because of the Root effect. If extracellular acidoses are transferred to the red blood cell in fish under stressful conditions, the reduction in pH and associated Root effect may therefore impose a serious limitation on the amount of O_2 that can be loaded into the arterial blood as it passes through the gills. Cossins and Richardson (1985) showed that adrenergic stimulation of rainbow trout red blood cells *in vitro* resulted in a 22–46 percent increase in the O_2-carrying capacity of the blood (Fig. 6). *In vivo* studies have also demonstrated that adrenergic regulation of red blood cell pH has an important role in preserving the O_2-carrying capacity of the blood during stressful conditions such as following

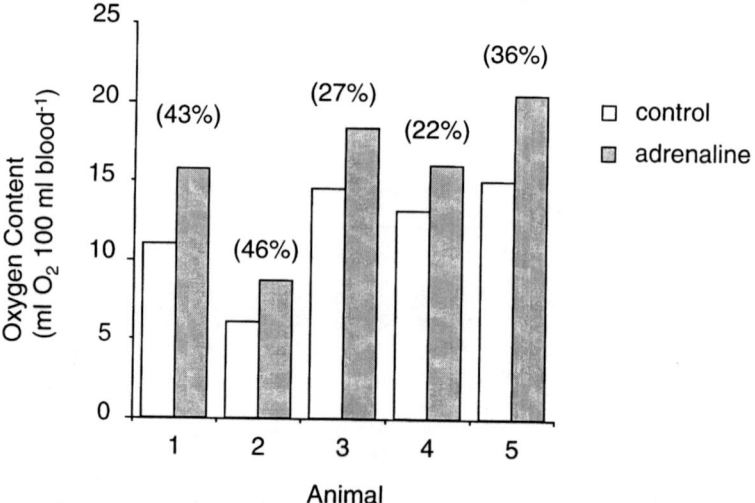

Fig. 6. Effects of adrenaline (10^{-4} mol/l) upon the O_2-carrying capacity of the red blood cells from five individual trout. All values assume 40 percent hematocrit. Values in parentheses indicate the percentage increase from the control condition. Drawn from data presented in Cossins and Richardson (1985).

exhaustive exercise (Nikinmaa, Cech & McEnroe, 1984; Primmett *et al.*, 1986).

Non-teleosts

Similar to those of most other vertebrates, elasmobranch red blood cells possess rapid anion exchange (Obaid, McElroy Critz & Crandall, 1979). The Jacobs–Stewart cycle will therefore equilibrate acid–base equivalents across the red blood cell membrane of elasmobranchs. In contrast to the situation in teleosts, red blood cell pH does not appear to be regulated by adrenergic mechanisms in elasmobranchs (Tufts & Randall, 1989). Indeed, according to Nikinmaa (1997), there is no published evidence that secondarily active transport mechanisms have a role in the regulation of pH_i in elasmobranch red blood cells. This may not be surprising because elasmobranch plasma contains significant levels of carbonic anhydrase (Wood *et al.*, 1994) and the buffer values of elasmobranch red blood cells are relatively high (Jensen, 1989). These features may preclude any possible regulation of pH_i by secondarily active transport mechanisms in elasmobranch red blood cells. Tufts *et al.* (1994) found that red blood cell pH in the bowfin, *Amia calva*, is also not affected by adrenergic stimulation. Thus, regulation of red blood cell pH *via* adrenergic mechanisms may be a feature that is restricted to teleosts, although further study in this area is certainly warranted.

Amphibians

Several studies have provided evidence that amphibian red blood cells possess secondarily active ion-transport mechanisms which may be capable of influencing red blood cell pH. Cala (1983, 1985) has shown that the red blood cells of the aquatic salamander, *Amphiuma*, possess both K^+/H^+ and Na^+/H^+ exchangers. Low levels of Na^+/H^+ exchange activity can be detected in these cells even under steady-state conditions (Tufts *et al.*, 1987a). Cala, Anderson and Cragoe (1988) have also shown that red blood cell Na^+/H^+ exchange in this species is stimulated by intracellular acidification. Although rapid anion exchange is also present in amphibian red blood cells, the secondarily active alkali metal–H^+ exchangers present in these cells can have a significant influence on the pH if their activity is greater than the rate of the uncatalyzed reaction forming CO_2 from HCO_3^- and protons in the plasma. A recent study by Jorgensen (1995) provides evidence that the Na^+/H^+ exchanger in the red blood cells of the frog, *Rana temporaria*, is capable of having a significant impact on red blood cell pH *in vitro* during recovery from intracellular acidification.

Rudolph and Greengard (1980) observed intracellular Na^+ accumulation and cell swelling in frog red blood cells following adrenergic stimulation, but only in the presence of a phosphodiesterase inhibitor that pharmacologically elevates intracellular cAMP levels. Thus, the Na^+/H^+ exchanger in amphibian red blood cells may be activated *via* a pathway that is similar to that in teleosts, although the physiological relevance of this mechanism in amphibians is questionable. In the absence of phosphodiesterase inhibition, adrenergic stimulation does not significantly influence red blood cell pH in the salamander, *Amphiuma tridactylum* (Tufts *et al.*, 1987a), or in the toad, *Bufo marinus* (Tufts, Mense & Randall, 1987b).

There is currently very limited information regarding the potential significance of red blood cell pH regulation in amphibians *in vivo*. Tufts *et al.* (1987b) observed that exhaustive exercise in the toad, *Bufo marinus*, caused relatively large decreases in both pH_e and red blood cell pH. Thus, secondarily active transport mechanisms in amphibian red blood cells may be less effective in regulating pH_i as compared to other vertebrate groups such as teleost fish. It is also noteworthy that the plasma of another amphibian, the mudpuppy (*Necturus maculosus*), appears to contain significant levels of carbonic anhydrase (B. Tufts, M. Gervais, A. Moss & R. Henry, unpublished data). As is probably the case in elasmobranchs, this feature may preclude any regulation of red blood cell pH *via* secondarily active transport mechanisms *in vivo* because the rate of the extracellular CO_2 reactions, and therefore equilibration of acid equivalents across the red blood cell membrane *via* the Jacobs–Stewart cycle, will be accelerated. To date, however, there is a paucity of information regarding the prevalence of extracellular carbonic anhydrase and/or plasma carbonic anhydrase inhibitors in amphibians. There have also been no attempts, either *in vitro* or *in vivo*, to determine whether secondarily active transport processes in amphibian red blood cells can ultimately have a significant impact on the gas-transport properties of the blood.

Birds and mammals

The red blood cells in all homeotherms examined to date, with the possible exception of the primitive ones in embryonic birds (Baumann & Engelke, 1987), possess high levels of anion exchange. Thus, the Jacobs–Stewart cycle has an important role in equilibrating acid equivalents across the red blood cell membrane of homeotherms.

Secondarily active ion-transport mechanisms have also been characterized in the red blood cells of a number of homeotherms. For example, Na^+/H^+ exchangers are present in the red blood cells of rabbits (Escobales & Figueroa, 1991), pigs (Sergeant, Sohn & Kim, 1989) and dogs (Parker, Coloclasure & McManus, 1991). According to Nikinmaa (1997), however, there is currently no conclusive evidence that secondarily active transport pathways have an important influence on red blood cell pH in homeotherms. Indeed, several factors might be expected to minimize its potential regulatory role in homeotherms. The rate of the uncatalysed formation of CO_2 from HCO_3^- and protons, for example, will be much faster in homeotherms simply because of the higher temperatures in these animals. Thus, the rate of any secondarily active mechanisms such as Na^+/H^+ exchange would also need to be much greater at these temperatures in order to have a significant impact on pH_i. In general, the hemoglobin of homeotherms also has relatively high buffer capacity, which will minimize the potential impact of secondarily active transport on red blood cell pH (Nikinmaa, 1997). Finally, the fact that mammalian red blood cells are essentially devoid of organelles such as mitochondria may place an important limitation on the potential energy that would be available to support secondarily active transporters with high enough activity levels to influence red blood cell pH significantly under these conditions.

Conclusions

The potential for red blood cell pH regulation *via* secondarily active transport in the red blood cells of vertebrates is influenced by a number of factors, including the activity of the transporter, the red blood cell buffer capacity, the rate of extracellular CO_2 reactions, and the presence or absence of rapid anion exchange. In some lower vertebrates (lamprey, teleost fish), an appropriate combination of these factors has provided circumstances in which secondarily active transport mechanims can have an important role in the regulation of red blood cell pH and therefore blood gas transport. In other vertebrates (elasmobranchs, homeotherms), these factors provide conditions that are probably less appropriate for the regulation of red blood cell pH *via* secondarily active ion-transport processes, and gas transport must be regulated *via* other mechanisms.

References

Albers, C. & Goetz, K.G. (1985). H^+ and Cl^- ion equilibrium across the red cell membrane in the carp. *Respiration Physiology* **61**, 209–19.

Baroin, A., Garcia-Romeu, F., Lamarre, T. & Motais, R. (1984). A transient sodium–hydrogen exchange system induced by catecholamines in erythrocytes of rainbow trout, *Salmo gairdneri*. *Journal of Physiology (London)* **356**, 21–31.

Baumann, R. & Engelke, M. (1987). The Cl^-/OH^- exchanger tributyltin depolarizes primitive red cells from chick embryo. In *Transport in cells and epithelia*, Book of Abstracts, p. 44. Ninth ESCPB Conference, Copenhagen, Denmark.

Berenbrink, M. & Bridges, C.R. (1994). Catecholamine-activated sodium/proton exchange in the red blood cells of the marine teleost *Gadus morhua*. *Journal of Experimental Biology* **192**, 253–67.

Borgese, F., Garcia-Romeu, F. & Motais, R. (1987). Ion movements and volume changes induced by catecholamines in erythrocytes of rainbow trout: effect of pH. *Journal of Physiology (London)* **382**, 145–57.

Borgese, F., Malapert, M., Fievet, B., Pouyssegur, J. & Motais, R. (1994). The cytoplasmic domain of the Na^+/H^+ exchangers (NHEs) dictates the nature of the hormonal response: behavior of a chimeric human NHE1/trout βNHE antiporter. *Proceedings of the National Academy of Science USA* **91**, 5431–5.

Borgese, F., Sardet, C., Cappadora, M., Pouyssegur, J. & Motais, R. (1992). Cloning and expression of a cyclic AMP-activated Na^+/H^+ exchanger: Evidence that cytoplasmic domain mediates hormonal regulation. *Proceedings of the National Academy of Science USA* **89**, 6765–9.

Boutilier, R.G. & Ferguson, R.A. (1989). Nucleated red cell function: metabolism and pH regulation. *Canadian Journal of Zoology* **67**, 2986–93.

Cala, P.M. (1983). Volume regulation by red blood cells: mechanisms of ion transport. *Molecular Physiology* **4**, 33–52.

Cala, P.M. (1985). Volume regulation by *Amphiuma* red blood cells: characteristics of volume sensitive K/H and Na/H exchange. *Molecular Physiology* **8**, 199–214.

Cala, P.M., Anderson, S.E. & Cragoe, E.J. Jr (1988). Na/H exchange-dependent cell volume and pH regulation and disturbances. *Comparative Biochemistry and Physiology A* **90**, 551–5.

Cameron, B.A. & Tufts, B.L. (1994). *In vitro* investigation of the factors contributing to the unique CO_2 transport properties of blood in the sea lamprey (*Petromyzon marinus*). *Journal of Experimental Biology* **197**, 337–48.

Cossins, A.R. & Kilbey, R.V. (1990). The temperature dependence of the adrenergic Na^+/H^+ exchanger of trout erythrocytes. *Journal of Experimental Biology* **148**, 303–12.

Cossins, A.R. & Richardson, P.A. (1985). Adrenaline-induced Na^+/H^+ exchange in trout erythrocytes and its effects upon oxygen-carrying capacity. *Journal of Experimental Biology* **118**, 229–46.

Ellory, J.C., Wolowyk, M.W. & Young, J.D. (1987). Hagfish (*Eptatretus stouti*) erythrocytes show minimal chloride transport activity. *Journal of Experimental Biology* **129**, 377–83.

Escobales, N. & Figueroa, J. (1991). Na^+/Na^+ exchange and Na^+/H^+ antiport in rabbit erythrocytes: two distinct transport systems. *Journal of Membrane Biology* **120**, 41–9.

Ferguson, R.A. & Boutilier, R.G. (1988). Metabolic energy production during adrenergic pH regulation in red cells of the Atlantic salmon, *Salmo salar*. *Respiration Physiology* **74**, 65–76.

Ferguson, R.A., Sehdev, N., Bagatto, B. & Tufts, B.L. (1992). *In vitro* interactions between oxygen and carbon dioxide transport in the blood of the sea lamprey (*Petromyzon marinus*). *Journal of Experimental Biology* **173**, 25–41.

Heisler, N. (1986). Acid–base regulation in fishes. In *Acid–base regulation in animals*, ed. N. Heisler, pp. 309–56. Amsterdam: Elsevier.

Heming, T.A., Randall, D.J., Boutilier, R.G., Iwama, G.K. & Primmett, D. (1986). Ion equilibria in red blood cells of rainbow trout (*Salmo gairdneri*). *Respiration Physiology* **51**, 303–18.

Hladky, S.B. & Rink, T.J. (1977). pH equilibrium across the red cell membrane. In *Membrane transport in red cells*, ed. J.C. Ellory & V.L. Lew, pp. 115–35. London: Academic Press.

Jensen, F.B. (1989). Hydrogen ion equilibria in fish haemoglobins. *Journal of Experimental Biology* **143**, 225–34.

Jorgensen, N.C. (1995). Amiloride-sensitive Na^+/H^+ exchange stimulated by cellular acidifcation or shrinkage in red blood cells of the frog, *Rana temporaria*. *Journal of Comparative Physiology B* **165**, 450–7.

Mahe, Y., Garcia-Romeu, F. & Motais, R. (1985). Inhibition by amiloride of both adenylate cyclase activity and the Na^+/H^+ antiporter in fish erythrocytes. *European Journal of Pharmacology* **116**, 199–206.

Milligan, C.L. & Wood, C.M. (1985). Intracellular pH transients in rainbow trout tissues measured by dimethadione distribution. *American Journal of Physiology* **248**, R668–73.

Motais, R., Garcia-Romeu, F. & Borgese, F. (1987). The control of Na^+/H^+ exchange by molecular oxygen in trout erythrocytes. A possible role of hemoglobin as a transducer. *Journal of General Physiology* **90**, 197–207.

Motais, R., Garcia-Romeu, F. & Thomas, S. (1989). Na^+/H^+ exchange and pH regulation in red blood cells: role of uncatalyzed H_2CO_3 dehydration. *American Journal of Physiology* **256**, C728–35.

Nikinmaa, M. (1986). Red cell pH of lamprey (*Lampetra fluviatilis*) is actively regulated. *Journal of Comparative Physiology B* **156**, 747–50.

Nikinmaa, M. (1990). *Vertebrate red blood cells*. Berlin, Heidelberg, New York: Springer.

Nikinmaa, M. (1992). Membrane transport and the control of haemo-globin oxygen-affinity in nucleated erythrocytes. *Physiological Reviews* **72**, 301–21.

Nikinmaa, M. (1993). Haemoglobin function in intact *Lampetra fluviatilis* erythrocytes. *Respiration Physiology* **91**, 283–93.

Nikinmaa, M. (1997). Oxygen and carbon dioxide transport in vertebrate erythrocytes: an evolutionary change in the role of membrane transport. *Journal of Experimental Biology* **200**, 369–80.

Nikinmaa, M., Cech, J.J. Jr. & McEnroe, M. (1984). Blood oxygen transport in stressed striped bass (*Morone saxitilis*): role of beta-adrenergic responses. *Journal of Comparative Physiology B* **154**, 365–9.

Nikinmaa, M. & Huestis, W.H. (1984). Adrenergic swelling in nucleated erythrocytes: cellular mechanisms in a bird, domestic goose, and two teleosts, striped bass and rainbow trout. *Journal of Experimental Biology* **113**, 215–24.

Nikinmaa, M., Kunnamo-Ojala, T. & Railo, A. (1986). Mechanisms of pH regulation in lamprey (*Lampetra fluviatilis*) red blood cells. *Journal of Experimental Biology* **122**, 355–67.

Nikinmaa, M. & Matsoff, L. (1992). Effects of oxygen saturation on the CO_2 transport properties of *Lampetra* red cells. *Respiration Physiology* **87**, 219–30.

Nikinmaa, M. & Railo, E. (1987). Anion movements across lamprey (*Lampetra fluviatilis*) red cell membrane. *Biochimica et Biophysica Acta* **899**, 134–6.

Nikinmaa, M., Steffensen, J.F., Tufts, B.L. & Randall, D.J. (1987). Control of red cell volume and pH in trout: effects of isoproter-enol, transport inhibitors and extracellular pH in bicarbonate/carbon dioxide-buffered media. *Journal of Experimental Zoology* **242**, 273–81.

Nikinmaa, M., Tiihonen, K. & Paajaste, M. (1990). Adrenergic control of red cell pH in salmonid fish: roles of the sodium/proton exchange, Jacobs–Stewart cycle and membrane potential. *Journal of Experimental Biology* **154**, 257–71.

Nikinmaa, M. & Tufts, B.L. (1989). Regulation of acid and ion transfer across the membrane of nucleated erythrocytes. *Canadian Journal of Zoology* **67**, 3039–45.

Nikinmaa, M., Tufts, B.L. & Boutilier, R.G. (1993). Volume and pH regulation in agnathan erythrocytes: comparisons between the hagfish, *Myxine glutinosa*, and the lampreys, *Petromyzon marinus* and *Lampetra fluviatilis*. *Journal of Comparative Physiology B* **163**, 608–13.

Obaid, A.L., McElroy Critz, A. & Crandall, E.D. (1979). Kinetics of bicarbonate/chloride exchange in dogfish erythrocytes. *American Journal of Physiology* **237**, R132–8.

Ohnishi, S.T. & Asai, H. (1985). Lamprey erythrocytes lack glycoproteins and anion transport. *Comparative Biochemistry and Physiology* **81B**, 405–7.

Parker, J.C., Coloclasure, G.C. & McManus, T.J. (1991). Coordinated regulation of shrinkage-induced Na/H exchange and swelling-induced [K-Cl] cotransport in dog red cells. *Journal of General Physiology* **98**, 869–80.

Primmett, D., Randall, D.J., Mazeaud, M. & Boutilier, R.G. (1986). The role of catecholamines in erythrocyte pH regulation and oxygen transport in rainbow trout (*Salmo gairdneri*) during exercise. *Journal of Experimental Biology* **123**, 139–48.

Roos, A. & Boron, W.F. (1981). Intracellular pH. *Physiological Reviews* **61**, 296–434.

Rudolph, S. & Greengard, P. (1980). Effect of catecholamines and prostaglandin E1 on cyclic AMP, cation fluxes and protein phosphorylation in the frog erythrocyte. *Journal of Biological Chemistry* **255**, 8534–40.

Salama, A. & Nikinmaa, M. (1988). The adrenergic responses of carp (*Cyprinus carpio*) red cells: effects of Po_2 and pH. *Journal of Experimental Biology* **136**, 405–16.

Salama, A. & Nikinmaa, M. (1989). Species differences in the adrenergic responses of fish red cells: studies on whitefish, pikeperch, trout and carp. *Fish Physiology and Biochemistry* **6**, 167–73.

Salama, A. & Nikinmaa, M. (1990). Effect of oxygen tension on catecholamine-induced formation of cyclic AMP and on swelling of carp red blood cells. *American Journal of Physiology* **259**, C723–6.

Sergeant, S., Sohn, D.H. & Kim, H.D. (1989). Volume-activated Na/H exchange activity in fetal and adult pig red cells: inhibition by cyclic AMP. *Journal of Membrane Biology* **109**, 209–20.

Tetens, V., Lykkeboe, G. & Christensen, N.J. (1988). Potency of adrenaline and noradrenaline for β-adrenergic proton extrusion from red cells of rainbow trout, *Salmo gairdneri*. *Journal of Experimental Biology* **134**, 267–80.

Tufts, B.L. (1991). Acid–base regulation and blood gas transport following exhaustive exercise in an agnathan, the sea lamprey, *Petromyzon marinus*. *Journal of Experimental Biology* **159**, 371–85.

Tufts, B.L. (1992). *In vitro* evidence for sodium-dependent pH regulation in sea lamprey (*Petromyzon marinus*) red blood cells. *Canadian Journal of Zoology* **70**, 411–16.

Tufts, B.L., Bagatto, B. & Cameron, B. (1992). *In vivo* analysis of gas transport in arterial and venous blood of the sea lamprey *Petromyzon marinus*. *Journal of Experimental Biology* **169**, 105–19.

Tufts, B.L. & Boutilier, R.G. (1989). The absence of rapid chloride/bicarbonate exchange in lamprey erythrocytes: implications for CO_2 transport and ion distributions between plasma and erythrocytes in the blood of *Petromyzon marinus*. *Journal of Experimental Biology* **144**, 565–76.

Tufts, B.L. & Boutilier, R.G. (1990a). CO_2 transport in agnathan blood: evidence of Cl^-/HCO_3^- exchange limitations. *Respiration Physiology* **80**, 335–48.

Tufts, B.L. & Boutilier, R.G. (1990b). CO_2 transport properties of the blood of a primitive vertebrate *Myxine glutinosa*. *Experimental Biology* **48**, 341–7.

Tufts, B.L. & Boutilier, R.G. (1991). Interactions between ion exchange and metabolism in erythrocytes of the rainbow trout *Oncorhynchus mykiss*. *Journal of Experimental Biology* **156**, 139–51.

Tufts, B.L., Drever, R.C., Bagatto, B. & Cameron, B.A. (1994). *In vitro* analysis of volume and pH regulation in the red blood cells of a primitive air-breathing fish, the bowfin, *Amia calva*. *Canadian Journal of Zoology* **72**, 280–6.

Tufts, B.L., Mense, D.C. & Randall, D.J. (1987b). The effects of forced activity on circulating catecholamines and pH and water content of erythrocytes in the toad. *Journal of Experimental Biology* **128**, 411–18.

Tufts, B.L., Nikinmaa, M., Steffensen, J.F. & Randall, D.J. (1987a). Ion exchange mechanisms on the erythrocyte membrane of the aquatic salamander, *Amphiuma tridactylum*. *Journal of Experimental Biology* **133**, 329–38.

Tufts, B.L. & Randall, D.J. (1988). The distribution of protons and chloride ions across the erythrocyte membrane of the toad, *Bufo marinus*. *Canadian Journal of Zoology* **66**(11), 2503–6.

Tufts, B.L. & Randall, D.J. (1989). The functional significance of adrenergic pH regulation in fish erythrocytes. *Canadian Journal of Zoology* **67**, 235–8.

Tufts, B.L., Vincent, C.J. & Currie, S. (1998). Different red blood cell characteristics in a primitive agnathan (*M. glutinosa*) and a more recent teleost (*O. mykiss*) influence their strategies for blood CO_2 transport. *Comparative Biochemistry and Physiology* **119A**(2), 533–41.

Virkki, L.V. & Nikinmaa, M. (1994). Activation and physiological role of Na^+/H^+ exchange in lamprey (*Lampetra fluviatilis*) erythrocytes. *Journal of Experimental Biology* **191**, 89–105.

Waddel, W.J. & Bates, R.G. (1969). Intracellular pH. *Physiological Reviews* **49**, 285–329.

Waddel, W.J. & Butler, T.C. (1959). Calculation of intracellular pH from the distribution of 5,5-dimethyl-2,4-oxazolidinedione (DMO). Application to skeletal muscle of the dog. *Journal of Clinical Investigation* **38**, 720–9.

Wood, C.M., Perry, S.F., Walsh, P.J. & Thomas, S. (1994). HCO_3^-

dehydration by the blood of an elasmobranch in the absence of a Haldane effect. *Respiration Physiology* **98**, 319–37.

Zeidler, R. & Kim, H.D. (1977). Preferential hemolysis of post-natal calf red cells induced by internal alkalinization. *Journal of General Physiology* **70**, 385–401.

ANDRÉ MALAN

Acid–base regulation in hibernation and aestivation

Introduction: from water to air breathing

Transition from water to land has freed animals from the constraints of water breathing: due to the low O_2 concentration in water, a water-breather is obliged to achieve a high ventilatory flow rate of a medium with a high capacitance both for CO_2 and for heat (Dejours, 1981). The animal thus has very little control over any of these. By contrast, the high O_2 concentration in air has enabled P_{CO_2} to be set at a higher level and air-breathers can therefore partly devote the ventilatory system to its control. Two major benefits accrue from this in terms of acid–base regulation. The first is a much higher open system CO_2 buffer value, and thereby a more efficient ventilatory control of pH (this buffer value is proportional to P_{CO_2} at a given pH). The second stems from the high diffusivity of CO_2. Krogh's diffusion coefficient for CO_2 in frog muscle at 21 °C is 37-fold higher than the same coefficient for O_2 (Dejours, 1981). In a mammal breathing air at sea level, the maximal possible P_{O_2} difference between arterial blood and mitochondria is 13.3 kPa. The corresponding difference for CO_2 is 0.36 kPa, less than 7 per cent of arterial P_{CO_2}. This percentage would be much higher in a water-breather. As a consequence, the ventilatory control sets the value of P_{CO_2}, not only in the arterial blood, but in all intracellular compartments (except for a few ill-perfused areas in which CO_2 may accumulate). Differing from water-breathers, air-breathers are thus endowed with an efficient centralised control of the major factor of pH in nearly all body compartments, extracellular and intracellular. Another independent benefit of air breathing is the capability to maintain a thermal gradient between the body core and the ambient medium, which was a prerequisite for the development of endothermy. However, terrestrial biotopes are often characterised by a high amplitude of seasonal changes in temperature or humidity. Cold and drought may be directly responsible for food shortage, or they may reduce the animal's ability to get food.

Terrestrial, air-breathing species are found in many zoological phyla: best known are gastropods, arthropods, fish (lungfish), amphibians, reptiles, birds and mammals. In all of these, at least a certain number of species respond to environmental conditions leading to reduced food availability on a seasonal or daily basis by entering hypometabolic or torpid states which have received various names: hibernation, aestivation, torpor, dormancy, overwintering. This chapter considers only the most common situation, in which the major determinant of the hypometabolic state is the shortage of substrates, not of O_2. The adaptations to overwintering in hypoxic water, such as are present in aquatic turtles (see Jackson, this volume), as well as the adaptations to freezing are therefore left aside. Because most available information has been obtained in pulmonate molluscs and mammals, this chapter concentrates on these two groups.

Reduction of metabolic rate

Hypometabolic states of air-breathers share some common features. First, whenever possible, body temperature is reduced. In an ectotherm such as the green lizard, this will be achieved by choosing a colder environment (Rismiller & Heldmaier, 1982, 1988). In endotherms, the apparent set-point of thermal regulation will be lowered (Heller et al., 1978). In both cases, metabolic rate is diminished via the Arrhenius effect of temperature on enzymatic reaction rates (the so-called Q_{10} effect). One can thus estimate that by reducing its body temperature from 37 °C to 5 °C, a hibernating mammal will reduce its resting metabolic rate by a factor of 10 to 12. To this, one should add the benefit of suppressing cold thermogenesis, a further factor of two to three in a small mammal (Malan, 1986, 1993; Geiser, 1988).

Whether or not the animal is able to lower its temperature, a further mechanism comes into play. This is called metabolic depression: to designate a reduction of energy expenditure below the normal resting metabolic rate at a given body temperature. For example, when aestivation is reversibly induced in the snail, *Oreohelix*, by exposing the animals to dry air for a few days, O_2 consumption is reduced by over 70 per cent within a few hours (Fig. 1; Rees & Hand, 1990). In the field, metabolic rate of aestivating snails may become extremely low (Schmidt-Nielsen, Taylor & Shkolnik, 1971). In aestivating African bullfrogs, O_2 consumption is reduced by 75 per cent (Loveridge & Withers, 1981).

Metabolic depression is much more difficult to demonstrate in mammals and birds, because the resulting reduction of metabolic rate (less than 75 per cent, and probably closer to 50 per cent, at least in steady-state hibernation) would be much smaller than that achieved concomi-

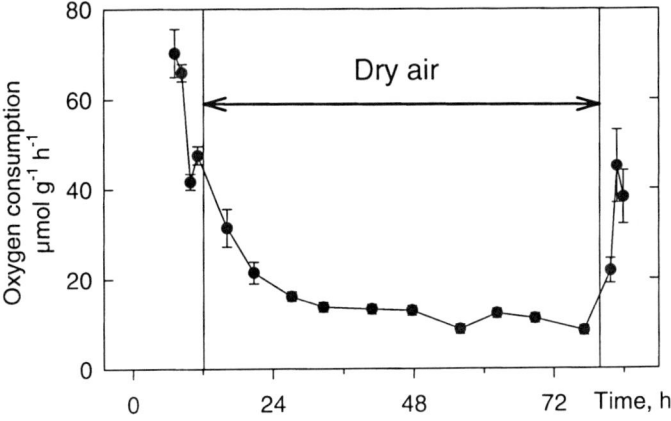

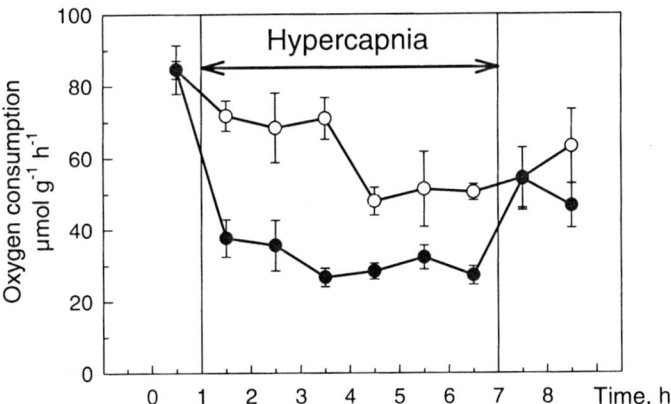

Fig. 1. *Upper panel.* O$_2$ consumption of snails (*Oreohelix*) during a cycle of activity, aestivation and arousal (mean ± SEM; $n = 8$). Aestivation was induced by exposure to dry air and arousal from aestivation by exposure to water-vapour-saturated air. *Lower panel.* The effect of imposed hypercapnia on the O$_2$ consumption of non-aestivating *Oreohelix.* Snails were exposed to ambient air (closed symbols) or to 4.7 kPa CO$_2$ in air (open symbols). The relative humidity of inspired air was 100 per cent throughout (means ± SEM; $n = 4$). Redrawn from Rees & Hand (1990).

tantly through the Arrhenius effect (90 per cent or more, see above) as a consequence of the decrease of body temperature. Simultaneous recording of O_2 consumption and body temperature during the entry into hibernation or torpor has unambiguously shown that the drop of metabolic rate may precede the decline in body temperature and cannot therefore be a mere consequence of it (Ruf & Heldmaier, 1992). Other arguments have been presented elsewhere (Malan, 1993).

Respiratory acidosis

Another common characteristic of torpid states in air-breathers is the presence of a respiratory acidosis. Ventilatory responses to changes in inspired air composition are conserved, in snails (Barnhart & McMahon, 1987; Barnhart, 1986b, 1989, 1992) as well as in mammals (McArthur & Milsom, 1991). Correspondingly, the variability of blood pH and P_{CO_2} is not increased in most hibernating mammals (Malan, 1982). The steady-state changes in blood pH and P_{CO_2} observed in torpor thus correspond to a resetting of the control of ventilation, similar to that observed for thermal regulation.

Methods applicable to hibernation and aestivation

In principle, any standard method could be used. Two major difficulties are often encountered, however, especially in higher vertebrates. First, torpid states may be fairly labile: a hibernating mammal may react to the slightest touch with an arousal characterised by a hyperventilation resulting in a rapid alkalinisation (see below). This will often impose a delay of at least several days between the surgery or injection and the restoration of steady hibernation. Second, changes in body temperature complicate the interpretation of changes in pH and P_{CO_2}.

Two methods have been successfully applied to the continuous monitoring of acid–base regulation in hibernating mammals or birds. One is the measurement of intracellular pH (pH_i) by ^{31}P-nuclear magnetic resonance (^{31}P-NMR; McArthur et al., 1990; see also Kinsey & Moerland, this volume). It suffers, however, from the constraint of requiring the strict conservation of spatial relationships between the antenna and the volume under study. In addition, the spatial resolution is not very good. This restricts its use to very quiet species, and to abundant and relatively homogeneous tissues such as skeletal muscles or liver. The second method is whole-body plethymography, which permits ventilatory flow rate and O_2 consumption (or CO_2 production) to be measured continuously (Malan, 1973; Jacky, 1978, 1980). From these measurements, relative ventilation, which sets arterial P_{CO_2}, can be calculated (relative ventilation is the ratio of ventilatory flow rate over CO_2 production).

Isothermal acid–base changes

In animals in which no changes in body temperature accompany the transition from active state to torpor, the interpretation of acid–base changes is straightforward. For example, when aestivation is induced in snails (*Otala lactea*) by exposure to dry air, haemolymph pH is decreased from 7.86 ± 0.05 to 7.65 ± 0.06 after three days and to 7.42 ± 0.12 after one month, all at 25 °C. By the same times, respectively, P_{CO_2} increases from 1.79 ± 0.27 kPa to 3.60 ± 0.70 kPa and to 6.0 ± 1.7 kPa (Barnhart, 1986a, 1986b). Similar results have been found in *Oreohelix* (Rees & Hand, 1990). These changes are reversed on return to humid air. They correspond to an uncompensated respiratory acidosis. This acidosis results from the lowering of relative ventilation, with less frequent openings of the pneumostome at the entry into aestivation, leading to an accumulation of metabolic CO_2 (Barnhart & McMahon, 1987). The reverse takes place on arousal (Rees & Hand, 1990). In snails which have been aestivating for more than one month, a metabolic acidosis adds to the respiratory acidosis (Barnhart, 1986a). A respiratory acidosis is also present in aestivating lungfish, *Protopterus aethiopicus* (DeLaney *et al.*, 1977), and African bullfrog, *Pyxicephalus adspersus* (Loveridge & Withers, 1981). Both species avoid dehydration by becoming surrounded by a cocoon, which limits the diffusion of CO_2.

Alkalinity relative to water

When the transitions from active state to torpor are not isothermal, which is generally the case in endotherms, the acid–base situation becomes more complicated due to the effects of temperature on the physicochemical constants involved in acid–base reactions: the ionisation constants of water and weak electrolytes, and the solubilities of CO_2 and NH_3. For example, due to the temperature dependency of its ionisation constant, the pH of pure water, pH_{nw}, varies from 7.0 at 25 °C to 6.8 at 37 °C or 7.5 at 0 °C. This occurs without any deviation from neutrality, the difference ($[OH^-] - [H^+]$) remaining zero throughout. The same pH of 7.0, which is neutral at 25 °C (Fig. 2B), is acidic below this temperature (A), and alkaline above it (C). Instead of using pH as an acid–base parameter of biological fluids, one should therefore use the relative alkalinity (Rahn, 1967; Rahn & Reeves, 1979); that is, the difference between pH and pH_{nw} at any given temperature. An example

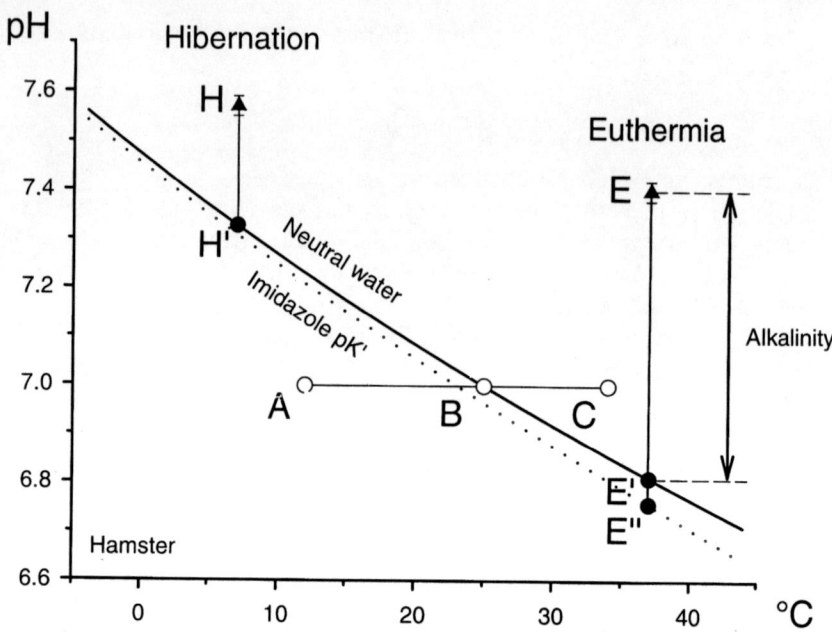

Fig. 2. *Solid line.* pH of neutral water (pH$_{nw}$) *versus* temperature. At any temperature, alkalinity (relative to water) is measured by the vertical distance above this curve, such as EE′ (Rahn, 1967). It can be modified only by adding or withdrawing acidic equivalents to the solution. Thus, when temperature varies, a constant pH is no longer equivalent to a constant alkalinity: A is more acidic and C more alkaline than B. *Dotted line*: the mean pK′ of histidyl imidazole moieties in a protein (Reeves, 1972). The curve is very close to that of pH$_{nw}$ (E′E″). Alkalinity relative to neutral water is thus a good index of the dissociation ratio of protein imidazole buffer groups (Reeves, 1977).

At 37 °C, the arterial pH of a euthermic hamster (*Cricetus cricetus*) is 7.4 while pH$_{nw}$ is 6.8, giving an alkalinity of 0.6 (EE′). In hibernation, arterial pH is nominally higher, but alkalinity is reduced (HH′). An acidification has taken place, as evidenced by the acid titration of both water and protein buffers.

is given in Figure 2 by the vertical distance EE′ from point E to the curve representing pH$_{nw}$ = f(T). The relative alkalinity is zero at E′, B and H′. At any temperature, alkalinity can be modified only by adding (or subtracting) acidic equivalents to the solution. This analysis may now be applied to arterial pH data from a hibernating mammal, the European hamster (*Cricetus cricetus*). From euthermia (E) to hibernation (H), the relative alkalinity decreases, i.e. there is an acidification.

The dissociation of protein imidazole groups

The limitation of this approach is that water itself is such a weak acid that its dissociation has very little biological relevance, except to facilitate diffusion of protons and to mediate their exchange between titratable substances. In fact, the functional significance of acid–base regulation rests on the control of the electric charge of the ionisable groups of proteins. About 80 per cent of protein titration properties at physiological pH are due to the imidazole groups of histidyl residues. Imidazole and equivalent groups are thus involved in the pH sensitivity of enzymes, in extracellular and intracellular non-bicarbonate buffers, and at the pH-sensitive sites of ion exchangers controlling pH_e and pH_i (Reeves, 1972, 1977; Somero, 1981; Aronson, 1985). Because pH_{nw} is both very close to pK' of protein imidazole groups and more accurately known, relative alkalinity may be used as a substitute index of protein titration for practical purposes.

The closed system approach

Acid–base physiologists also need to assess the respective contributions of respiratory and metabolic (ionic) factors in acid–base changes occurring at a variable temperature. Temperature affects CO_2 solubility (or capacitance) and therefore Pco_2 is no longer a good index because it may change in the absence of any change of relative ventilation. This difficulty may be circumvented by the closed system approach (Malan, 1977, 1978; Rodeau & Malan, 1979). In brief, a reference temperature is selected, 37 °C for example, at which pH and Pco_2 will be measured. Any acid–base change occurring at a variable temperature may then be considered equivalent to the sum of a temperature change in closed system ('anaerobic') conditions, from the actual animal temperature to 37 °C, followed by a conventional acid–base change at this temperature. In practice, it suffices to warm (or cool) the blood sample in an air-tight syringe up to 37 °C (no acid equivalent is gained or lost to the medium in this process) and then measure pH and Pco_2 as usual. When this is not possible, e.g. for intracellular fluids, approximate results can be obtained from model calculations. Both methods result in similar temperature-corrected values of pH, Pco_2 and $[HCO_3^-]$, respectively pH^*, P^*co_2, and $[HCO3_3^-]^*$, which can then be subjected to standard methods of analysis of acid–base data (Malan, 1977, 1978).

Extracellular acid–base data in hibernation

Once the blood pH and Pco_2 data collected from a hibernating mammal have been corrected to 37 °C, they can be directly compared to euthermic data (Fig. 3). In all mammals studied, whether in deep hibernation or in daily torpor, there is a marked respiratory acidosis (Malan, 1982; Bickler, 1984b; Nestler, 1990a; Szewczak & Jackson, 1992b). For example, in a hibernating hamster (*Cricetus cricetus*), arterial pH* is 7.01 ± 0.01 against 7.40 ± 0.01 in euthermy, and $P*_{CO_2}$ is 21.3 ± 0.5 kPa against 6.0 ± 0.3 kPa in euthermy (Malan, Mioskowski & Calgari, 1988). In this species, there is no metabolic compensation, whereas there is a partial compensation in others (Malan, 1982). Production of urine virtually stops in hibernation (Zatzman & South, 1975), and therefore kidney acidification cannot contribute much to acid–base control.

Intermittent breathing

A number of hibernating mammals present intermittent breathing (Malan, 1982; McArthur and Milsom, 1991; Szewczak & Jackson,

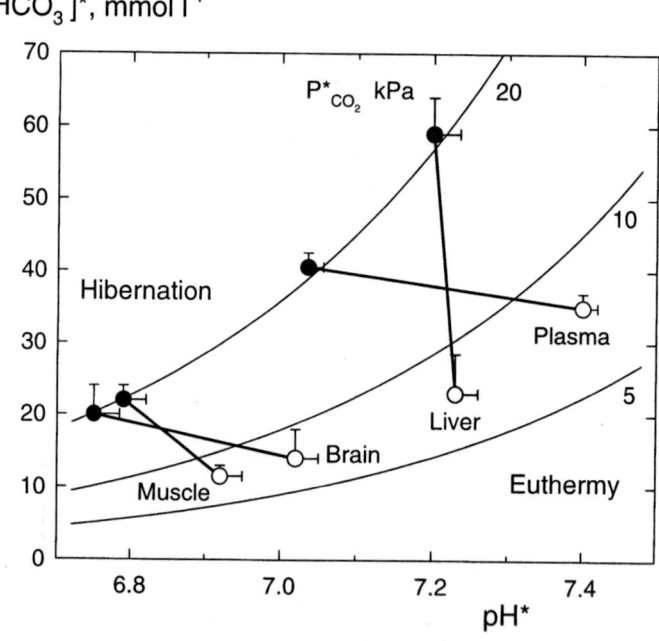

Fig. 3. Extracellular (venous) and intracellular acid–base data of hamsters (*Cricetus cricetus*), active (open symbols) or hibernating (closed symbols). Data measured at hibernation body temperature (about 9 °C) have been corrected to 37 °C (see text). Notice the respiratory acidosis, and the lack of metabolic compensation in blood, skeletal muscles and brain. Partial compensation took place in liver, and to a lesser extent in heart (not shown). Adapted from Malan, Rodeau & Daull (1985).

1992b). In the hedgehog, *Erinaceus europaeus*, for example, apnoeas lasting 60–70 minutes are separated by bursts of 40–50 ventilatory cycles (Kristoffersson & Soivio, 1964). From the beginning to the end of an apnoea, arterial P_{O_2} drops from 16.0 ± 3.6 kPa to 1.4 ± 0.5 kPa – a 91 per cent reduction. Yet due to the much higher capacitance (solubility) for CO_2 than for O_2, and to protein buffering of protons, arterial pH drops only by 0.04 unit while P_{CO_2} is increased by 30 per cent (Tähti & Soivio, 1975). In torpid bats (*Eptesicus fuscus*), the fluctuations of pH and P_{CO_2} are wider, 0.15 unit and 59 per cent respectively (Szewczak & Jackson, 1992b). The incidence of these fluctuations at the intracellular level has not been determined.

Changes in relative ventilation

To develop the acidosis, metabolic CO_2 needs to be accumulated during the entry into hibernation, which requires a decrease of relative ventilation. Conversely, restoration of euthermic acid–base conditions on arousal from hibernation requires a relative hyperventilation. Measuring these variables provides an independent confirmation of the existence of a respiratory acidosis in hibernation. It is technically difficult, and the methods are open to a number of artifacts. Alternatively, one can measure the respiratory exchange ratio – that is, the ratio of CO_2 elimination over O_2 uptake in the lungs. In the presence of a relative hypoventilation or hyperventilation, this ratio is respectively lower or higher than the metabolic respiratory quotient (ratio of metabolic CO_2 production over O_2 consumption). Hibernating mammals depend mostly on lipid reserves as metabolic fuels, except in the second half of the arousal process when carbohydrate consumption is restored in the shivering muscles. Whether or not these assumptions are fully correct, respiratory exchange ratios lower than 0.7 or higher than 1.0 result from changes in relative ventilation.

Relative ventilation is depressed during entry into hibernation, especially at the beginning, as indicated both by direct measurements of ventilatory flow rate and CO_2 production and by measurements of respiratory exchange ratio (Snapp & Heller, 1981; Bickler, 1984a; Nestler, 1990b). Conversely, the beginning of arousal is generally characterised by an intense relative hyperventilation (Snapp & Heller, 1981; Bickler, 1984a; Malan, 1982; Malan *et al.*, 1988; Nestler, 1990b). In the hamster, *Cricetus cricetus*, arterial P_{CO_2} is reduced by half within the first 30 minutes of arousal, while the rise in body (cheek pouch) temperature is still very slow (Malan *et al.*, 1988). This partial reversal of the respiratory acidosis may be very important in view of the inhibitory effects of low pH (see below). Similar results have been obtained during the entry into and arousal from torpor in birds: poorwills (Withers, 1977) and hummingbirds (Bucher and Chappell, 1992). In the absence of direct acid–base measurements, this is the only evidence of respiratory acidosis in avian torpor.

Intracellular compartments

As explained above, all intracellular compartments are necessarily affected by the rise in P_{CO_2}. However, numerous cell types are endowed with powerful mechanisms of regulation of pH_i, especially by Na^+/H^+ exchange (Aronson, 1985). Direct measurements of pH_i in aestivating snails, by the 5,5-dimethyl-oxazolidine-2,4-dione (DMO) method or by ^{31}P-NMR, confirm that intracellular fluids are acidotic (Barnhart & McMahon, 1988; Rees et al., 1991). In hibernating hamsters, the respiratory acidosis is present in all compartments, as determined by the DMO method (Fig. 3; Malan et al., 1985). This has been confirmed for muscle by ^{31}P-NMR (McArthur et al., 1990). In skeletal muscles and brain, which together with extracellular fluids represent a majority of total body water, the respiratory acidosis is uncompensated. Compensation is present in liver, and to a lesser degree in heart. Similar results, but with a less pronounced acidosis, have been found in the daily torpor of the deer mouse, Peromyscus maniculatus (body temperature 24 °C) (Nestler, 1990a). Notice, however, that the two methods give only a global average of tissue pH_i, which may hide circumscribed heterogeneities.

The absence of metabolic compensation in brain and muscle is somewhat surprising in view of the powerful mechanisms of ionic regulation of pH_i in euthermic conditions, especially in brain. This requires an inhibitory mechanism, either active or by a passive effect of cold. The latter would compromise the intense heat production necessary for arousal, and is therefore less likely. In addition, the acidic equivalents rejected from pH_i-regulating cells acidify the extracellular compartment, which is smaller than the intracellular compartment. In the probable absence of urinary acidification, they might be buffered by bones. Morphological evidence of bone decalcification in hibernation has been provided (Kayser & Franck, 1963). The capacity of such a mechanism is probably limited, which would impose an upper limit on pH_i regulation in hibernation. Localised exceptions to an overall inhibition of pH_i regulation in the brain might have remained unnoticed, however, owing to the poor spatial resolution of the DMO method. Potential candidates are the few brain areas which keep active in hibernation (Kilduff et al., 1990).

Effects of acidosis

The effects of intracellular acidosis in the downregulation of cellular metabolism in aquatic invertebrates have been extensively studied, especially in connection with the cryptobiosis of shrimp larvae. Glycolysis and protein synthesis are the major sites of pH sensitivity (Hand, 1991; Hand & Hardewig, 1996). In air-breathers, the hypothesis of an inhibitory role of acidosis in mammalian hibernation has long been supported mainly by correlative evidence (Malan, 1982). Acute exposure to hypercapnia depresses O_2 consumption in non-aestivating snails (see Fig. 1; Barnhart & McMahon, 1988; Rees & Hand, 1990, 1991). In active, euthermic mammals, some attempts at depressing metabolic rate or thermogenesis by acute exposure to 10 per cent CO_2 have been unsuccessful (Kuhnen, Wloch & Wünnenberg, 1987). The failure probably resulted from the stimulation by hypercapnia of catecholamine release, which counteracts depressing effects of acidosis even in anaesthetised animals (Willford *et al.*, 1990). This notwithstanding, a depression of O_2 consumption, linearly related to arterial pH decrease, was obtained in unanaesthetised ground squirrels by exposure to 10 per cent CO_2 (Bharma & Milsom, 1993). Hypercapnia also depresses regulatory thermogenesis and facilitates induction of hypothermia in the rat (Morita *et al.*, 1993). Notice that the decrease in blood pH produced by breathing 10 per cent CO_2 is about half that observed in mammalian or snail hibernation.

In mammals, the high pH sensitivity of phosphofructokinase, a major regulating enzyme of glycolysis, is well known (Hand & Somero, 1983). Another important site of pH effects is brown adipose tissue. This tissue is the major contributor to cold-induced thermogenesis in small mammals (Nicholls & Locke, 1984). When brown adipocytes isolated from hibernating hamsters are exposed to a bicarbonate buffer with the pH and P_{CO_2} of hibernating blood, their thermogenic response to norepinephrine is reduced by 50 per cent compared to control conditions (Malan, 1989). The hyperventilation of the beginning of arousal will thus partly reverse this inhibition, at a time when brown adipose tissue thermogenesis is essential for rapid arousal.

In addition to these metabolic effects, inhalation of a hypercapnic mixture depresses the thermal threshold for eliciting cold thermogenesis in guinea-pigs, thus mimicking what occurs in the entry into hibernation (Schaefer & Wünnenberg, 1976). In itself, such a depression may be sufficient to start a cumulative process: as soon as thermogenesis is reduced, body temperature will start declining, and then the Arrhenius effect of temperature will depress metabolic rate further. This may therefore represent a major factor of the entry into hibernation. At the cellular level, the effect of hypercapnia is likely to involve the depression of firing rate of thermosensitive pre-optic neurons. This has been evidenced in euthermic hamsters *in vivo* as well as in brain slices (Wünnenberg & Baltruschat, 1982; Matsumura, Nakayama & Kaminaga, 1987).

It should be noted, however, that when experimental hypercapnia is applied to active animals, snails or mammals, acidosis alone is not sufficient to depress metabolic rate to the same extent as in aestivation or hibernation. The mechanisms involved in the effects of low pH on enzymes in the aestivating animal may differ from those observed in experimental hypercapnia on active animals (Rees & Hand, 1991; Michaelidis & Pardalidis, 1994), or they may operate in conjunction with other factors such as protein phosphorylation (Whitwam & Storey, 1990, 1991).

Conclusions

Air breathing has endowed terrestrial animals with two major assets: the possibility to control their body temperature, and an efficient, centralised ventilatory control of pH, extracellular and intracellular. The conquest of land has also exposed these animals to seasonal hardships – cold or drought. To survive these hardships, the animals have developed various kinds of hypometabolic states – hibernation, aestivation, daily torpor or others. All these processes share a common feature: they take advantage of the acquisition of ventilatory control of pH and, when present, of thermal regulation to achieve hypometabolism in an efficient and rapidly reversible way. From snails to mammals, respiratory acidosis seems to be a common feature. It is therefore proposed to regroup all these hypometabolic processes of air-breathing animals under the common denomination of *acidotic torpidity*.

References

Aronson, P.S. (1985). Kinetic properties of the plasma membrane Na$^+$-H$^+$ exchanger. *Annual Review of Physiology* **47**, 545–60.

Barnhart, M.C. (1986a). Control of acid–base status in active and dormant land snails, *Otala lactea* (Pulmonata, Helicidae). *Journal of Comparative Physiology B* **156**, 347–54.

Barnhart, M.C. (1986b). Respiratory gas tensions and gas exchange in active and dormant land snails *Otala lactea*. *Physiological Zoology* **59**, 733–45.

Barnhart, M.C. (1989). Respiratory acidosis and metabolic depression in dormant invertebrates. In *Living in the cold/La vie au froid*, ed. A. Malan & B. Canguilhem, pp. 321–31. Paris: John Libbey Eurotext/INSERM.

Barnhart, M.C. (1992). Acid–base regulation in pulmonate molluscs. *Journal of Experimental Zoology* **263**, 120–6.

Barnhart, M.C. & McMahon, B.R. (1987). Discontinuous carbon dioxide release and metabolic depression in dormant land snails. *Journal of Experimental Biology* **128**, 123–38.

Barnhart, M.C. & McMahon, B.R. (1988). Depression of aerobic metabolism and intracellular pH by hypercapnia in land snails, *Otala lactea*. *Journal of Experimental Biology* **138**, 289–99.

Bharma, S. & Milsom, W. (1993). Acidosis and metabolic rate in golden mantled ground squirrels (*Spermophilus lateralis*). *Respiration Physiology* **94**, 337–51.

Bickler, P.E. (1984a). CO_2 balance of a heterothermic rodent: comparison of sleep, torpor, and awake states. *American Journal of Physiology* **246**, R49–55.

Bickler, P. E. (1984b). Blood acid–base status of an awake heterothermic rodent, *Spermophilus tereticaudus*. *Respiration Physiology* **57**, 307–16.

Bucher, T.L. & Chappell, M.A. (1992). Ventilatory and metabolic dynamics during entry into and arousal from torpor in *Selasphorus* hummingbirds. *Physiological Zoology* **65**, 978–93.

Dejours, P. (1981). *Principles of comparative respiratory physiology*, 2nd edition. Amsterdam: Elsevier/North-Holland.

DeLaney, R.G., Lahiri, S., Hamilton, R. & Fishman, A.P. (1977). Acid–base balance and plasma composition in the aestivating lungfish (*Protopterus*). *American Journal of Physiology* **232**, R10–17.

Geiser, F. (1988). Reduction of metabolism during hibernation and daily torpor in mammals and birds: temperature effect or physiological inhibition? *Journal of Comparative Physiology B* **158**, 25–38.

Hand, S. C. (1991). Metabolic dormancy in aquatic invertebrates. *Advances in Comparative and Environmental Physiology* **8**, 1–50.

Hand, S.C. & Hardewig, I. (1996). Downregulation of cellular metabolism during environmental stress: mechanisms and implications. *Annual Review of Physiology* **58**, 539–63.

Hand, S.C. & Somero, G.N. (1983). Phosphofructokinase of the hibernator *Citellus beecheyi*: temperature and pH regulation of activity *via* influences on the tetramer–dimer equilibrium. *Physiological Zoology* **56**, 380–8.

Heller, H.C., Walker, J.M., Florant, G.L., Glotzbach, S.F. & Berger, R.J. (1978). Sleep and hibernation: electrophysiological and thermoregulatory homologies. In *Strategies in cold: natural torpidity and thermogenesis*, ed. L.C.H. Wang & J.W. Hudson, pp. 225–65. New York: Academic Press.

Jacky, J.P. (1978). A plethysmograph for long term measurements of ventilation in unrestrained animals. *Journal of Applied Physiology* **45**, 644–7.

Jacky, J.P. (1980). Barometric measurement of tidal volume: effects of pattern and nasal temperature. *Journal of Applied Physiology* **49**, 319–25.

Kayser, C. & Franck, R.M. (1963). Comportement des tissus calcifiés du hamster d'Europe *Cricetus cricetus* au cours de l'hibernation. *Archives of Oral Biology* **8**, 703–18.

Kilduff, T.S., Miller, J.D., Radeke, C.M., Sharp, F.R. & Heller, H.C. (1990). ^{14}C-2-Deoxyglucose uptake in the ground squirrel brain during entrance to and arousal from hibernation. *Journal of Neuroscience* **10**, 2463–75.

Kristoffersson, R. & Soivio, A. (1964). Hibernation in the hedgehog (*Erinaceus europaeus* L.). Changes of respiratory pattern, heart rate and body temperature in response to gradually decreasing or increasing ambient temperature. *Annales Academiae Scientiarum Fennicae, Series A, IV* **82**, 3–17.

Kuhnen, G., Wloch, B. & Wünnenberg, W. (1987). Effects of acute hypoxia and/or hypercapnia on body temperatures and cold induced thermogenesis in the golden hamster. *Journal of Thermal Biology* **12**, 103–7.

Loveridge, J.P. & Withers, P.C. (1981). Metabolism and water balance of active and cocooned African bullfrogs *Pyxicephalus adspersus*. *Physiological Zoology* **54**, 203–14.

Malan, A. (1973). Ventilation measured by body plethysmography in hibernating mammals and poikilotherms. *Respiration Physiology* **17**, 32–44.

Malan, A. (1977). Blood acid–base state at a variable temperature. A graphical representation. *Respiration Physiology* **31**, 259–75.

Malan, A. (1978). Intracellular acid–base state at a variable temperature in air-breathing vertebrates and its representation. *Respiration Physiology* **33**, 115–19.

Malan, A. (1982). Respiration and acid–base state in hibernation. In *Hibernation and torpor in mammals and birds*, ed. C.P. Lyman, J.S. Willis, A. Malan & L.C.H. Wang, pp. 237–82. New York: Academic Press.

Malan, A. (1986). pH as a control factor in hibernation. In *Living in the cold. Physiological and biochemical adaptations*, ed. H.C. Heller, X.J. Musacchia & L.C.H. Wang, pp. 61–70. New York: Elsevier.

Malan, A. (1989). pH as a control factor of cell function in hibernation: the case of brown adipose tissue thermogenesis. In *Living in the cold/La vie au froid*, ed. A. Malan & B. Canguilhem, pp. 333–41. Paris: John Libbey Eurotext/INSERM.

Malan, A. (1993). Temperature regulation, enzyme kinetics and metabolic depression in mammalian hibernation. In *Life in the cold III: Ecological, physiological and molecular mechanisms*, ed. C. Carey, G.L. Florant, B.A. Wunder & B. Horwitz, pp. 241–52. Boulder, USA: Westview Press.

Malan, A., Mioskowski, E. & Calgari, C. (1988). Time-course of blood acid–base state during arousal from hibernation in the European hamster. *Journal of Comparative Physiology B* **158**, 495–500.

Malan, A., Rodeau, J.L. & Daull, F. (1985). Intracellular pH in hibernation and respiratory acidosis in the European hamster. *Journal of Comparative Physiology B* **156**, 251–8.

Matsumura, K., Nakayama, T. & Kaminaga, T. (1987). Effects of carbon dioxide on preoptic thermosensitive neurons *in vitro*. *Pflügers Archiv – European Journal of Physiology* **408**, 120–3.

McArthur, M.D., Hanstock, C.C., Malan, A., Wang, L.C.H. & Allen, P.S. (1990). Skeletal muscle pH dynamics during arousal from hibernation measured by ^{31}P NMR spectroscopy. *Journal of Comparative Physiology B* **160**, 339–47.

McArthur, M.D. & Milsom, W.K. (1991). Changes in ventilation and respiratory sensitivity associated with hibernation in Columbian (*Spermophilus columbianus*) and golden-mantled (*Spermophilus lateralis*) ground squirrels. *Physiological Zoology* **64**, 940–59.

Michaelidis, B. & Pardalidis, T. (1994). Regulation of pyruvate kinase (PK) from the ventricle of the land snail *Helix lucorum* L. during early and prolonged estivation and hibernation. *Comparative Physiology and Biochemistry B Comparative Biochemistry and Molecular Biology* **107**, 585–91.

Morita, T., Konaka, K., Kawasaki, Y., Kawai, F., Kanamori, M. & Mitsuda, H. (1993). Effects of moderate hypercapnia on hypothermia induced by cold He–O_2 in rats. *Comparative Biochemistry and Physiology A Comparative Physiology* **104**, 215–18.

Nestler, J.R. (1990a). Intracellular pH during daily torpor in *Peromyscus maniculatus*. *Journal of Comparative Physiology B* **159**, 661–6.

Nestler, J.R. (1990b). Relationships between respiratory quotient and metabolic rate during entry to and arousal from daily torpor in deer mice (*Peromyscus maniculatus*). *Physiological Zoology* **63**, 504–15.

Nicholls, D.G. & Locke, R.M. (1984). Thermogenic mechanisms in brown fat. *Physiological Reviews* **64**, 1–64.

Rahn, H. (1967). Gas transport from the environment to the cell. In *Development of the lung*, ed. A.V.S. de Reuck & R. Porter, pp. 3–23. London: Churchill.

Rahn, H. & Reeves, R.B. (1979). Patterns in vertebrate acid–base regulation. In *Evolution of respiratory processes*, ed. S.C. Wood & C. Lenfant, pp. 225–52. New York: Dekker.

Rees, B.B. & Hand, S.C. (1990). Heat dissipation, gas exchange and

acid–base status in the land snail *Oreohelix* during short-term aestivation. *Journal of Experimental Biology* **152**, 77–92.

Rees, B.B. & Hand, S.C. (1991). Regulation of glycolysis in the land
snail *Oreohelix* during aestivation and artificial hypercapnia. *Journal of Comparative Physiology B* **161**, 237–46.

Rees, B.B., Malhotra, D., Shapiro, J.I. & Hand, S.C. (1991). Intracellular pH decreases during entry into estivation in the land snail
Oreohelix strigosa. *Journal of Experimental Biology* **159**, 525–30.

Reeves, R.B. (1972). An imidazole alphastat hypothesis for vertebrate
acid–base regulation: tissue carbon dioxide content and body temperature in bullfrogs. *Respiration Physiology* **14**, 219–36.

Reeves, R.B. (1977). The interaction of body temperature and acid–
base balance in ectothermic vertebrates. *Annual Review of Physiology* **39**, 559–86.

Rismiller, P.D. & Heldmaier, G. (1982). The effect of photoperiod on
temperature selection in the European green lizard, *Lacerta viridis*.
Oecologia **53**, 222–6.

Rismiller, P.D. & Heldmaier, G. (1988). How photoperiod influences
body temperature selection in *Lacerta viridis*. *Oecologia* **75**, 125–
31.

Rodeau, J.L. & Malan, A. (1979). A two compartment model of blood
acid–base state at constant or variable temperature. *Respiration
Physiology* **37**, 5–30.

Ruf, T. & Heldmaier, G. (1992). The impact of daily torpor on energy
requirements in the Djungarian hamster, *Phodopus sungorus*.
Physiological Zoology **65**, 994–1010.

Schaefer, K.E. & Wünnenberg, W. (1976). Threshold temperatures
for shivering in acute and chronic hypercapnia. *Journal of Applied
Physiology* **41**, 67–70.

Schmidt-Nielsen, K., Taylor, C.R. & Shkolnik, A. (1971). Desert
snails: problems of heat, water and food. *Journal of Experimental
Biology* **55**, 385–98.

Snapp, B.D. & Heller, H.C. (1981). Suppression of metabolism during
hibernation in ground squirrels (*Citellus lateralis*). *Physiological
Zoology* **54**, 297–307.

Somero, G.N. (1981). pH-temperature interactions on proteins: principles of optimal pH and buffer system design. *Marine Biology
Letters* **2**, 163–78.

Szewczak, J.M. & Jackson, D.C. (1992a). Ventilatory response to
hypoxia and hypercapnia in the torpid bat, *Eptesicus fuscus*. *Respiration Physiology* **88**, 217–32.

Szewczak, J.M. & Jackson, D.C. (1992b). Acid–base state and intermittent breathing in the torpid bat, *Eptesicus fuscus*. *Respiration
Physiology* **88**, 205–15.

Tähti, H. & Soivio, A. (1975). Blood gas concentrations, acid–base
balance and blood pressure in hedgehogs in the active state and in

hibernation with periodic respiration. *Annales Zoologici Fennici* **12**, 188–92.

Whitwam, R.E. & Storey, K.B. (1990). Pyruvate kinase from the land snail *Otala lactea*: regulation by reversible phosphorylation during aestivation and anoxia. *Journal of Experimental Biology* **154**, 321–38.

Whitwam, R.E. & Storey, K.B. (1991). Regulation of phosphofructo-kinase during aestivation and anoxia in the land snail, *Otala lactea*. *Physiological Zoology* **64**, 595–610.

Willford, D.C., Moores, W.Y., Ji, S.Y., Chen, Z.T., Palencia, A. & Daily, P.O. (1990). Importance of acid–base strategy in reducing myocardial and whole-body O_2-consumption during perfusion hypothermia. *Journal of Thoracic and Cardiovascular Surgery* **100**, 699–707.

Withers, P.C. (1977). Respiration, metabolism, and heat exchange of euthermic and torpid poorwills and hummingbirds. *Physiological Zoology* **50**, 43–52.

Wünnenberg, W. & Baltruschat, D. (1982). Temperature regulation of golden hamsters during acute hypercapnia. *Journal of Thermal Biology* **7**, 83–6.

Zatzman, M.L. & South, F.E. (1975). Concentration of urine by the hibernating marmot. *American Journal of Physiology* **228**, 1326–40.

VICTOR A. ZAMMIT

Hepatic metabolism and pH in starvation and refeeding

Introduction

Metabolic activity results in the net generation of protons. While this property is utilised by cells to maintain function, they are equipped with mechanisms that restrict the changes in intracellular pH (pH_i) to a range compatible with cell viability. These mechanisms also protect the cell from changes in pH_i that arise from fluctuations in the extracellular pH (pH_e) induced by systemic alkalosis or acidosis. For example, in rat isolated liver cells there is a close correlation between variation in pH_e and pH_i (Kashiwagura *et al.*, 1984). However, although pH_i is a linear function of pH_e, it only changes by 0.45 units for every 1.0 unit change in pH_e, the two parameters being equal at approximately pH 7.1. This is due not only to intracellular buffering capacity but also to the activity of proton and other ion exchangers present in the plasma membrane. These ion transporters are also important in determining cell volume and the response of cells (regulatory volume increase/decrease) to conditions that induce cell swelling or shrinking (see below). Consequently, there is a close relationship between the regulation of pH_i and cell volume.

Liver cells contain a Na^+–H^+ antiporter in the basolateral plasma membrane (Arias & Forgag, 1984; Moseley *et al.*, 1986). This exchanger, together with the Na^+–HCO_3^- co-transport, regulates pH_i in isolated hepatocytes (Anderson, Graf & Boyer, 1987; Gleeson, Smith & Boyer, 1989; Renner *et al.*, 1989). The Na^+–H^+ antiporter exchanges extracellular Na^+ for intracellular protons and it plays an important role in the regulation of both pH_i and cell volume in hepatocytes (Mahnenesmith & Aronson, 1985; Grinstein & Rothstein, 1986; Moolenar, 1986). The driving force for the Na^+–H^+ antiporter is the Na^+ gradient across the plasma membrane. The system is secondarily dependent on ATP hydrolysis by the Na^+/K^+-ATPase. Under physiological conditions, the system functions to keep the pH_i high. The antiporter is quiescent above pH 7.5 and becomes (sigmoidally) more active as the pH_i is lowered (Frelin *et al.*, 1988). Therefore, it is activated by a decrease in pH_i. It is also activated by hormones and growth factors, as well as by hyperosmotic cell shrinking, all of which give changes in pH_i of 0.1 to 0.4 units. The mechanism involves a shift in the pH-dependence curve of its activity towards higher pH (i.e. an alkaline shift) (Frelin *et al.*, 1988). Conversely, the fall in pH_i observed upon cell swelling is mediated by inhibition of Na^+–H^+ exchange in rat hepatocytes (Gleeson, Corasanti & Boyer, 1990) and Hep G2 cells (Madshus *et al.*, 1987). The effect appears to be due to a lowering of the V_{max} of Na^+–H^+ exchange in both directions, without altering K_m (Green *et al.*, 1988). The volume changes that occur during the return of cells to isosmotic medium are accompanied by reciprocal changes in pH_i of similar magnitude (Gleeson *et al.*, 1990) and by the activation or inhibition, respectively, of the activity of the Na^+–H^+ antiporter (Green *et al.*, 1988).

Insulin causes cell swelling but increases pH_i through its activation of the Na^+–H^+ antiporter – as well as the Na^+/K^+-ATPase and the Na^+–K^+–$2Cl^-$ co-transporter (Haussinger, 1996). Therefore, the differences between the metabolic effects of insulin and of hyposmotic swelling can be explained by the inverse pH changes caused by insulin and hyposmotic swelling (Peak, Al-Habori & Agius, 1992).

Plasma membrane HCO_3^- transporters, of which three types are known to exist (Frelin *et al.*, 1988), play an essential role in the regulation of pH_i in vertebrate cells (Thomas, 1976; Roos & Boron, 1981; Boron, 1983). Bicarbonate transport into the cell counteracts the effects of proton extrusion. The regulation of HCO_3^- transporter activity is analogous to that of the Na^+–H^+ antiporter, namely through alteration of its pH dependence (Olsnes, Tonnessen & Sandvig, 1986). Consequently, hormonal effects due to acidification of the cytosol that are observed in isolated cells in the absence of HCO_3^- (e.g. due to mitogens) may not be observed in the presence of extracellular HCO_3^- as the change in pH_i is counteracted. Depending on the cell type, the Na^+–H^+ and HCO_3^- transporter systems could be differentially dominant with respect to their importance in regulating pH_i.

These mechanisms for stabilising pH_i might suggest that cells do not make use of changes in pH_i as a mechanism to regulate metabolic pathways. However, although not numerous (Frelin *et al.*, 1988), examples are known whereby changes in pH_i appear to be used as mechanisms through which metabolic control is exerted. In this chapter, emphasis is placed on the potential for the regulation of hepatic mitochondrial fatty acid oxidation by changes in pH_i, and particularly for the regulation of the activity of the first step, catalysed by mitochohdrial outer-membrane carnitine palmitoyltransferase (CPT I), that commits long-chain fatty acids to β-oxidation in the mitochondrial matrix.

Hormonal action, volume change and pH_i in hepatocytes

A relatively recent development in the field of the study of hepatic metabolism has been the appreciation that changes in the hydration state (effectively cell volume) of hepatocytes (and other cells) can affect the activity of multiple metabolic processes. In particular, Na^+-dependent uptake of amino acids (through systems A or N), and incubation of cells with certain hormones are able to affect cell hydration state and pH_i through their acute regulation of the activities of plasma membrane ion transporters (Baquet, Lavoinne & Hue, 1991; Haussinger & Lang, 1992; Haussinger, 1993, 1996; Haussinger *et al.*, 1994; Haussinger & Schliess, 1995). The question as to whether these hormones exert their effects through changes in cell volume and/or pH_i, or whether the latter are parallel consequences of the activation of hormonal signal transduction systems, has been addressed in relatively few instances. Of particular note are the studies of Agius and co-workers (Agius, Peak & Alberti, 1990; Agius, Peak & Al-Habori, 1991; Al-Habori *et al.*, 1992; Peak *et al.*, 1992; Agius *et al.*, 1994) in which the effects on insulin action of either preventing cell volume changes or manipulating the pH_i were studied in rat cultured hepatocytes in which the rates of glycolysis and glycogen synthesis were measured. It was concluded from these studies that both the increase in cell volume induced by insulin and the associated alkalinisation of the cytosol probably play a role in the stimulation of glycolysis by the hormone, i.e. that both effects form part of the insulin signal transduction system (Peak *et al.*, 1992). Similarly, some of the effects of amino acid-induced changes in cell metabolism can be well correlated to the degree of cell swelling induced by their Na^+-dependent uptake (Baquet *et al.*, 1991), whereas other effects appear to depend on the intracellular accumulation or depletion of specific anions and/or metabolic intermediates that affect protein phosphatase activity (Baquet *et al.*, 1993).

A reasonably comprehensive description of the ion movements involved in the mediation of changes in cell volume and pH_i induced by aniso-osmotic media, volume-active amino acids and hormones is outwith the scope of this chapter. It is to be noted, however, that the actions of the hormones and growth factors known to affect cell volume and pH_i are mediated through protein phosphorylation signalling cascades and that aniso-osmotic shock of cells is itself associated with the activation of two stress-activated signalling pathways (Galcheva-Gargova et al., 1994; Matsuda et al., 1995; Meier et al., 1996) as well as the activation of signal transduction pathways used by hormones and growth factors themselves (Agius et al., 1994; Schliess, Schreiber & Haussinger, 1995; Krause, Rider & Hue, 1996; Schliess et al., 1996). Consequently, the scope for cross-talk between the two types of cell stimulation is extensive.

The relevance of interactions between insulin action, cell volume and pH_i in the liver

All the elements of these interactions between hormonal and cell volume effects on pH_i are present for hepatocytes in the intact liver during the absorptive phase after food intake. The hepatic portal circulation channels to the liver simultaneously elevated concentrations of insulin and dietary constituents, particularly osmolyte amino acids and inorganic ions. The latter include K^+, which is a major dietary constituent and the single most important intracellular cation and, thus, a major determinant of cell volume. As described above, insulin increases hepatocyte volume by increasing net K^+ uptake secondary to activation of the Na^+–H^+ exchange (Fehlman & Freychet, 1981), thus resulting in the alkalinisation of the cytosol. The increased concentration of amino acids (particularly the most abundant ones such as glutamine and alanine) and their Na^+-driven uptake will also tend to increase cell volume, but the consequent counter-regulatory volume decrease will tend to induce an acidification of the cytosol, as it is accompanied by efflux of HCO_3^- (Haussinger, 1996).

Therefore, in the normal absorptive phase, the overall effect of the delivery of high concentrations of insulin and osmolytes is to increase cell volume through an increase in cell hydration. The increase in liver weight that occurs upon refeeding has been documented for a long time (Fenn, 1939). The increase in liver water and K^+ contents coincides with the increase in liver protein content and the net deposition of glycogen. Agius and co-workers (1991), using the data of Fenn (1939), have argued persuasively in favour of a causative relationship between insulin-mediated net K^+ influx, increased cell volume and glycogen deposition during the absorptive phase. In addition, they have provided experimental evidence, obtained on rat cultured hepatocytes, that glycogen deposition and glycolysis are both activated by alkalinisation of the cytosolic compartment (Al-Habori et al., 1992; Agius et al., 1994) and that cell swelling induced by hypo-osmotic medium tends to blunt this response because of its ability to lower pH_i. Moreover, they have provided evidence that the effects of changes in pH_i on a particular metabolic process are highly dependent on whether the effects of changes in cell volume and other signal transduction mechanisms act in the same or opposite directions. For example, it is possible to demonstrate experimentally that the apparent lack of effect of hypo-osmotic cell swelling on either glycogen synthesis or glycolysis in cultured rat hepatocytes is due to the opposite effects of the resultant acidification of the cytosol and the swelling *per se* (Al-Habori *et al.*, 1992). Similarly, it would appear from experimental observations on the perfused liver and freshly isolated rat hepatocytes that although the Na^+-dependent uptake of amino acids would be expected to lower pH_i, and thus antagonise the effects of insulin on glycogen synthase, its net effect is to stimulate glycogen synthesis. Evidently, the other changes associated with the accumulation of high concentrations of intracellular amino acids (Baquet *et al.*, 1993) and of the activation of signalling protein phosphorylation cascades (Krause *et al.*, 1996) override any effects on pH_i.

Therefore, in the *in vivo* situation, the resultant of all the effects of changes in the portal concentrations of hormones and osmolytes may depend on the physiological condition of the animal. In particular, because of the opposite effects of insulin-mediated cell swelling and that mediated by Na^+-co-transported amino acids on pH_i, the response of the liver will depend on the circulating concentration of, and tissue sensitivity to, insulin.

Examples of effects of pH_i on energy metabolism in hepatocytes

Discussion of the likely effects on metabolism of changes in pH_i on metabolism will be restricted to glycogen synthesis, glycolysis, gluconeogenesis and mitochondrial fatty acid oxidation. It has already been mentioned that a fall in pH_i inhibits both glycogen synthesis and glycolysis (Peak *et al.*, 1992) in hepatocytes. Similarly, gluconeogenesis is inhibited by acidification of the cytosol (Hems *et al.*, 1966; Iles *et al.*, 1977; Kashiwagura *et al.*, 1984). This has been postulated to occur secondarily to the inhibition of pyruvate carboxylase by a lowered intra-mitochondrial pH (Iles *et al.*, 1977), owing to the known sensitivity of the activity of this enzyme to lower pH values (Scrutton & Utter, 1967). A possible complementary mechanism is described below, based on the dependence of the activity of the enzyme on the intra-mitochondrial concentration of acetyl-CoA.

The relationship between gluconeogenesis and fatty acid oxidation in the absorptive phase is particularly important. When starved rats are refed a high-carbohydrate diet, a large proportion (of the order of 60 to 70 per cent) of the high rate of glycogen synthesis that occurs during the first few hours of refeeding uses lactate rather than glucose as the substrate, even though glucose is abundantly available in the hepatic portal circulation (Sugden *et al.*, 1983; Newgard *et al.*, 1984). This 'indirect' pathway necessitates (i) the provision of an abundant supply of lactate and gluconeogenic amino acids in the portal circulation, and (ii) the continued operation of gluconeogenesis. The first of these two conditions is met by virtue of the fact that dietary glucosyl units are rapidly converted into lactate by the gut (Katz & McGarry, 1984). Continued gluconeogenic flux also requires the maintenance of the activity of pyruvate carboxylase which catalyses a reaction at which substantial control over gluconeogenic flux is exerted (Quinlan & Halestrap, 1986). In the starved state, the intra-mitochondrial acetyl-CoA required for the activity of the enzyme is thought to be provided by the β-oxidation of fatty acids (Zammit, 1994, 1996). It has been shown by experiments *in vivo*, in which the fate of intrahepatic fatty acids was monitored in awake, unrestrained rats, that the metabolism of fatty acids continues to favour their oxidation for as long as gluconeogenesis is required for replenishment of liver glycogen (Moir & Zammit, 1993a, 1993b; Zammit, 1994; Zammit & Moir, 1994; Moir, Park & Zammit, 1995).

Inferences about changes in pH_i in the liver during the absorptive phase

The fact that fatty acids continue to be preferentially oxidised during the first three to four hours after refeeding suggests that pH_i is not decreased in the normal rat (i.e. in the presence of a normal insulin secretory response to a meal). This is because of the fact that CPT I would need to remain active. CPT I is extremely sensitive to inhibition by micromolar concentrations of malonyl-CoA (McGarry, Mannaerts & Foster, 1977), i.e within the range of concentrations that occur physiologically in the liver (see Zammit, 1984). Moreover, this sensitivity is extremely dependent on pH. A shift in pH from 7.4 to 6.8 increases the degree of inhibition of the enzyme by malonyl-CoA by several-fold (Stephens, Cook & Harris, 1983). This property has been suggested to limit ketogenic rate under ketoacidotic conditions and may account for the inhibition of ketogenesis in rat isolated hepatocytes by a decrease in pH_e (Fafournoux, Demigne & Remesey, 1987). Consequently, any decrease in pH_i of hepatocytes during refeeding of rats would be expected to inhibit CPT I activity, especially as the hepatic malonyl-

CoA concentration is rapidly doubled during this period (although it falls short of reaching the 'fed' level) (Moir & Zammit, 1993b). Evidently, the effect of the surge in hepatic portal insulin concentration (on which the rise in malonyl-CoA also depends – see Zammit, 1984) to alkalinise the cytosol overcomes the effects of any tendency of cell swelling *per se*, or of Na^+-dependent amino acid uptake, to acidify the cytosol.

However, an interesting situation arises in insulin-dependent, streptozotocin-diabetic rats. In these animals, the insulin secretory response to a meal is markedly blunted. *In vivo* experiments (Moir & Zammit, 1995) have shown that, when these animals are refed after 24 hours' starvation, the switching off of fatty acid oxidation is much more rapid than in normal rats. In addition, the rapid inhibition of fatty acids precedes the transient and modest rise in hepatic malonyl-CoA concentration (Moir & Zammit, 1995). Consequently, Moir and Zammit have suggested that CPT I becomes inhibited through the increased effectiveness of the concentrations of malonyl-CoA already present in the cytosol, through the acute lowering of pH_i. Evidence in favour of this suggestion is provided by the observation that the inhibition of fatty acid oxidation is accompanied by a rapid increase in the absolute concentration of lactate in the liver and in the lactate : pyruvate concentration ratio (Moir & Zammit, 1995). Both these changes would be expected to be associated with increased cytosolic $[H^+]$ concentration (Alberti & Cuthbert, 1982; Poole & Halestrap, 1993).

As mentioned above, lactate is produced at high rates in the gut. Diabetic rats consume a much larger amount of food upon refeeding (Moir & Zammit, 1995) and, because of their hyperphagic adaptation, absorption from their gut may be enhanced. Hepatic concentrations of lactate rise rapidly from 1 mM to 5 mM (Moir & Zammit, 1995). This marked increase in lactate concentration may be causatively associated with the putative decrease in pH_i. Lactate is in fact transported as lactic acid by the hepatocyte plasma membrane monocarboxylate transporter (Poole & Halestrap, 1993). However, at physiological pH, the acid is dissociated within the cell, such that uptake involves the generation of protons and cytosolic acidification. Under normal conditions, no significant drop in pH_i is thought to occur because lactate acts as a substrate for gluconeogenesis, which is a net consumer of H^+ (Alberti & Cuthbert, 1982). Cohen and co-workers (Beech *et al.*, 1989) have suggested that this is even more important in protecting pH_i from acidotic conditions in fed diabetic rats. However, during refeeding of starved diabetic rats, it would appear that the rate of gluconeogenesis does not keep pace with the rate of entry of lactate into the hepatocyte, giving rise to the marked increase in hepatic lactate content observed experimentally. (It has been pointed out (Poole & Halestrap, 1993) that the rates of lactic acid transport measured by Beech *et al.* (1989) were underestimates.) It is known that both starvation and diabetes, separately, result in the upregulation of lactate entry into isolated rat hepatocytes (Metcalf *et al.*, 1988). In addition, the Na^+-dependent transport of amino acids is also activated (McGivan & Pastor-Anglada, 1994). The liver of the starved–refed insulin-dependent rat may thus represent a condition in which the acidification induced by cell swelling, uptake of amino acids such as glutamine, and rapid lactic acid uptake all combine to lower pH_i, whereas the countering (alkalinising) effects of the ion movements normally induced by insulin are absent. This combination may also occur in insulin-resistant states normally associated with such conditions as obesity and non-insulin-dependent diabetes in humans.

Any pH-induced inhibition of CPT I would have the effect of lowering intra-mitochondrial acetyl-CoA, which will suppress the rate of gluconeogenesis, and thus of the utilisation of H^+. Under these conditions, the high concentrations of lactate (and, presumably, those of H^+) are not dissipated until the absorptive phase declines. It is suggested, therefore, that the strong pH dependence of CPT I inhibition by malonyl-CoA plays a central role in rapidly limiting the rate of hepatic fatty acid oxidation in insulin-dependent or insulin-resistant states.

Acknowledgements

Work in the author's laboratory was supported by the British Diabetic Association, The Leverhume Trust and the Scottish Office Agriculture and Fisheries Department.

References

Agius, L., Peak, M. & Alberti, K.G.M.M. (1990). Regulation of glycogen-synthesis from glucose and gluconeogenic precursors by insulin in periportal and perivenous rat hepatocytes. *Biochemical Journal* **266**, 91–102.

Agius, L., Peak, M. & Al-Habori, M. (1991). What determines the increase in liver-cell volume in the fasted-to-fed transition – glycogen or insulin? *Biochemical Journal* **276**, 843–5.

Agius, L., Peak, M., Beresford, G., Al-Habori, M. & Thomas, T.H. (1994). Activation of MAPkinase by cell swelling. *Biochemical Society Transactions* **22**, 516–21.

Alberti, K.G.M.M. & Cuthbert, C. (1982). The hydrogen ion in normal metabolism: a review. *CIBA Foundation Symposium* **87**, 1–19.

Al-Habori, M., Peak, M., Thomas, T.H. & Agius, L. (1992). The role of cell swelling in the stimulation of glycogen-synthesis by insulin. *Biochemical Journal* **282**, 789–96.

Anderson, R.M., Graf, J. & Boyer, J.L. (1987). Na⁺–H⁺ exchanger regulates intracellular pH in isolated rat hepatocyte couplets. *American Journal of Physiology* **252**, G109–13.

Arias, I.M. & Forgag, M. (1984). The sinusoidal domain of the plasma membrane of rat hepatocytes contains an amiloride-sensitive Na⁺/H⁺ antiport. *Journal of Biological Chemistry* **259**, 5406–8.

Baquet, A., Gaussin, V., Bollen, M., Stalmans, W. & Hue, L. (1993). Mechanism of activation of liver acetyl-CoA carboxylase by cell swelling. *European Journal of Biochemistry* **217**, 1083–9.

Baquet, A., Levoinne, A. & Hue, L. (1991). Comparison of the effects of various amino acids on glycogen synthesis, lipogenesis and keto-genesis in isolated rat hepatocytes. *Biochemical Journal* **273**, 57–62.

Beech, J.S., Williams, S.R., Cohen, R.D. & Iles, R.A. (1989). Gluco-neogenesis and the protection of hepatic intracellular pH during diabetic ketoacidosis in rats. *Biochemical Journal* **263**, 737–44.

Boron, W.F. (1983). Transport of H⁺ and ionic weak acids and bases. *Journal of Membrane Biology* **72**, 1–16.

Fafournoux, P., Demigne, C. & Remesey, C. (1987). Mechanisms involved in ketone body release by rat liver cells: influence of pH and bicarbonate. *American Journal of Physiology* **252**, G200–8.

Fehlman, M. & Freychet, P. (1981). Insulin and glucagon stimulation of (Na⁺–K⁺)-ATPase transport activity in isolated hepatocytes. *Journal of Biological Chemistry* **256**, 7449–53.

Fenn, W.O. (1939). The deposition of potassium and phosphate with glycogen in rat livers. *Journal of Biological Chemistry* **128**, 297–307.

Frelin, C., Vigne, P., Ladoux, A. & Lazdunski, M. (1988). The regulation of the intracellular pH in cells from vertebrates. *European Journal of Biochemistry* **174**, 3–14.

Galcheva-Gargova, Z., Derijard, B., Wu, I.-H. & Davis, R.J. (1994). An osmosensing signal transduction pathway in mammalian cells. *Science* **265**, 806–8.

Gleeson, D., Corasanti, J.G. & Boyer, J.L. (1990). Effects of osmotic stress on isolated rat hepatocytes II. Modulation of intracellular pH. *American Journal of Physiology* **258**, G299–307.

Gleeson, D., Smith, N.D. & Boyer, J.L. (1989). Bicarbonate-dependent and -independent intracellular pH regulatory mechanisms in rat hepatocytes. Evidence for Na–HCO₃ transport. *Journal of Clinical Investigation* **84**, 312–21.

Green, J., Yamguchi, D.T., Kleeman, C.R. & Muallem, S. (1988). Selective modification of the kinetic properties of Na⁺/H⁺ exchange in both directions without altering K_m. *Journal of Biological Chemistry* **263**, 5012–15.

Grinstein, S. & Rothstein, A. (1986). Mechanisms of regulation of the Na^+/H^+ exchanger. *Journal of Membrane Biology* **90**, 1–12.

Haussinger, D. (1993). Control of protein-turnover by the cellular hydration state. *Italian Journal of Gastroenterology* **25**, 42–8.

Haussinger, D. (1996). The role of cellular hydration in the regulation of cell-function. *Biochemical Journal* **313**, 697–710.

Haussinger, D. & Lang, F. (1992). Cell-volume and hormone action. *Trends in Pharmacological Sciences* **13**, 371–3.

Haussinger, D., Newsome, W., Dahl, S.V. *et al.* (1994). Control of liver-cell function by the hydration state. *Biochemical Society Transactions* **22**, 497–502.

Haussinger, D. & Schliess, F. (1995). Cell-volume and hepatocellular function. *Journal of Hepatology* **22**, 94–100.

Hems, R., Ross, B.D., Berry, M.N. & Krebs, H.A. (1966). Gluconeogenesis in the perfused liver. *Biochemical Journal* **101**, 284–92.

Iles, R.A., Cohen, R.D., Rist, A.H. & Baron, P.G. (1977). The mechanism of inhibition by acidosis of gluconeogenesis from lactate in rat liver. *Biochemical Journal* **164**, 185–91.

Kashiwagura, T., Deutsch, C.J., Taylor, J., Erecinska, M. & Wilson, D.F. (1984). Dependence of gluconeogenesis, urea synthesis, and energy metabolism of hepatocytes on intracellular pH. *Journal of Biological Chemistry* **259**, 237–43.

Katz, J. & McGarry, J.D. (1984). The glucose paradox: is glucose a substrate for liver metabolism? *Journal of Clinical Investigation* **74**, 1901–9.

Krause, U., Rider, M.H. & Hue, L. (1996). Protein kinase signalling pathway triggered by cell swelling and involved in the activation of glycogen synthase and acetyl-CoA carboxylase in isolated rat hepatocytes. *Journal of Biological Chemistry* **271**, 16668–73.

Madshus, I.H., Tonnessen, T.I., Olsen, S. & Sandvig, K. (1987). Effect of potassium depletion of Hep 2 cells on intracellular pH and on chloride uptake by anion antiport. *Journal of Cell Physiology* **131**, 6–13.

Mahnenesmith, R.L. & Aronson, P.S. (1985). The plasma membrane sodium–hydrogen exchanger and its role in physiological and pathological processes. *Circulation Research* **56**, 773–88.

Matsuda, S., Kawasaki, H., Moriguchi, T., Gotoh, Y. & Nishida, E. (1995). Activation of protein kinase cascades by osmotic shock. *Journal of Biological Chemistry* **270**, 12781–6.

McGarry, J.D., Mannaerts, G.P. & Foster, D.W. (1977). A possible role for malonyl-CoA in the regulation of hepatic fatty acid oxidation and ketogenesis. *Journal of Clinical Investigation* **60**, 265–70.

McGivan, J.D. & Pastor-Anglada, M. (1994). Regulatory and molecular aspects of mammalian amino acid transport. *Biochemical Journal* **299**, 321–34.

Meier, R., Rouse, J., Cuenda, A., Nebreda, A.R. & Cohen, P. (1996). Cellular stress and cytokines activate multiple mitogen-activated protein kinase kinase homologues in PC12 and KB cells. *European Journal of Biochemistry* **236**, 796–805.

Metcalf, H.K., Monson, J.P., Cohen, R.D. & Padgham, C. (1988). Enhanced carrier-mediated lactate entry into isolated hepatocytes from starved and diabetic rats. *Journal of Biological Chemistry* **263**, 19505–9.

Moir, A.M.B., Park, S.-B. & Zammit, V.A. (1995). Quantification *in vivo* of the effects of different types of dietary fat on the loci of control involved in hepatic triacylglycerol secretion. *Biochemical Journal* **308**, 537–42.

Moir, A.M.B. & Zammit, V.A. (1993a). Monitoring of changes in hepatic fatty acid and glycerolipid metabolism during the starved-to-refed transition *in vivo*. *Biochemical Journal* **289**, 49–55.

Moir, A.M.B. & Zammit, V.A. (1993b). Rapid switch of hepatic fatty acid metabolism from oxidation to esterification during diurnal feeding of meal-fed rats correlates with changes in the properties of acetyl-CoA carboxylase, but not of carnitine palmitoyltransferase I. *Biochemical Journal* **291**, 241–6.

Moir, A.M.B. & Zammit, V.A. (1995). Acute meal-induced changes in hepatic glycerolipid metabolism are unimpaired in severely diabetic rats: implications for the role of insulin. *FEBS Letters* **370**, 255–8.

Moolenar, W.H. (1986). Effects of growth factors on intracellular pH regulation. *Annual Review of Physiology* **48**, 363–76.

Moseley, R.H., Meier, P.J., Aronson, P.S. & Boyer, J.L. (1986). Na^+–H^+ exchange in rat liver basolateral but not canalicular plasma membrane vesicles. *American Journal of Physiology* **250**, G35–43.

Newgard, C.B., Moore, S.V., Foster, D.W. & McGarry, J.D. (1984). Efficient hepatic glycogen synthesis in refeeding rats requires continued carbon flow through the gluconeogenic pathway. *Journal of Biological Chemistry* **259**, 6958–63.

Olsnes, S., Tonnessen, T.I. & Sandvig, K. (1986). pH-regulated anion antiport in nucleated mammalian cells. *Journal of Cell Biology* **102**, 1967–71.

Peak, M., Al-Habori, M. & Agius, L. (1992). Regulation of glycogen synthesis and glycolysis by insulin, pH and cell volume – interactions between swelling and alkalinization in mediating the effects of insulin. *Biochemical Journal* **282**, 797–805.

Poole, R.C. & Halestrap, A.P. (1993). Transport of lactate and other monocarboxylates across mammalian plasma membranes. *American Journal of Physiology* **264**, C761–82.

Quinlan, P.T. & Halestrap, A.P. (1986). The mechanism of the hormonal activation of respiration in isolated hepatocytes and its importance in the regulation of gluconeogenesis. *Biochemical Journal* **236**, 789–800.

Renner, E.L., Lake, J.R., Pensco, M. & Scharschmidt, B.F. (1989) Na^+–H^+ exchange activity in rat hepatocytes: role in regulation of intracellular pH. *American Journal of Physiology* **256**, G44–52.

Roos, A. & Boron, W.F. (1981). Intracellular pH. *Physiological Reviews* **61**, 296–334.

Schliess, F., Schreiber, R. & Haussinger, D. (1995). Activation of extracellular signal-regulated kinases Erk-1 and Erk-2 by cell swelling in H4IIe hepatoma-cells. *Biochemical Journal* **309**, 13–17.

Schliess, F., Sinning, R., Fischer, R., Schmalenbach, C. & Haussinger, D. (1996). Calcium-dependent activation of Erk-1 and Erk-2 after hypoosmotic astrocyte swelling. *Biochemical Journal* **320**, 167–71.

Scrutton, M.C. & Utter, M.F. (1967). Pyruvate carboxylase IX. Some properties of the activation by certain acyl derivatives of coenzyme A. *Journal of Biological Chemistry* **242**, 1723–35.

Stephens, T.W., Cook, G.A. & Harris, R.A. (1983). Effect of pH on malonyl-CoA inhibition of carnitine palmitoyltransferase I. *Biochemical Journal* **212**, 521–4.

Sugden, M.C., Watts, D.I., Palmer, T.N. & Myles, D.D. (1983). Direction of carbon flux in starvation and after refeeding – *in vivo* and *in vitro* effects of 3-mercaptopicolinate. *Biochemistry International* **7**, 329–37.

Thomas, R.C. (1976). The effect of carbon dioxide on the intracellular pH and buffering power of snail neurons. *Journal of Physiology* **255**, 715–35.

Zammit, V.A. (1984). Time-dependence of inhibition of carnitine palmitoyltransferase I by malonyl-CoA in mitochondria from livers of fed or starved rats. Evidence for transition of the enzyme between states of low and high affinity for malonyl-CoA. *Biochemical Journal* **218**, 379–86.

Zammit, V.A. (1994). Regulation of ketone body metabolism: a cellular perspective. *Diabetes Reviews* **2**, 132–55.

Zammit, V.A. (1996). The role of insulin in hepatic fatty acid partitioning: emerging concepts. *Biochemical Journal* **314**, 1–14.

Zammit, V.A. & Moir, A.M.B. (1994). Monitoring the partitioning of hepatic fatty acids *in vivo*: keeping track of control. *Trends in Biochemical Sciences* **19**, 313–17.

R. TYLER-JONES and E. W. TAYLOR

Back to basics: a plea for a fundamental reappraisal of the representation of acidity and basicity in biological solutions

Introduction

The acid–base status of an aqueous solution is reflected by the relative activities of hydrogen ($\{H^+\}$) and hydroxide ($\{OH^-\}$) ions. These two quantities can be altered by changes in temperature, the concentrations of weak and strong electrolytes, and P_{CO_2} (Stewart, 1978). In biological solutions, $\{H^+\}$ is determined by the strong ion difference (SID) (defined as the sum of all strong base cations minus the sum of all strong acid anions), P_{CO_2}, and the total weak acids present. At constant temperature, acid–base status can be changed only by changes in one or more of these independent variables (Stewart, 1978).

The maintenance of a stable acid–base status of the internal body fluids is thought to be of importance to animals primarily because of the need to maintain protein function. Protein structure (and hence function) is dependent on the degree of ionisation of the component amino acids and of the α-imidazole group in particular (Reeves, 1972). The degree of ionisation of amino acids can be altered by changes in the acid–base conditions of the medium. To preserve protein function and maintain the physiological processes to which it contributes, animals must regulate acid–base status and may, for example when faced by changes in temperature, either maintain a constant relative alkalinity (see below), or regulate pH independently of temperature, in order to depress or enhance rates of metabolism (see Whiteley, this volume).

To understand fully how these processes are affected by acid–base conditions and how acid–base status is maintained in living organisms requires a suitable means for representing the acid–base status of a solution. Because the range of values for $\{H^+\}$ and $\{OH-\}$ can be extremely large, Sørenson (1909) devised the logarithmic pH scale as a convenience for calculations. Since then, pH has become the standard measure of acidity and basicity and is used by both chemists and physiologists for describing the acid–base chemistry of solutions. However, as a representative of a solution's acidity or basicity, pH has been criticised as being misleading and indirect (Stewart, 1978) and for distorting the physiological reality (Davenport, 1974).

In addition to the problems associated with the way in which pH represents acid–base status, there is some confusion over the correct statistical treatment of replicated pH measurements. Sometimes average pH and other statistics are calculated directly from measured values of pH. Other, possibly more enlightened, researchers convert pH values to {H$^+$} before statistics are calculated. The results may then either be reported as {H$^+$} or they may be reconverted to pH.

This chapter considers the representation of acid–base data and what may be the proper use of pH or {H$^+$} as a means of representing such data. The study examines this issue on a theoretical basis, on a statistical basis and by a re-examination of published data to show how the use of the two scales can affect the results of analysis.

Theoretical considerations

One of the principal laws of chemistry, the Law of Mass Action, states that 'at a given temperature the rate of a chemical reaction is directly proportional to the "active masses" of the reactants'. This law governs the rates of all chemical reactions, including reactions between solutes, in which the rate will be directly proportional to their activities.

As stated above, the acid–base status of a biological solution is determined by three independent variables: SID, $P\text{CO}_2$ and the total concentration of weak acids (Stewart, 1978). These independent variables determine the acid–base status through three fundamental physical laws: the need to conserve mass, to maintain electrical neutrality and to maintain the dissociation equilibrium of any weak electrolytes present. Because the Law of Mass Action will, in part, determine the activities at equilibrium, it is the activities of the individual acid–base variables, including H$^+$ and OH$^-$, which are determined, not a logarithmic function of these quantities.

For physiologists, one of the most fundamental of relationships determining acid–base status can be described by the Henderson–Hasselbalch equation. This enables the calculation of the concentration of a key variable, such as [HCO$_3^-$], in a biological fluid from measurements of related variables, such as $P\text{CO}_2$ or total CO$_2$ and pH. The calculation requires prior knowledge of the solubility coefficient for CO$_2$ in the fluid and the first dissociation constant for the hydration reaction of CO$_2$ with the fluid (pK$_i$).

The Henderson–Hasselbalch equation is derived from the equation for the dissociation of carbonic acid: $H_2CO_3 \rightleftharpoons H^+ + HCO_3^-$. Application of the Law of Mass Action gives:

$$\frac{[H^+][HCO_3^-]}{[H_2CO_3]} = k'$$

which, by appropriate rearrangement and substitution and by taking the negative logarithm of both sides, can be transformed to the Henderson–Hasselbalch equation:

$$pH = pK + \log \frac{[\text{total CO}_2] - \alpha P, CO_2}{\alpha P, CO_2}$$

the full derivation of which is given in Davenport (1974). It is the artifice of taking the negative logarithms to obtain a pH term which confounds our understanding of the true nature of this physiologically important relationship. It conceals the fact that, as it is derived from the Law of Mass Action, the relationships between the variables must change in a linear fashion. The equation indicates that $\{H^+\}$ and $\{HCO_3^-\}$ change in response to changes in any of the independent variables, including P_{CO_2}, and that their product must be proportional to P_{CO_2}. It does not imply an interdependency between $\{H^+\}$ and $\{HCO_3^-\}$ (Stewart, 1978).

The distribution of pH and $\{H^+\}$ in biological solutions

Prompted by the confusion over the statistical treatment of pH data, Boutilier and Shelton (1980) investigated whether pH or $[H^+]$ is the natural quantity of biological importance. They suggested that receptor systems in animals monitor $[H^+]$ on a logarithmic basis and that the concentrations would therefore be regulated on the same basis. If a logarithmic function of the concentrations is the regulated quantity, then it would be expected to show a normal distribution in replicated samples.

To investigate which quantity is biologically significant, Boutilier and Shelton examined the distribution of blood pH and $[H^+]$ values in both humans and the toad, *Bufo marinus*. In neither species did the test statistic, the Kolmogorov–Smirnov test, reveal any departure from normality for the distributions of either pH or $[H^+]$ values.

Because the study of Boutilier and Shelton (1980) was inconclusive, the authors examined the frequency distributions of haemolymph pH and {H⁺} values in two species (humans and the lobster *Homarus vulgaris*), but used a different test statistic. Values for blood pH in samples taken from newborn infants were kindly supplied by Dr H. Gee and staff at the Sorrento Maternity Hospital, Moseley, Birmingham. The pH of these blood samples was measured using a Corning 178 pH/ Blood Gas Analyzer, calibrated with pH 6.838 and pH 7.382 buffers. The values for haemolymph pH in *H. vulgaris* were supplied by Dr N.M. Whiteley (School of Biological Sciences, University of Birmingham). All the lobster samples had been taken from settled, submerged animals at 15 °C. The pH of the lobster samples was measured using a Radiometer PHM73 pH meter with a G279/G2 glass capillary micro-electrode and a K497 calomel electrode thermostatted to the experimental temperature and calibrated with the Radiometer precision buffers S1500 and S1510. For each pH value, {H⁺} was calculated by a 10^{-pH} transformation.

The mean, standard deviation and coefficient of skewness (Snedecor & Cochran, 1980) were estimated for each set of data and the results are given in Table 1. The data are shown in Figure 1 as frequency plots together with theoretical normal distribution curves having the same mean, standard deviation and maximum frequency as the samples (see Boutilier & Shelton, 1980). The coefficient of skewness showed that the distribution of blood pH values in each species departs significantly from normality. However, it was unable to dis-

Table 1. *Summary statistics for the distribution of pH and {H⁺} values in blood from humans and lobster (*Homarus vulgaris*)*

	Sample data Human (*n* = 167)		Lobster (*n* = 99)	
	pH	{H⁺}	pH	{H⁺}
x̄	7.39	41.4	7.79	16.6
M	7.38	41.7	7.77	17.0
SD	0.06	5.44	0.09	3.34
√b₁	0.397	0.018	0.619	0.062
	*		**	

The statistics include the mean (x̄), median (M), and standard deviation (SD) and the test for normality, the coefficient of skewness (√b₁). Significant values are denoted by askerisks * = 5% level; ** = 1% level).

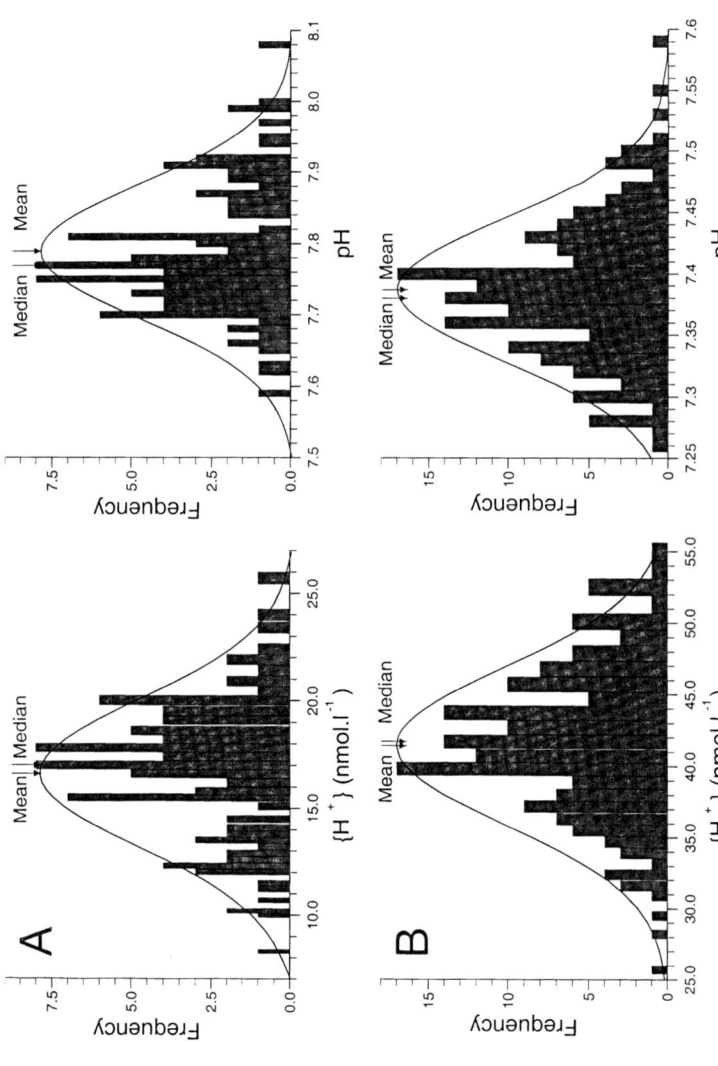

Fig. 1. The frequency distributions of pH and $\{H^+\}$ values from the blood of lobsters (**A**) and humans (**B**). The distributions are each shown with a theoretical normal curve calculated for a distribution with the same mean, standard deviation and maximum frequency as the sample distributions. The mean and median values for each distribution are indicated by arrows.

tinguish either of the {H$^+$} distributions from normal distributions. The analysis also indicated that the non-normality displayed by the pH distributions was due to the acidic values being grouped close to the mean and the basic values extending far above it, which accords with the fact that it is a logarithmic transform.

This contrasts with the findings of Boutilier and Shelton (1980), who were unable to show any departure from a normal distribution for either the linear or logarithmic data in the two species they examined. The coefficient of skewness provides a good measure of non-normality against both highly skewed and long-tailed distributions, whereas the Kolmogorov–Smirnov test is generally insensitive to departures from normality (Shapiro, Wilk & Chen 1968). Thus, the test statistic used in the present study is more sensitive to detecting non-normal distributions than the test used in the earlier study. The present results indicate that {H$^+$} and not pH is normally distributed in the blood or haemolymph and support the hypothesis that it is {H$^+$}, or some related variable, that is regulated. This is consistent with the fact that in animals, acid–base regulation is achieved primarily by adjustments of P_{CO_2} and SID. Although the non-normality displayed by the distributions of the blood pH values is not great, it would seem to be statistically exact and biologically more informative to calculate and report mean values for {H$^+$}.

The modulation of haemocyanin–O$_2$ affinity

As an example of how differing conclusions may be reached depending on how acid–base data are expressed and analysed, the authors re-examined some data on the modulation of the O$_2$ affinity of the crustacean respiratory pigment, haemocyanin. In each case, the data were re-analysed using {H$^+$} calculated by a 10^{-pH} transformation, to represent the acid–base status of the medium.

The O$_2$ affinity of haemocyanin in crustaceans is modulated by a number of factors, including temperature (Truchot, 1973), inorganic ions (Truchot, 1975; Morris, Tyler-Jones & Taylor, 1986) and organic ions (Mangum, 1983; Morris, Bridges & Grieshaber, 1985). The extent of this modulation is thought to be regulated to ensure an adequate supply of O$_2$ to the tissues under varying conditions (Truchot, 1975).

The acid–base status of the haemolymph is one of the inorganic modulators of haemocyanin–O$_2$ affinity. For most species examined, an increase in acidity decreases O$_2$ affinity, an effect known as the Bohr effect (Mangum, 1983). The Bohr effect or factor (ϕ) is quantified as $\Delta \log P_{50}/\Delta pH$, and in some species $\log P_{50}$ has been found to vary linearly with pH. The Bohr factor often appears to be independent of the levels of other modulators of haemocyanin–O_2 affinity, but in some species the size of the Bohr factor may be influenced by other modulators such as Mg^{2+} (Truchot, 1975) and lactate (Lac⁻) (Morris *et al.*, 1985).

Morris *et al.* (1986) investigated the modulation of haemocyanin–O_2 affinity in the crayfish, *Austropotamobius pallipes*, by pH and $[Ca^{2+}]$ and [Lac⁻]. The Bohr factor was independent of both $[Ca^{2+}]$ and [Lac⁻] and the mean magnitude was −0.455. Both $[Ca^{2+}]$ and [Lac⁻] were found to affect O_2 affinity, with $logP_{50}$ varying linearly with $log[Ca^{2+}]$ and $log[Lac^-]$. The effects of $[Ca^{2+}]$ and [Lac⁻] on P_{50} were found to be interdependent and mutually extenuating.

The results of re-analysis of these data are shown in Figure 2, for comparison with Figure 3 in Morris *et al.* (1986). At each concentration of Ca^{2+} and Lac⁻, P_{50} varied linearly with $\{H^+\}$ (Table 2). This modulation of O_2 affinity by acid–base status is the Bohr effect, and the Bohr factor can be defined, for present purposes, as $\Delta P_{50}/\Delta\{H^+\}$. The Bohr

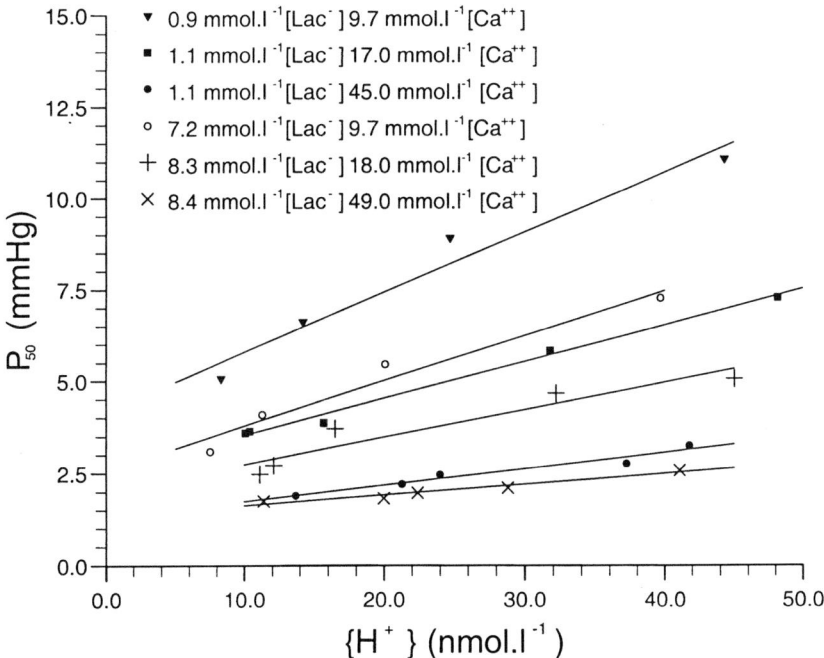

Fig. 2. The relationship between $\{H^+\}$, derived as a 10^{-pH} transformation, and P_{50} at 15 °C in dialysed *Autropotamobius pallipes* haemolymph containing different concentrations of L-lactate and Ca^{2+}. The lines for the regression of P_{50} with $\{H^+\}$ (see Table 2) are included.

Table 2. *Regression equations calculated for the effect of*
{H⁺}, calculated as a 10^{-pH} transmission, on the O_2 affinity
(P_{50}) of haemocyanin from the crayfish A. pallipes at different
concentrations of Ca^{2+} and Lac^-

[Lac⁻] (mmol/l)	[Ca²⁺] (mmol/l)	Regression equation
0.9	9.7	$P_{50} = 4.19 + 0.16\{H^+\}$
1.1	17.0	$P_{50} = 2.53 + 0.10\{H^+\}$
1.1	45.0	$P_{50} = 1.30 + 0.04\{H^+\}$
7.2	9.7	$P_{50} = 2.55 + 0.12\{H^+\}$
8.3	18.0	$P_{50} = 1.00 + 0.07\{H^+\}$
8.4	49.0	$P_{50} = 1.33 + 0.03\{H^+\}$

factor ranged from +0.163 ($p<0.025$) at the lowest concentrations of
Ca^{2+} and Lac^- to +0.029 ($p<0.001$) at the highest levels of [Ca²⁺] and
[Lac⁻] used. Comparison of the regression lines by analysis of covariance indicated that increasing [Lac⁻] from 1 to 8 mmol/l at each [Ca²⁺]
significantly decreased O_2 affinity but did not significantly affect the
Bohr factor. Increasing [Ca²⁺] from 9 to 17 mmol/l and from 17 to 45
mmol/l at 1 mmol/l [Lac⁻] significantly decreased both O_2 affinity and
the Bohr effect. However, at 8 mmol/l [Lac⁻], increasing [Ca²⁺] from 9
to 18 mmol/l only significantly decreased O_2 affinity and did not have
a significant affect on the Bohr factor.

The modulation of the Bohr effect by Ca^{2+} was examined at both
[Lac⁻] by analysis of the regression of ϕ with [Ca²⁺], log[Ca²⁺], ln[Ca²⁺],
1/[Ca²⁺] and $\sqrt{[Ca^{2+}]}$. At both concentrations, it was found that only
the regression of 1/[Ca²⁺] with ϕ was significant; the two regression
coefficients were 1.47 ($p < 0.05$) at 1 mmol/l Lac⁻ and 1.14 ($p < 0.05$)
at 8 mmol/l Lac⁻. The regression lines are plotted in Figure 3 and show
how increases in [Ca²⁺] decrease the Bohr effect. Analysis of covariance
showed that the slopes and elevations of the two lines differed significantly ($p < 0.05$). Thus, although it was not possible to quantify the
effect of Lac⁻, these data indicate that increasing Lac⁻ concentrations
decrease the Bohr factor and also decrease modulation of the Bohr
effect by Ca^{2+}. At the two Lac⁻ concentrations, the modulation of O_2
affinity by both {H⁺} and [Ca²⁺] can be described by the equations:

$$P_{50} = -2.01 + 0.11\{H^+\} + 73.0/[Ca^{2+}] \text{ at 1 mmol/l Lac}^-$$

$$P_{50} = -0.59 + 0.08\{H^+\} + 40.2/[Ca^{2+}] \text{ at 8 mmol/l Lac}^-$$

(calculated by multiple regression analysis)

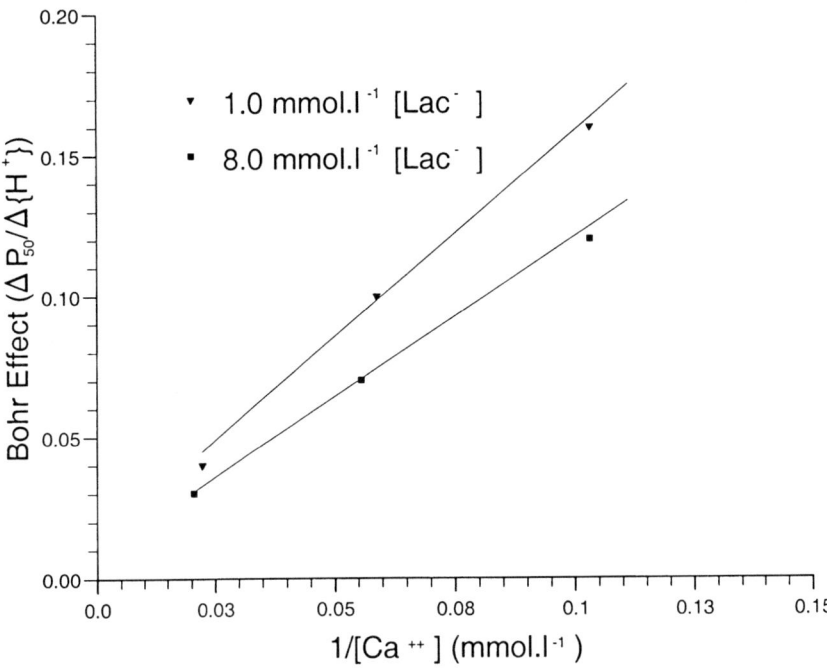

Fig. 3. The effect, *in vitro*, of $[Ca^{2+}]$ on the Bohr effect in haemo-cyanin from the crayfish *A. pallipes*. The Ca^{2+} effect was found at both concentrations (1.0 and 8.0 mmol/l) of Lac^- used. The lines calculated from regression analysis of the data are included; the regression coefficients for the two lines are 1.47 and 1.14 at 1.0 mmol/l and 8.0 mmol/l Lac^- respectively.

The effect of Lac^- on the modulation of haemocyanin O_2 affinity has been studied in more detail in the prawn *Palaemon elegans* (Bridges, Morris & Grieshaber, 1984). Increasing $[Lac^-]$ was found to increase O_2 affinity, but the Bohr effect was independent of $[Lac^-]$. The mean value for the Bohr factor, as $\Delta logP_{50}/\Delta pH$, was -0.18 for blood sampled in the summer. Re-examination of these data, using $\{H^+\}$, showed that the Bohr factor (calculated as $\Delta P_{50}/\Delta\{H^+\}$) varied from 2.22 at 0 mmol/l Lac^- to 0.61 at 9.1 mmol/l Lac^- and was significant ($p < 0.001$) in each case. The slopes of contiguous regression lines were not significantly different, so each incremental increase in $[Lac^-]$ did not significantly affect the Bohr factor. However, there was a significant difference between the regression coefficients at the lowest and highest concentrations, indicating that, overall, Lac^- does modulate the Bohr effect.

This effect was examined by regression analysis of ϕ against untransformed data ($[Lac^-]$) and against transformed data ($\log[Lac^-]$, $\ln[Lac^-]$, $1/[Lac^-]$ and $\sqrt{[Lac^-]}$). The most significant regression was for untransformed data, for which the coefficient of regression was -0.184 ($p < 0.01$), indicating that increases in $[Lac^-]$ decrease the Bohr factor.

Uric acid has also been identified as a modulator which enhances the O_2 affinity of haemocyanin in the crayfish *A. pallipes* (Morris *et al.*, 1985). However, this study was unable to demonstrate any interaction between the effects of urate and the Bohr effect. The effect of urate was re-analysed, using $\{H^+\}$ in place of pH. P_{50} varied linearly with $\{H^+\}$, with the Bohr factor ranging from 0.141 ($p < 0.001$) at 0.19 mmol/l urate to 0.09 ($p < 0.001$) at 0.84 mmol/l urate. In this case, the most significant regression obtained was for the regression of $\sqrt{[urate]}$ with the Bohr factor, with a coefficient of -0.02 ($p < 0.005$).

In each of the examples used, analysis of the data using $\{H^+\}$ revealed interactions not found by analysis based on pH. In contrast to the original studies, both lactate and urate were found to modulate haemocyanin–O_2 affinity and also to modify the Bohr effect. Increases in the concentration of either ion increased P_{50} but decreased the Bohr factor. Re-analysis of the data of Morris *et al.* (1986) indicated that not only do increases in $[Ca^{2+}]$ and $[Lac^-]$ increase O_2 affinity, but that increases in these ions decrease the Bohr effect. Additionally, there is an interaction between the effects of these two modulators and their effect on the Bohr factor: an increase in $[Lac^-]$ decreases the modulation of the Bohr factor by Ca^{2+}. These interactions, revealed by analysis of the data using $\{H^+\}$ as the measure of acid–base status, were concealed in the original analyses because of the way the pH scale distorts variations in $\{H^+\}$, and, in doing so, loses significant trends in the data.

The regulation of acid–base status

The way in which the acid–base status of a solution is represented must be important in studies of acid–base regulation. In particular, the results obtained and conclusions reached in comparisons or analyses of acid–base data will differ, depending on which scale, logarithmic or linear, is used.

In animals, the acid–base status of the internal body fluids is maintained primarily by adjustments in either P_{CO_2} or SID, with only an insignificant contribution from weak acids or bases. Regulation of P_{CO_2}, achieved by alterations in ventilation rate, is a mechanism for acid–base control of primary importance to air-breathers and of less importance to water-breathers, because the solubility of CO_2 in water is so much greater than that of O_2 (Dejours, 1978). Changes in SID, available to both air-breathing and water-breathing animals, are achieved by ion exchanges between body compartments and between the body and the external environment (Cameron, 1985). In the normal physiological range of acid–base conditions, lactate, an important end-product of anaerobiosis in many species, acts as a strong acid.

As an example of how comparisons are affected by the scale used, Boutilier and Shelton (1980) concluded, on the basis of a relatively small range of measured pH (and $[H^+]$) values, that the regulation of acid–base status is precise in humans while, because the range of pH values was much larger, it is less precise in the toad. Although this was apparently the case on the basis of their pH values, the range of $[H^+]$ values was almost identical in the two species, suggesting that they can regulate acid–base status with the same degree of precision.

In a study on the effects of environmental hypoxia in the marine worm *Sipunculus nudus*, Pörtner, Grieshaber and Heisler (1984) found that after 24 hours of anaerobiosis extracellular pH (pH_e) in the coelomic fluid had dropped by 0.38 units, while intracellular pH (pH_i) had dropped by 0.30 units in the body wall and by 0.31 units in the introvert retractors. It was concluded that pH_i and pH_e are affected by hypoxia to the same extent. However, if the data are converted by a 10^{-pH} transformation, then the respective increases in $\{H^+\}$ are 47.6 nmol/l for the body wall, 41.5 nmol/l for the retractors and only 11.4 nmol/l for the coelomic fluid. When the regulated variable $\{H^+\}$ is considered rather than pH, then it can be seen that in *S. nudus* the intracellular acid–base status of the relatively acidic tissues examined is affected substantially more by environmental hypoxia than the acid–base status of the more alkaline extracellular fluid.

A similar discrepancy between observed changes in pH and $\{H^+\}$ has been reported in a study of the acid–base status of different tissue compartments in the crayfish, *A. pallipes*, acclimated to a range of ambient temperatures (see Whiteley, this volume). Similarly, during aerial exposure of the crayfish, the decrease in haemolymph pH was greater than the decrease in pH_i of the abdominal muscles (Tyler-Jones & Taylor, 1988). As the intracellular compartment is more acidic than the haemolymph, any distortion introduced by the logarithmic function of pH is greater for the intracellular than for the extracellular compartment. Thus, although the changes in pH_i were smaller than those in pH_e, they actually corresponded to much greater changes in $\{H^+\}$ in the more acidic intracellular compartment.

The changes in acid–base status of a solution can be plotted on a pH-HCO$_3^-$ diagram together with CO$_2$ isopleths (Davenport, 1974). The causes of initial disturbances to acid–base status and the mechanisms of subsequent regulation can be analysed using the method described by Wood, McMahon and McDonald (1977). By plotting measured changes in [Lac$^-$] as changes in [HCO$_3^-$] equivalents, this analysis partitions the total change in acid–base status to changes in PCO$_2$ and [Lac$^-$], both of which can be measured experimentally.

In keeping with convention, the Davenport diagram and the analysis of Wood *et al.* (1977) are based on representing acid–base conditions using pH. The data presented in Figure 4 for acid–base changes in a single animal during recovery from exercise (Wood *et al.*, 1977) were re-analysed to determine if using {H$^+$} as the measure of acid–base conditions would make any substantial differences to the conclusions reached. The same values for pK and αPCO$_2$ were used for the analysis. The contribution of the increase in [Lac$^-$] to the changes in acid–base

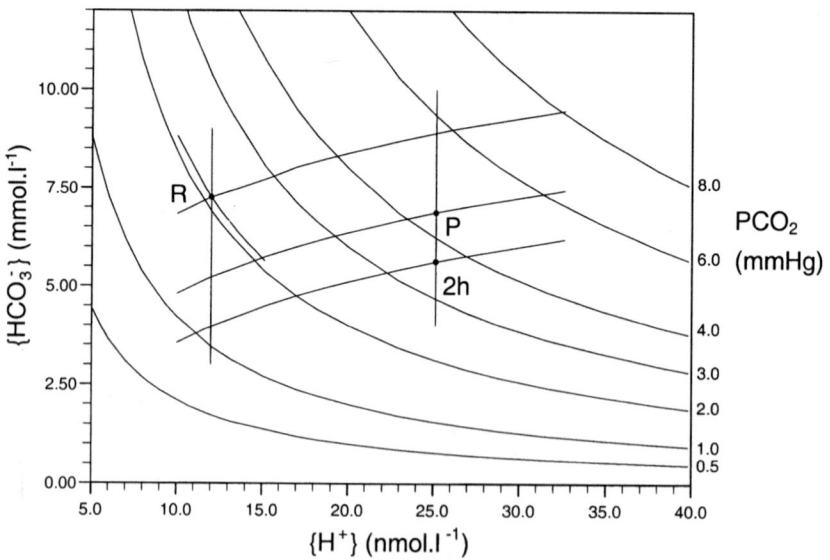

Fig. 4. Davenport diagram to show the changes in venous blood acid–base status of a single flounder at rest (point R) and following 2-hour (point 2h) recovery after 10 minutes of exhausting activity. The point predicted by analysis of the changes in [Lac$^-$] is indicated by P. Data were obtained from Figure 4 in Wood, McMahon and McDonald (1977).

status was analysed from the diagram by the method of Wood *et al.* (1977). Using pH, the contribution of lactate at 20 minutes and 1, 2 and 4 hours post exercise was approximately 12, 11, 30 and 52 per cent respectively. Analysed using {H$^+$}, the contribution at the same times was 7, 10, 24 and 41 per cent. Thus, performing the analysis using pH as the representative of acid–base status overestimates the contribution of lactic acid to the post-exercise acidosis by as much as 70 per cent.

Another popular index of acid–base relationships in poikilothermic animals has been the pH/temperature plot, which examines the degree to which the animal maintains so-called 'constant relative alkalinity'. This is based on a supposedly fundamental observation that blood pH varies predictably with body temperature in many ectothermic animals, so as to maintain a constant relationship with the change in pH at neutrality (pN). This relationship was first expounded by Hermann Rahn and his co-workers at New York State University in Buffalo (Howell *et al.*, 1970). Its graphical interpretation was later dignified as the 'Buffalo curve' (Jackson, 1982). What the Buffalo curve actually showed was that, as environmental temperature rose, blood pH for a wide range of species dropped. The progressive fall in pH was apparently parallel to, but at a more alkaline pH than, the change in pH arising from the progressive dissociation of water (i.e. the pH at neutrality, or pN). The supposed selective advantage of this relationship was that it maintained an unchanging ionisation state on cellular proteins, and in particular their α-imidazole groups, in the face of changing environmental temperatures. This explanation was termed the alphastat hypothesis (Reeves, 1972), later amended to the Z-stat model, based on the near-linear relationship between pH and net protein charge (Cameron, 1989).

Recently, this relationship has been questioned. It was founded on laboratory-based studies of the effects of acute temperature change. It often does not hold true for the blood of animals seasonally acclimatised to temperature change, nor for the various tissue compartments in animals in which maintenance of acid–base status is likely to be most critical (e.g. see Whiteley, this volume). More fundamentally, however, the apparent relationship is heavily dependent on the expression of acid–base status as pH. When the Buffalo curve proposed by Rahn's laboratory is replotted with [H$^+$] substituted for pH, the slope changes direction. More importantly, the changes in [H$^+$] in the animal body fluids are over a narrow range and are no longer remotely in parallel with the changes due to dissociation of water (Fig. 5). Once again, interpretation of acid–base data seems to have been distorted by the pH scale.

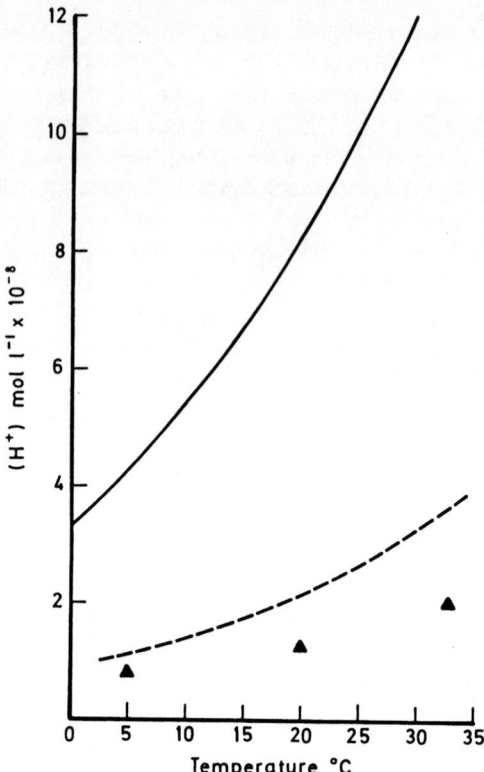

Fig. 5. Data from Rahn and Howell (1978), replotted against [H⁺] rather than pH. The slopes of the lines are now negative and clearly different, implying that there is no basis for the hypothesis that animals sustain a constant relationship between the acid–base status of their tissues and the change in ionisation state of water, as it changes with temperature. The solid line is change in [H⁺] in water; the broken line traces the lower limit of the pH range plotted by Rahn and Howell; and the triangular data points are an approximate re-positioning of some of their data, as they were plotted against pH.

A scale for representing acid–base status

An essential aid to understanding acid–base chemistry and physiological processes affected by it is a useful scale for measuring the degree of acidity or basicity of a solution. To be useful, a scale should provide two key elements. First, a scale should provide a direct measure of the chemically and physiologically important quantities, the activities of the entities involved in acid–base chemistry. Second, as the dissociation of water and consequently {H$^+$} and {OH$^-$} at neutrality are highly dependent on the temperature, ionic strength and osmolarity of a solution, a scale should ideally include some information about the degree of acidity or basicity relative to neutrality.

The pH scale fails to meet either of these criteria. As an inverse logarithmic scale, it substantially distorts the significant quantities by condensing order of magnitude changes in quantity into single-unit increments. The pH scale also conceals the fact that in most biological solutions, H$^+$ is present in only minute quantities. Neither does the pH scale measure the degree of acidity or basicity because it does not include any adjustment for the variation of pH at neutrality (pN), which can vary markedly with changes in ambient conditions such as temperature.

Almost 80 years ago, Benjamin Moore suggested that the [OH$^-$] : [H$^+$] ratio, or its inverse, is the most useful scale for measurement of the degree of acidity or basicity in solutions (Moore, 1919). This view has received some contemporary support (e.g. Stewart, 1978; Truchot, 1983). This ratio certainly provides a temperature-independent definition of neutrality, with a value of 1.0 indicating a neutral solution. However, it does not provide a direct measure of important quantities such as {H$^+$}. Using k_w=[H$^+$][OH$^-$] to substitute for [H$^+$], the ratio can be changed to [OH$^-$]^2/k_w. Thus the [OH$^-$] : [H$^+$] ratio is an exponential and hence curvilinear function. In this respect, the [OH$^-$] : [H$^+$] ratio would not be much improvement over pH for representing variables which change in a linear fashion.

In the absence of a more detailed and informed examination of this subject, acid–base status would probably best be represented by {H$^+$}. When [SID] is negative and greater than about 10^{-3} mol/l, {H$^+$} varies proportionally and inversely with [SID] (Stewart, 1978). Over this range, {H$^+$} would at least provide a linear scale. However, it would not provide a linear measure in fluids where [SID] is less than 10^{-3} mol/l and it could not provide a complete description of acid–base status because it does not include any information about {H$^+$} at neutrality.

If the value of {H$^+$} at neutrality ($\sqrt{K_w}$) can be determined, then a measure of acid–base status independent of temperature, osmolarity and ionic strength could be used. {H$^+$}, like {OH$^-$} and [HCO$_3^-$], is dependent on [SID], P_{CO_2} and the total concentration of weak acids and bases, and so only provides a measure of how these independent factors interact (Stewart, 1978). Additionally, in solutions in which [SID] is less than 10^{-3} mol/l, {H$^+$} does not vary proportionally with [SID] but will equal $\sqrt{K_w}$ when [SID] is zero and will tend towards zero as [SID] becomes increasingly positive. If $\sqrt{K_w}$ for a solution is known, it would be possible, using the equations given by Stewart (1978), to calculate the quantity of strong acid or base which would be required to achieve the observed $\{H^+\}$. This quantity, the [SID] equivalents, might provide the most suitable scale for measuring acid–base status: a linear scale in terms of one of the acid–base-determining independent variables which is independent of the effects of temperature, osmolarity and ionic strength.

Conclusions

The study of acid–base regulation in animals has advanced enormously over the last few decades. With the tools and techniques now available to physiologists, it is possible to study and analyse in detail the causes of acid–base disturbances and the mechanisms of regulation. In particular, accurate measurements of acid–base variables, including $\{H^+\}$, in small quantities of liquid are possible using ion-selective electrodes. To complement the accurate and sensitive tools and techniques available, a scale which will aid the analysis and understanding of processes involving acid–base chemistry is needed.

The pH scale was devised nearly a century ago (Sørenson, 1909), principally as a convenient means of rationalising the very large range of possible values of $\{H^+\}$. It has since been adopted as the standard scale for representing acidity or basicity in physiological studies. However, with the development of modern numerical and computing methods, this rationalisation is now unnecessary.

The frequency plots of replicated blood pH and $\{H^+\}$ values provided evidence that it is $\{H^+\}$ values, not pH values, that are normally distributed. This would suggest that it is $\{H^+\}$, or some other related linear quantity, which is monitored and regulated by animals and is, therefore, of biological significance.

That a linear, not logarithmic, acid–base quantity is regulated would be consistent with the mechanisms of acid–base regulation. In animals, acid–base regulation is achieved either by the ventilatory control of blood CO_2 levels (Wheatly, Burggren & McMahon, 1984) or by ion-exchange mechanisms operating either between different body compartments (Cameron, 1981) or between the body and the environment (Booth et al., 1984). Control of blood CO_2 depends on the P_{CO_2} in the blood and the air and is therefore a linear function. Ion-exchange mechanisms often involve the coupled exchange of acid–base equivalents for other ions such as Na^+/H^+ (or Na^+/NH_4^+) or Cl^-/HCO_3^- (Truchot, 1983) and will therefore depend on the activities of the ions involved.

However, recent studies on ionoregulation over fish gills have identified a specific role for an electrogenic proton pump in regulating the relative acidity of blood plasma and the boundary layer of water at the gill surface (Lin & Randall, 1991), which would seem to provide uncompromised evidence for the fact that it is {H^+} which is regulated.

If linear quantities are of chemical and biological significance, it would be appropriate to represent acid–base data using a linear expression both for the presentation and analysis of data. Our re-analysis of some published data using a 10^{-pH} transformation to derive {H^+} indicates that the distortion introduced by pH can conceal physiologically significant effects or introduce misleading trends into data which are only truly revealed when the data are presented using a linear scale.

This remains a complex issue, and further investigation into the most appropriate scale for presenting acid–base data would not be amiss. The pH scale was a convenient means for expressing the wide range of values possible for {H^+} before modern techniques were developed and before any detailed analysis of acid–base status was attempted. However, on the basis of the work presented here, it is perhaps pertinent to reiterate a point made by Stewart (1978), that 'whatever the justifications may have been for the use of pH . . . they are no match for the serious inherent dangers in its use'. It could aptly be said of pH that 'conceptions from the past blind us to facts which almost slap us in the face' (William S. Halstead, 1852–1922).

Acknowledgements

The authors wish to acknowledge gratefully the helpful comments and suggestions of Dr S. Egginton and Dr N.M. Whiteley during the preparation of this manuscript.

References

Booth, C.E., McMahon, B.R., De Fur, P.L. & Wilkes, P.R.H. (1984). Acid–base regulation during exercise and recovery in the blue crab, *Callinectes sapidus*. *Respiration Physiology* **58**, 359–76.

Boutilier, R.G. & Shelton, G. (1980). The statistical treatment of hydrogen ion concentration and pH. *Journal of Experimental Biology* **84**, 335–9.

Bridges C.R., Morris, S. & Grieshaber, M.K. (1984). Modulation of haemocyanin O_2 affinity in the intertidal prawn *Palaemon elegans* (Rathke). *Respiration Physiology* **57**, 189–200.

Cameron, J.N. (1981). Acid–base responses to changes in CO_2 in two Pacific crabs: the coconut crab, *Birgus latro*, and a mangrove crab, *Cardisoma carnifex*. *Journal of Experimental Zoology* **218**, 65–73.

Cameron, J.N. (1985). Compensation of hypercapnic acidosis in the aquatic blue crab, *Callinectes sapidus*: the predominance of external sea water over carapace carbonate as a proton sink. *Journal of Experimental Biology* **114**, 197–206.

Cameron, J.N. (1989). Acid–base homeostasis: past and present perspectives. *Physiological Zoology* **62**, 845–65.

Davenport, H.W. (1974). *The ABC of acid–base chemistry*, 6th edition, Chicago: The University of Chicago Press.

Dejours, P. (1978). Carbon dioxide in water- and air-breathers. *Respiration Physiology* **33**, 121–8.

Howell, B.J., Baumgardner, F.W., Bondi, K. & Rahn, H. (1970). Acid–base balance in cold-blooded vertebrates as a function of body temperature. *American Journal of Physiology* **218**, 600–6.

Jackson, D.C. (1982). Strategies of blood acid–base control in ectothermic vertebrates. In *A companion to animal physiology*, ed. C.R. Taylor, K. Johansen & L. Bolis, pp. 73–90. Cambridge: Cambridge University Press.

Lin, H & Randall, D.J. (1991). Evidence for the presence of an electrogenic proton pump on the trout gill epithelium. *Journal of Experimental Biology* **161**, 119–34.

Mangum, C.R. (1983). On the distribution of lactate sensitivity among the hemocyanins. *Marine Biology Letters* **4**, 139–49.

Moore, B. (1919). The cause of the exquisite sensitivity of living cells to changes in hydrogen- and hydroxyl-ion concentration. *Journal of Physiology* **53**, LVII–LVIII.

Morris, S., Bridges, C.R. & Grieshaber, M.K. (1985). A new role for uric acid: modulator of haemocyanin oxygen affinity in crustaceans. *Journal of Experimental Biology* **235**, 135–9.

Morris, S., Tyler-Jones, R. & Taylor, E.W. (1986). The regulation of haemocyanin oxygen affinity during emersion of the crayfish *Austropotamobius pallipes*. I. An *in vitro* investigation of the interactive effects of calcium and L-lactate on oxygen affinity. *Journal of Experimental Biology* **121**, 315–26.

Pörtner, H.O., Grieshaber, M.K. & Heisler, N. (1984). Anaerobiosis and acid–base status in marine invertebrates: effect of environmental hypoxia on extracellular and intracellular pH in *Sipunculus nudus* L. *Journal of Comparative Physiology B* **155**, 13–20.

Rahn, H. & Howell, B.J. (1978). The OH⁻/H⁺ concept of acid–base balance: historical development. *Respiration Physiology* **33**, 91–7.

Reeves, R.B. (1972). An imidazole alphastat hypothesis for vertebrate acid–base regulation: tissue carbon dioxide content and body temperature in bullfrogs. *Respiration Physiology* **14**, 219–36.

Shapiro, S.S., Wilk, M.B. & Chen, H.J. (1968). A comparative study of various tests for normality. *Journal of the American Statistical Society* **63**, 1343–72.

Snedecor, G.W. & Cochran, W.G. (1980). *Statistical methods*, 7th edition. Ames: The Iowa State University Press.

Sørenson, S.P.L. (1909). Enzymstudien. II. Mitteilung. Uber die Messung und die Bedeutung der Wasserstoffionenkonzentration bei enzymatischen Prozessen. *Biochemical Zoology* **21**, 131–304.

Stewart, P.A. (1978). Independent and dependent variables of acid–base control. *Respiration Physiology* **33**, 9–26.

Truchot, J.P. (1973). Temperature and acid–base regulation in the shore crab *Carcinus maenas* (L.). *Respiration Physiology* **17**, 11–20.

Truchot, J.P. (1975). Factors controlling the *in vitro* and *in vivo* oxygen affinity of hemocyanin in the crab *Carcinus maenas* (L.). *Respiration Physiology* **24**, 173–89.

Truchot, J.P. (1983). Regulation of acid–base balance. In *Biology of the Crustacea*, Vol. II, ed. L. Mantel & D.C. Bliss, pp. 431–57. New York: Academic Press.

Tyler-Jones, R. & Taylor, E.W. (1988). Analysis of haemolymph and muscle acid–base status during aerial exposure in the crayfish *Austropotamobius pallipes*. *Journal of Experimental Biology* **134**, 409–22.

Wheatly, M.G., Burggren, W.W. & McMahon, B.R. (1984). The effects of temperature and water availability on ion and acid–base balance in hemolymph of the land hermit crab *Coenobita clypeatus*. *Biological Bulletin* **166**, 427–45.

Wood, C.M., McMahon, B.R. & McDonald D.G. (1977). An analysis of changes in blood pH following exhausting activity in the starry flounder *Platichthys stellaris*. *Journal of Experimental Biology* **69**, 173–85.

Index

Page numbers in bold print indicate figures or tables.

Printed in the United Kingdom
by Lightning Source UK Ltd.
129364UK00001B/481/A